T0187615

Errata

Activities in Navigation
Marine Navigation and Safety of Sea Transportation
Editor: Adam Weintrit
ISBN: 978-1-138-02858-6

Page 250:
Due to an error during the typesetting stage of this volume, the following corrections should be applied to the equations:

Eqn (4) $\sigma_{x,z,y} = \sigma_T * c * PDOP$

Eqn (5) $\sigma_{x,y} = \sigma_T * c * HDOP$

Eqn (6) $\sigma_{z} = \sigma_T * c * VDOP$

We sincerely apologize for these errors.

The publishers

ACTIVITIES IN NAVIGATION

Activities in Navigation

Marine Navigation and Safety of Sea Transportation

Editor

Adam Weintrit
Gdynia Maritime University, Gdynia, Poland

CRC Press
Taylor & Francis Group
Boca Raton London New York

CRC Press is an imprint of the
Taylor & Francis Group, an **informa** business

A BALKEMA BOOK

Published by:
CRC Press/Balkema
P.O. Box 447, 2300 AK Leiden, The Netherlands
e-mail: Pub.NL@taylorandfrancis.com
www.crcpress.com – www.taylorandfrancis.com

First issued in paperback 2020

Typeset by V Publishing Solutions Pvt Ltd., Chennai, India

ISBN 13: 978-0-367-73823-5 (pbk)
ISBN 13: 978-1-138-02858-6 (hbk)

**Visit the Taylor & Francis Web site at
http://www.taylorandfrancis.com**

**and the CRC Press Web site at
http://www.crcpress.com**

Contents

List of reviewers

Prof. Michael **Baldauf**, Word Maritime University, Malmö, Sweden
Prof. Andrzej **Banachowicz**, West Pomeranian University of Technology, Szczecin, Poland
Prof. Eugen **Barsan**, Constanta Maritime University, Romania
Prof. Knud **Benedict**, University of Wismar, University of Technology, Business and Design, Germany
Prof. Heinz Peter **Berg**, Bundesamt für Strahlenschutz, Salzgitter, Germany
Prof. Tor Einar **Berg**, Norwegian Marine Technology Research Institute, Trondheim, Norway
Prof. Jarosław **Bosy**, Wroclaw University of Environmental and Life Sciences, Wroclaw, Poland
Prof. Krzysztof **Czaplewski**, Gdynia Maritime University, Poland
Prof. Eamonn **Doyle**, National Maritime College of Ireland, Cork Institute of Technology, Cork, Ireland
Prof. Branislav **Dragović**, University of Montenegro, Kotor, Montenegro
Prof. Daniel **Duda**, Polish Naval Academy, Polish Nautological Society, Poland
Prof. Milan **Džunda**, Technical University of Košice, Slovakia
Prof. Bernd **Eisfeller**, University of FAF, Munich, Germany
Prof. Ahmed **El-Rabbany**, University of New Brunswick; Ryerson University in Toronto, Ontario, Canada
Prof. Naser **El-Sheimy**, The University of Calgary, Canada
Prof. Akram **Elentably**, King Abdulaziz University (KAU), Jeddah, Saudi Arabia
Prof. Włodzimierz **Filipowicz**, Gdynia Maritime University, Gdynia, Poland
Prof. Renato **Filjar**, University College of Applied Sciences, Bjelovar, Croatia
Prof. Börje **Forssell**, Norwegian University of Science and Technology, Trondheim, Norway
Prof. Alberto **Francescutto**, University of Trieste, Trieste, Italy
Prof. Masao **Furusho**, Kobe University, Japan
Prof. Wiesław **Galor**, Maritime University of Szczecin, Poland
Prof. Yang **Gao**, University of Calgary, Canada
Prof. Péter **Gáspár**, Computer and Automation Research Institute, Hungarian Academy of Sciences, Budapest, Hungary
Prof. Aleksandrs **Gasparjans**, Latvian Maritime Academy, Latvia
Prof. Jerzy **Gaździcki**, President of the Polish Association for Spatial Information; Warsaw, Poland
Prof. Mirosław **Gerigk**, Gdańsk University of Technology, Gdańsk, Poland
Prof. Witold **Gierusz**, Gdynia Maritime University, Poland
Prof. Dariusz **Gotlib**, Warsaw University of Technology, Warsaw, Poland
Prof. Hugh **Griffiths**, University College London (UCL), the United Kingdom
Prof. Lucjan **Gucma**, Maritime University of Szczecin, Poland
Prof. Jerzy **Hajduk**, Maritime University of Szczecin, Poland
Prof. Esa **Hämäläinen**, University of Turku, Finland
Prof. Qinyou **Hu**, Shanghai Maritime University, Shanghai, China
Prof. Jacek **Januszewski**, Gdynia Maritime University, Poland
Prof. Mirosław **Jurdziński**, Gdynia Maritime University, Poland
Prof. Tadeusz **Kaczorek**, Warsaw University of Technology, Poland
Prof. Kalin **Kalinov**, Nikola Y. Vaptsarov Naval Academy, Varna, Bulgaria
Prof. John **Kemp**, Royal Institute of Navigation, London, UK
Prof. Eiichi **Kobayashi**, Kobe University, Japan
Prof. Serdjo **Kos**, University of Rijeka, Croatia
Prof. Pentti **Kujala**, Helsinki University of Technology, Helsinki, Finland
Prof. Krzysztof **Kulpa**, Warsaw University of Technology, Warsaw, Poland
Prof. Shashi **Kumar**, U.S. Merchant Marine Academy, New York
Prof. Bogumił **Łączyński**, Gdynia Maritime University, Poland
Prof. David **Last**, The Royal Institute of Navigation, London, the United Kingdom
Prof. Nadav **Levanon**, Tel Aviv University, Tel Aviv, Israel
Prof. Józef **Lisowski**, Gdynia Maritime University, Poland
Prof. Evgeniy **Lushnikov**, Maritime University of Szczecin, Poland
Prof. Francesc Xavier **Martinez de Oses**, Polytechnical University of Catalonia, Barcelona, Spain
Prof. Boyan **Mednikarov**, Nikola Y. Vaptsarov Naval Academy,Varna, Bulgaria
Prof. Sergey **Moiseenko**, Kaliningrad State Technical University, Kaliningrad, Russian Federation
Prof. Wacław **Morgaś**, Polish Naval Academy, Gdynia, Poland
Prof. Janusz **Narkiewicz**, Warsaw University of Technology, Poland
Prof. Nikitas **Nikitakos**, University of the Aegean, Chios, Greece
Prof. Andy **Norris**, The Royal Institute of Navigation, University of Nottingham, UK
Prof. Stanisław **Oszczak**, University of Warmia and Mazury in Olsztyn, Poland
Prof. Gyei-Kark **Park**, Mokpo National Maritime University, Mokpo, Korea
Prof. Jin-Soo **Park**, Korea Maritime University, Pusan, Korea
Mr. David **Patraiko**, The Nautical Institute, UK
Prof. Jan **Pawelski**, Gdynia Maritime University, Poland
Prof. Zbigniew **Pietrzykowski**, Maritime University of Szczecin, Poland
Prof. Tomasz **Praczyk**, Polish Naval Academy, Gdynia, Poland
Prof. Jerzy B. **Rogowski**, Gdynia Maritime University, Poland
Prof. Hermann **Rohling**, Hamburg University of Technology, Hamburg, Germany
Prof. Wojciech **Ślączka**, Maritime University of Szczecin, Poland
Prof. Roman **Śmierzchalski**, Gdańsk University of Technology, Poland

Activities in Navigation
Introduction

A. Weintrit
Gdynia Maritime University, Gdynia, Poland
Polish Branch of the Nautical Institute
Chairman of TransNav

The contents of the book are partitioned into seven separate chapters: Safety of navigation (covering the subchapters 1.1 through 1.10), Navigational Bridge Equipment (covering the chapters 2.1 through 2.4), Automatic Identification System AIS (covering the chapters 3.1 through 3.5), Route planning (covering the chapters 4.1 through 4.5), Anti-collision and collision avoidance (covering the chapters 5.1 through 5.4), GNSS - Global Navigation Satellite Systems (covering the chapters 6.1 through 6.3), and Aviation (covering the chapters 7.1 and 7.2).

In each of them readers can find a few subchapters. Subchapters collected in the first chapter, titled 'Safety of navigation', concerning conceptual foundation of safety of navigation, question of vessels safety ensuring in the emergency situations, safety of dynamic positioning, safety of offshore supply operations, risk evaluation model for management of navigation safety in an entire ship route area, identification of typical hazards and limitations to the commercial shipping safety, created by offshore activity and crew transfer high speed crafts, operating in the vicinity of the intensive traffic flow areas, physical characteristics of virtual aids to navigation, analysis and decision-making for control of extreme situation of fishing vessels on the base dynamic model of catastrophe, and case study in flawed accident investigation as well as some results by real time simulation study.

In the second chapter there are described problems related to navigational bridge equipment: radar detection in duct situations in maritime environment, the vital influence of the radar antenna height, signal processing optimization in the FMCW navigational radars, autopilot using the nonlinear inverse ship model, and reliability and exploitation analysis of navigational system consisting of ECDIS and ECDIS back-up systems.

The third chapter deals with Automatic Identification System AIS problems. The contents of the third chapter are partitioned into five subchapters: an analysis of ship behaviour induced by the Great East Japan Earthquake tsunami based on AIS; onboard AIS reception performance advances for a small boat; a subject of class B AIS for small trawler; research on real movement of container ship between China, Japan and South Korea using AIS data; and modelling of observed ship domain in coastal sea area based on AIS data.

In the fourth chapter there are described problems related to route planning: development of the method of safe ship trajectory planning based on the ant algorithm, ship evolutionary trajectory planning method with application of polynomial interpolation, supply and demand of transit cargo along the Northern Sea Route, Northern labyrinths as navigation network elements, and ship's route planning in ice infested areas of Northern Svalbard following ice charts made by remote sensing methods.

The fifth chapter deals with anti-collision and collision avoidance. The contents of the fifth chapter are partitioned into four subchapters: apprisal of the coordinability of the vessels for collision avoidance manoeuvres by course alteration, comparison of anti-collision game trajectories of ship in good and restricted visibility at sea, concretization of the concept "nearly reciprocal course" in rule 14 of Colreg-72 and visualization of holes and relationships between holes and latent conditions.

In the sixth chapter there are described problems related to GNSS - Global Navigation Satellite Systems: nominal unique BeiDou satellite constellation, its advantages and disadvantages, experiments with reception of IRNSS satellite navigation signals in the S and C frequency bands, and evaluation of a low cost tactical grade MEMS IMU for maritime navigation.

The seventh chapter deals with aviation aspects of navigation. The contents of the seventh chapter concerns use of passive surveillance systems in aviation and the concept of the SWIM system in air traffic management.

Each chapter was reviewed at least by three independent reviewers. The Editor would like to express his gratitude to distinguished authors and reviewers of chapters for their great contribution for expected success of the publication. He congratulates the authors and reviewers for their excellent work.

Safety of Navigation

Safety of Navigation – Conceptual Foundation

V.G. Torskiy, V.P. Topalov & M.V. Chesnokova
Odessa National Maritime Academy, Ukraine

ABSTRACT: The most important global problem being solved by the whole world community nowadays is to provide sustainable mankind development. Recent research in the field of sustainable development states that civilization safety is impossible without transfer sustainable development. At the same time, sustainable development (i.e. preservation of human culture and biosphere) is impossible as a system that serves to meet economical, cultural, scientific, recreational and other human needs and plays an important role in sustainable development goals achievement. An essential condition of effective navigation functioning is to provide its safety. The "prescriptive" approach to the navigation safety, which is currently used in the world maritime field, is based on long-term experience and ship accidents investigation results. Thus this approach acted as an important fact in reduction of number of accidents at sea.

Having adopted the International Safety Management Code all the activities connected with navigation safety problems solution were transferred to the higher qualitative level. Search and development of new approaches and methods of ship accidents prevention during their operation have obtained greater importance. However, the maritime safety concept (i.e. the different points on ways, means and methods that should be used to achieve this goal) hasn't been formed and described yet. The article contains a brief review of the main provisions of Navigation Safety Conventions, which contribute to the number of accidents and incidents at sea reduction.

1 INTRODUCTION

The symposium dedicated to the problems of sustainable maritime transportation system creation was held under the auspices of the IMO in London on 13 September, 2013. Such perspective efforts of shipping industry subjects' correspond with the policy of conversion to the mankind sustainable development declared on the UN conference in Rio de Janeiro (1992), which should be followed by all fields of the world economy. Navigation safety is named among the main features of sustainable maritime transportation system in the concluding document approved at the symposium. Due to this fact elaboration of the generally accepted navigation safety concept is supposed to conform to successful creation of sustainable maritime transportation system. The abovementioned concept implies organizational and technical strategy of safety at sea with the use of means and methods complex which meets modern requirements and conditions of shipping industry. This article is concerned with certain fragments of such concept, which could serve as the basis for its formation and development.

Sustainable mankind development has been much paid attention to recently. It concerns our civilization transfer to a new development strategy which should provide survival and further constant society development. Sustainable development is the world community response to the crisis phenomena in biosphere, economy, social sphere, foreign affairs which may cause the global disaster and apocalypse some day. The term "sustainable development" was implemented for the first time at the UN conference on environment and development (Rio de Janeiro, 3-14 July 1992) as an alternative to the prior "unsteady" course followed by the mankind. At this conference the concept of sustainable development was put across and the policy paper "Agenda 21" was approved. Since then numerous international events concerning the realization of different provisions of the Agenda were held in different countries. The IMO as the specialized agency of UN was fully engaged in this work, aiming its activity at maritime safety improving and environmental pollution protection while vessel operation.

Consequently, all IMO regulatory documents have formed the single complex, and each of these documents fulfills an important protection function of vessels, people and environment from current harmful factors. These documents are being constantly modified, improved and amended according to the changes in construction, equipment, ships operating conditions and arising requirements of their industrial use.

Intentional efforts of the IMO, classification societies, flag states, ship-owners resulted in merchant fleet accident reduction. The World's Oceans pollution from ships has also considerably decreased.

Some outstanding tendencies of maritime safety improvement and environmental protection enhancement can be observed. It is necessary to support steadiness of these tendencies with universal application of either traditional standards or by means of new approaches and methods development, corresponding to the international shipping needs and requirements.

2 STATEMENT OF THE PROBLEM

The regular UN conference on sustainable development, known as the Rio +20, was held in Rio de Janeiro in June, 2012, where results of the work carried out during the last 20 years were summed up and tasks for the future were established. The idea of the sustainable maritime transportation system creation aimed to develop action program was chosen as the subject of discussion during the celebration of the Day of the Seafarer in 2013.

The symposium on this topic was held on the initiative of Koji Sekimizu, the IMO Secretary-General, at the IMO Headquarters in London on 26 September, 2013. It was further agreed that the main goal of such system creation is to ensure that «this system will offer the entire planet a safe, efficient and reliable means of transportation of goods, all the while reducing pollution, maximizing energy, efficiency and ensuring the conservation of resources». The fundamental principles of the sustainable maritime transportation system creation are: development of «global standards that support level playing fields across the world, supporting global safety and environmental standards, addressing technical and operational requirements for ships as well as the appropriate education and training of crews». In addition: «All actors will need to collaborate with the aim of achieving the three dimensions – the economic, social and environmental dimensions – but with the safety of navigation always being the overriding priority». The safety culture is needed to be formed on global

standards should be promoted and developed in the sustainable maritime transportation system. The principle «Safety and environmental awareness should be the priorities» should be followed under professional seafarer's training.

Thus, conceptual grounds of international navigation safety development could contribute to successful achievement of these goals.

It should be noted that issues on people on-board and vessel safety were paid much attention to during the merchant shipping developmental period. System of principles about ways, means and methods to be applied to achieve safety (i.e. maritime safety concept) wasn't described and formulated in appropriate way.

The number of essential propositions, which should be used while creating of such concept are discussed below. Offered recommendations, methods and means are fragmentary, but together they provide a clear idea of its possible structure and separate components.

3 SAFETY OF NAVIGATION – CONCEPTUAL FOUNDATION

It is commonly known that to a wide extent a concept is a system of views on any fact, way of understanding and explaining of any facts, the main idea of any theory. Therefore, maritime safety concept is a system of views on providing safety of the operations at sea, description of its facts and processes. However, «navigation safety» is defined as a relatively constant state, clear of dangers provided by the system of international and national technical, organizational, economical, social, and juridical standards, which are aimed at the reduction and prevention of accidents at sea to provide safety of life and property at sea and marine environmental protection.

The Maritime Safety Concept, formed under the influence of numerous and contradictory factors, according to the needs of the world economy in accordance with prevailing political, economical and ecological conditions, shouldn't be presented as a still form of the defined content. As shipping is an activity which maintains mankind varying needs under environmental protection provision, its conditions require flexibility of safety concept as well as allowance to make necessary modifications in existing principles and methods according to the changes in the maritime transportation system and its functional environment. The navigation safety concept is aimed to form a comprehensive picture about ways, principles, methods, means of vessels and other maritime objects safety provision as well as its staff and environment under modern conditions of industrial activity. It is aimed to reduce vessel accident and injury rate by means of concept provisions to the lowest possible level.

It defines goals, objectives, principles, key activities undertaken to reduce accidents and incidents at sea. It establishes a single conceptual and technological system in this field and aimed at formation of global system which provides navigation safety.

The currently scientifically grounded maritime safety concept allows:
1 undertaking effective measures to ensure protection of vessels, people and environment from dangerous factors;
2 responding quickly to the changes in fleet operation, application of newest achievements in the field of ship navigation, marine engineering and technology;
3 developing appropriate methods of vessel operation safety improvement;
4 providing crew training conforming to further requirements;

5 ensuring appropriate support of transfer to mankind sustainable development policy implemented by UN.

Subjects of navigation safety are: the member states of corresponding international conventions and intergovernmental agreements, represented by its bodies of legislative, executive and juridical power as well as ship-owners, masters and crew members. Objects of navigation safety include vessels, people on board, shipping ways with appropriate navigation means of navigation equipment and natural environment influenced by the vessel operation. Navigation safety should provide vessels' operation condition which is responsible for its use as intended. Negative impacts on the vessels are dangerous factors for navigation. Level of maritime operation safety depends on the interaction of favorable and unfavorable factors in following areas related to the navigation safety:
1 Technical and technological;
2 Organizational and managerial ;
   Anti-terrorist (protection of maritime objects from illegal actions).- Fig. 1

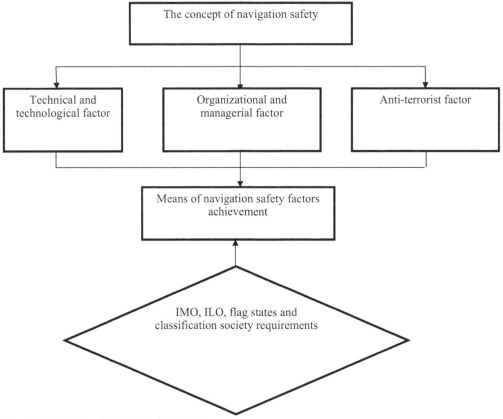

Figure 1 The structure of navigation safety concept

The key issues of technical and technological factor of the navigation safety are:
- technical condition and applicability of vessels, its equipment, port infrastructures, coastal and maritime facilities of navigation equipment, etc.
- ensuring the functioning reliability of the "Man-machine-environment" systems.

Principal means for achieving the objectives of this navigation safety field include:
- providing of all the parties involved in navigation with necessary technical standards and norms in time;
- systematic monitoring of vessels and other objects technical condition;
- maintenance and repair of vessels and fleet serving objects;
- scientific and technical progress ensuring in the maritime industry;
- corresponding staffing and resources provision of enterprises and organizations engaged in maritime operations.

There are numerous methods related to this area of navigation safety. They are usually subdivided into the following groups:
1 Technical;
2 Organizational;
3 Ergonomic;
4 Informational;
5 Juridical;
6 Social.

Each of these methods can't be characterized separately because of the limited article length. It should be noted that they have different degrees of development and implementation in the safety providing practice. Technical methods are more commonly used in practice, including the stages of shipbuilding and operation. Ergonomic and social methods are less developed and implemented restrictedly.

The issues of organizational and managerial factor of navigation safety are:
- organization and maintenance of navigation safety services at the national level, shipping companies and ports;
- development and application of standards aimed at proper organization and control of staff of on-board, enterprises and fleet serving organizations, safety management systems (SMS);
- improvement of navigation safety management bodies and systems at all levels – according to the conditions and requirements of fleet operation practice.

There are the following means of the above mentioned objectives achievement:
- creation and improvement of methods of vessel operation safety management and environmental pollution prevention;
- organizational and managerial provision of search and rescue of persons under distress;

- application of the assessment methodology and risk management on the objects of maritime safety;
- formation of the safety culture elements at shipping companies and on vessels;
- organization and maintenance of safety international standards monitoring on vessels, at enterprises and in maritime institutions.

Provision of the navigation safety anti-terrorist factor covers an ample amount of problems, connected with planning and adoption of protective measures from terroristic acts and other illegal actions against vessels and port facilities, enforceability of onshore personnel and other persons involved in activities on prevention and protection of maritime objects from terroristic acts and pirate attacks, elimination of its consequences for vessels, people and environment.

Main means to achieve the anti-terrorist safety objectives include: international treaties and national legislative acts against terrorism at sea, creation of accepted legal framework related to self protection of vessels and maritime navigation from armed attacks, etc. (Fig. 2)

1	Definition of the general structure of navigation safety concept
2	Information collection and analysis on every constituent element of the Safety of Navigation Convention
3	Correspondence assessment of the parts of Navigation Safety Conventions with update and perspective requirements of navigation industry
4	Definition of main research and development fields within the framework of the structure of navigation safety conventions
5	Measures elaboration of enhancement of informative and regulatory provision of navigation safety

Figure 2. Stages of safety of navigation concepts development

4 CONCLUSIONS

Practical realization of maritime safety concept provisions should be carried out under support and control of states possessing fleet by means of necessary application of corresponding standards, rules and methods regulated by the international and national sea safety documents at the maritime field objects.

Therefore, systematic approach to the navigation as the marine transportation system functioning aimed to meet needs of the world community in cargo shipment and passengers transportation should be the basis of the modern navigation safety concept.

The «prescriptive» approach to the navigation safety provision, adopted by the IMO, and legal documents developed for its practical implementation should remain at the centre of the maritime operations safety concept. The existing standards and regulations of vessel safe operation should be supplemented by the rules reflecting modern navigation conditions, new means and methods of vessel and environmental protection from unfavorable and dangerous factors.

A message from Koji Sekimizu, the IMO Secretary-General (IMO News #1, 2013) noted that: «The world relies on the safe, secure, efficient and clear international shipping industry. And the comprehensive regulatory framework developed and maintained by IMO creates the conditions in which shipping can achieve those objectives». In our opinion, the navigation safety concept should be considered as one of the necessary IMO regulatory instruments.

REFERENCES

Kuo Ch.1998. Managing Ship Safety.LLP London – Hong-Kong, -189 p.
Lutzhoft M. 2005.. Maritime Technology and Human Integration on the Ship's Bridge. «Seaways», NI.
Joykody N., Zhengiang Liu. Modern Technology. «Seaways», NI.
T. Crowch, 2013. Navigating the Human Element, MLB Publishing.
Torskiy V.G., Topalov V.P. 2011. Ship's crew management .Second ed. – Odessa, Astroprint, - 244 p.
Torskiy V.G., Topalov V.P., Pozolotin L.A. The hazards for navigation. 2009, Odessa, Astroprint, -240 p.
Guidelines for the Investigation of Human Factors in Marine Casualties and Incidents. Res.IMO A.884(21), 1999.

To the Question of Vessels Safety Ensuring in the Emergency Situations

E.P. Burakovskiy & P.E. Burakovskiy
Kaliningrad State Technical University, Kaliningrad, Russia

ABSTRACT: In this article there are considered new mechanisms of ships loss in the sea storm conditions based on the washing down over the bow. Herewith loss of ships may arise from both the destruction of a ship and the loss of transverse roll metacentric stability. It is presented hull modernization variant decreasing the catastrophe realization possibility in the forward end.

1 CATASTROPHE OF TANKER "NAKHODKA" AT STORM SEA CONDITION

Despite the constant hull design development, average operating conditions, regulatory documents ensuring the seafaring safety [8], instrumental and navigation equipment severe accidents inherent in loss of ships, people and cargo, occur anyway that gives evidence of the existence the mechanisms of the hull and the environment interaction which were not paid adequate attention. In general, it's possible to analyze the emergency situations effectively only when there's enough source information. If the vessel within seconds gets into the emergency situation and it leads to the loss of a ship the source information may be not enough for catastrophe cause explanation. Only in rare investigations cases such information is provided which analysis allows to gain insight into the catastrophe cause or, at least, to work out the version that may be used for explaining the catastrophe cause. Such a catastrophe is the loss of the tanker "Nakhodka".

In 1996 the tanker "Nakhodka" has broken and sank at the shores of Japan. The architectural and constructional type, ratios of main dimensions and characteristics of this tanker are typical for the similar class vessels. (length of the vessel L=166 m, the width B=22.4 m, the draft T=9.38 m and the hull height H=12.32 m.) Wave slaps to the bow have lead to the strength loss and the loss of the ship. The measurements of the hull constructions condition have shown that the vessel was in the emergency condition because of the warn out hull and it didn't withstand real dynamic loads.

Let's consider the catastrophe physical dependency on the basis of the mathematical modeling of the vessel interaction dynamics with the environment executed in the works [3, 9]. At the moment of the catastrophe the sea storm force was 8, and the vessel speed was about 8 knots and was reducing to 3.5 knots. The process of the vessel oscillatory motion is presented in Fig. 1 in the form of subsequent fragments corresponding to different phases of the vessel with oncoming sea. On Fig. 1 it is seen that the pitch is leading to the periodical immersion of the bow into the wave and significant washing down of the bow. The main "hidden" interaction regularities follow from the most unfavorable phase which fits the last fragment of Fig. 1. The discussion of it will be carried out from the position of the hydrodynamic interaction and peculiar properties of the vessel strength condition.

The peculiar properties of the hydrodynamic interaction are to be considered further. The periodical bow immersion into water on the head sea resulted in sufficient changing of the regime of deck flow. In such conditions the dynamic interaction is characterized by the appearance of heavy loads caused by the asymmetric flow of the loaded deck which can be considered as the complex shape wing inclined under the attack angle to the liquid main flow [11]. Moreover, the load line loss (up to 50% and more) is caused by the periodical deterioration of longitudinal and transverse metacentric stability. As a result there appeared the phenomenon of getting the vessel into "potential well" [5] leading to the sufficient decreasing of the vessel resistance to external dynamic loads.

Such hydrodynamic task about nonlinear longitudinal pitching in the framework of a differential equation with periodical coefficients hasn't been considered at all in the works in the theory of pitching and the vessel metacentric stability while sea swell [11]. The existing theoretical and experimental investigations were about the transverse metacentric stability at the following sea, and the questions of pitching at longitudinal rolling were executed in the deterministic formulation on the basis of linear models. These "hidden" effects of the interaction have lead to the catastrophe and the loss of the ship. The evaluation of the metacentric stability at the given waterline zone loss has shown that its real characteristics may fall down comparing to the nominal values, and the hydrodynamic loads while flowing the deck exceeded ten times usual forces and moments working in the vessel forward end while pitching on the irregular sea.

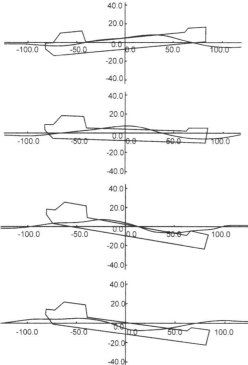

Figure 1. The movement of the vessel at storm sea condition

The vessel exploitation strength evaluation must be carried out with the regard to its forward end periodical immersion under water. If the hull of the vessel is worn off then the dynamics of the vessel and the environment interaction changes sufficiently. According to our preliminary estimation on the basis of the data [3, 9] the loads from the action of

the hydrodynamic force and the moment at the asymmetric flow of the deck have resulted in the intolerable tensions in the hull constructions. It sufficiently exceeded the nominal values. As a result the worn down hull was unable to take the acting forces and the vessel having lost the strength has broken and sank. Some regularity has appeared which was not been taken into consideration in the process of designing of the vessel with such architecture. So the appearance of the sufficient dynamic loads due to the flow of the deck sank into water was not considered at all in the strength practical tasks and at vessels designing.

The principal moment in it is that the hull destruction occurred not in the midship body but sufficiently further that approximately 1/3 of the vessel was torn away which does not fit in the ordinary scheme. It is necessary to find out the mechanism of the hull and water interaction which would let strictly prove the appearance of the bending moment extreme values which are sufficiently dislocated from the midship. The offered mechanism of creating additional bending moment based on the flow of the deck in the forward end is physical enough and it reveals the reasons of tearing away parts of vessels which have open flat surfaces in forward ends of decks.

In recent years severe accidents with tankers of such class remind constantly the designers about errors and disadvantages of architectural-constructional decisions evidence of which is in the periodical press [1, 4]. A special interest is given to offers of improving the forward end shape of tankers of such class shown in the work [4]. The variant of such a decision got the name "axe bow".

The authors offer a principally new approach to the architectural decision of the forward end which intend decreasing of the hydrodynamic loads and sufficient reduction of effects from the forward end wash over. The distinction of the offered architecture is the usage of mold line of the hull bow having a vertical stem post [1, 4].

As emergency situation conducted analysis result it was found out that the hull destruction of the tanker "Nakhodka" began from the upper deck then there began the bottom destruction and both cracks came together. For finding out the reasons having caused the deck destruction the vessel dynamics evaluation was carried out.

2 RESULTS OF THE MATHEMATICAL MODELING OF THE VESSEL AND THE ENVIRONMENT INTERACTION DYNAMICS

One of the main problems in the decision making systems is the necessity of producing a great deal of calculations. It is especially characteristic of interpreting complex situations by means of catastrophe theory which are to be dealt with when formalizing knowledge in intelligent system of ship strength control. When the number of attributes of investigated situation is large the use of conventional calculations methods brings about a sharp rise of calculations volume ("the curse of dimensions").

At the pitching while the forward end immersion into water the deck presents itself the wing of the finite extending. As a result of the flow there appears the perpendicular to the deck the press force P_y combining from two components (Fig. 2). The component P'_y is determined by the horizontal water flow of the deck as the fragment of the wing of the finite end with the speed

$$V_H = V_C + V_{WH}, \tag{1}$$

where V_C – the speed of the vessel, V_{WH} – the horizontal velocity of the wave.

The component P''_y is determined by the flow of the deck in the vertical direction with the speed V_V

$$V_V = V_{WV} + V_{OV}, \tag{2}$$

where V_{WV} – vertical speed of water particles, V_{OV} – vertical speed of the forward end caused by the vessel pitching.

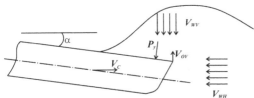

Figure 2. The scheme of the vessel forward end flow

For the approximate evaluation of P'_y and P''_y values the formulas for the wing of unlimited extending with some corrections may be used:

$$P'_y \cong k_1 k_2 k_3 c_y \frac{1}{2} \rho V_H^2 F, \tag{3}$$

where k_1 – proportionality coefficient connected with the flow of the deck close to the two environments division, k_2 – the coefficient regarding the finiteness of the wing extending, k_3 – the coefficient of superstructures influencing, ρ – water density. The component P''_y may be evaluated by formula:

$$P''_y \cong k'_1 k'_2 k'_3 c'_y \frac{1}{2} \rho V_V^2 F, \tag{4}$$

where F – the immersed deck area, k'_1, k'_2, k'_3 – the coefficients analogous to k_1, k_2, k_3 but connected to the flow in the vertical plane.

Gaining the strict task solution is connected with certain difficulties, however, the values of P'_y and

P''_y may be evaluated approximately, after that the vessel's dipping may be analyzed. It should be mentioned that this regime of the flow (the flow in two orthogonally related directions) is short and is estimated at seconds, however, this may be enough for developing the emergency situation.

Let's consider the longitudinal metacentric stability of the vessel (Fig. 3) while developing the trim under the moment action caused by force P_y applied to the deck in the vessel forward end

$$M_{dif} \cong P_y \cdot \left(\frac{L}{2} - \frac{a}{2} \right), \tag{5}$$

where a is the length of the wetted area.

As follows from Figure 3 the presented curves 1 and 2 cross in some point A where the trimming moment becomes equal the restoring one. In that case "the wash over" the forward end by water occurs and further the uncontrollable situation develops: the increasing of the bow-down pitch angle and the sharp rise of bending moments in the vessel hull up to the extreme values. Practically that means that the vessel will be destructed if it does not go out of the condition of "the wash over" the forward end. It should be noticed that the vessel position while moving to the point A is unstable due to the asymmetric process of the flow, the rolling etc, therefore, the P_y force is not applied to the vessel centerline plane strictly. If to take into consideration that at dipping and partial aft outlet from water the cross metacentric height sharply drops, then the displacement of the pressure centre to the deck P_y from the centerline plane to the value ε (ε – the resultant displacement from the centerline plane) gives the inclining moment $P_y \cdot \varepsilon$ which will turn the vessel under some angle θ to the vertical surface – the centerline plane.

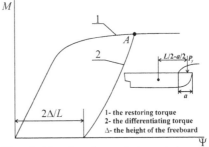

Figure 3. The dependence of the restoring and trimming moments from the bow-down pitch angle

At that process will occur the cross-axis forces which will sharply turn the hull beam-on to the oncoming sea [2]. The angular velocities of the vessel turning will be two next-higher orders than the angular velocities while rotating.

The analysis shows that only the displacement of the resultant of pressure forces to the ε value from the centerline plane may prove to be enough to overturn the vessel as the sufficient decreasing of the metacentric height does not ensure the necessary transverse metacentric stability. The curve of the metacentric height changing is shown at Figure 4.

Let's evaluate the value of the additional bending moments caused by the water pressure to the dipped part of the deck. The task will be narrowed down to the simplest scheme of the beam lying on the resilient bed. It is evident that ≈(1/5÷1/4) part of the hull dipped into water doesn't lie on such bed. Then the analytic model will be seen according to Figure 5a, where there is a hull part dipped into water, P – the total pressure to the deck achieved while the horizontal and vertical flow. The trim angle was α=30° in the horizontal surface which corresponds to a real vessel pitching.

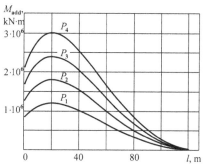

Figure 4. Changing the transverse metacentric height depending on the vessel trim

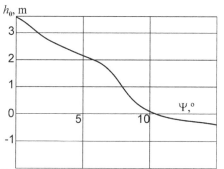

Figure 5. The analytic models of the hull

Cast aside segment a in the analytical model and replacing its action to the beam lying on the resilient bed with the equivalent forces system we shall get the analytical model at Figure 5a, where $M_a = P \times a / 2$ (provided that the pressure total force is in the midspan). In this case the hull bending moment may be determined by formula

$$M(z) = \frac{d^2 w(z)}{dz^2} \cdot EI. \qquad (6)$$

Inserting the particular values into the bending moment equation we obtain the dependences presented at Figure 6. It is apparent that the bending

moment diagram has the distinctive extreme, the position of which is determined mainly by the value of the dipped part of the hull (the value a), and the extreme value itself is dependent on the P force and the length of the water-washed part of the hull a. So with a =20 m and the overall length 170 m the extreme is at 45m distance from the forward end but with a =42 m the extreme is at 62 m distance which is in the range from 1/4 to 1/3 of the vessel length.

According to Figure 6 the values of the additional bending moments obtained as a result of "washing over" forward end of the the vessel have rather big quantities and while certain conditions may achieve values almost two times bigger than the hydrostatic and wave bending moments taken together. The attention should be paid to the fact that the sign of the additional moment is different from the sign of the hydrostatic and wave bending moment that leads after all to the hull strain and in certain conditions to its destruction. While big P forces to the deck in the forward end the situation of going the vessel aft out of water may be observed that leads to the appearance of the second extreme in the vessel aft. In this case the breakaway of the vessel aft or the hull destruction into three parts is possible.

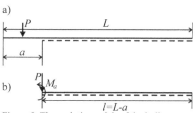

Figure 6. Changes of the additional bending moments

That is why there appears the question how to ensure the ship survivability in this situation. On the one hand there is the craft loss from the turnover; on the other hand there is the appearance of additional bending moments able in the sum with traditional components of bending moments on still water and the roll to tear the vessel into pieces. To ensure the vessel strength is the simplest task, however, not everything is clear here as the torn away bow part of the tanker "Nakhodka" allowed the vessel go out of the wash over and gave the crew 2-2,5 hours for conducting rescue operations. Therefore the hull strengthening is not the best solution of this task. Moreover the analysis of the strength of particular tankers has shown that their strength was enough but that made worse the problem of their survivability due to the possible loss of the transverse metacentric stability.

3 SOLUTION FOR NAVIGATION SAFETY INCREASING

Therefore the problem of navigation safety increasing is to be solved excluding the wash over forward end of the vessel for account of creating the optimal conditions of the flow of the vessel forward end, namely, the reduction of the resultant of pressure forces and ensuring the conditions for eliminating the displacement of this force from the centerline plane [10]. For solving the assigned task the traditional construction of the vessel bow should be equipped additionally with the deck cowling (Fig. 7) which should be supplied with the wave breaker. At Figure 7 the following nomenclature is taken: 1– the stem post, 2 – the bow bulb, 3 – the bottom, 4 – the deck, 5 – the deck cowling, 6 – the wave breaker; L_1 – the deck cowling length; R_1, R_2, R_3 are curvature radiuses characterizing the deck cowling shape.

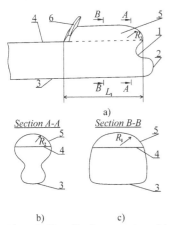

a)

Section A-A   Section B-B

b)   c)

Figure 7. The offered construction of the vessel forward end: a) the main view of the vessel forward end; b) section A-A; c) section B-B.

In the case of the offered constructions realization the fluid flow affects the deck cowling but not the flat surface of the deck due to what the hydrodynamic force acting on the bow decreases and the possibility of this force displacement from the centerline plane is excluded due to that the vessel turnover is excluded.

The vessel forward end acts the following way. When the vessel goes head sea periodically forward end can immerse into water and that leads to the appearance of a difficult regime of the flow of the deck. In the offered construction the liquid flow affects the deck cowling and due to that the resultant of pressure forces decreases and the possibility of this force displacement off the centerline plane is excluded, herewith, the cowling allows to decrease the deck wetness. In such conditions the dynamics of the vessel interaction with the forward end of the

traditional construction is characterized by the appearance of sufficient loads caused by the asymmetric flow of the dipped deck which can be considered as the complex shape wing placed under the angle of the trim to the liquid incoming flow. Herewith, the situation is redoubled in the presence of the gunwale increasing the load to the vessel hull at the flow of the forward end with the fluid stream.

The force affecting the forward end with a flat deck and the modernized forward end may be estimated with the help of [6]. So, the force affects the body being in the fluid flow

$$F = C \cdot S \cdot V^2 \cdot \frac{\rho}{2}, \qquad (7)$$

where C – the coefficient depending on the body shape, Reynolds number (for a plate $C_{pl} = 1.11$, for a cylinder $C_{cyl} = 0.4$); S – the biggest section area of the body in the plane perpendicular to the flow direction; ρ – the fluid density, V – the speed of the incoming fluid flow.

As the parameters S, V, ρ are equal for both constructions then the relation of forces affecting them while the flow of the forward end is equal the coefficient ratio C_{pl} and C_{cyl} that is

$$\frac{F_{t.c.}}{F_{o.c.}} = \frac{C_{pl}}{C_{cyl}} = \frac{1.11}{0.4} \approx 2.78, \qquad (8)$$

where $F_{t.c.}$ – the force affecting the traditional construction at the flow of the forward end; $F_{o.c.}$ – the force affecting the offered construction at the flow of the forward end.

Thus, the offered construction allows almost in 3 times to decrease the hydrodynamic force affecting the vessel forward end at its flow that allows to reduce the possibility of the vessel wash over by wave and its loss both for the reason of the turnover and the reason of the hull destruction and consequently to increase the navigation safety.

It should be mentioned that any structural and especially architectural changes in the forward end need to be considered not only from the positions of ensuring the hull strength but also from the positions of reducing the negative hydrodynamic phenomena appearing when wash over the vessel forward end by water.

The reasons of the vessel hull destruction shown in this work and also dangerous hydrodynamic effects [2] accompanying this phenomenon may be the probable reason of the loss of many ships.

4 CONCLUSIONS

The analysis of the emergency situation with the tanker "Nakhodka" have allowed to find out new facts and regularities of the vessel and the environment interaction dynamics. Theses regularities don not lay into the frames of traditional views about the vessel behavior at the roll as a complex dynamic system and demand more in-depth study at choosing the architecture of the vessels of such class and the necessity of the situation operative control in the exploitation conditions. This control can be realized on the basis of intellectual technologies and high-production computational facilities [7] in the frames of the concept of the modern catastrophe theory.

REFERENCES

1. Buckley Tork. The Axe Factor. Damen & Amels take a bow // The Yacht Report, issue 111, march 2010, p. 46–52.
2. Burakovskiy E.P., Burakovskiy P.E., Nechaev Yu. I., Prokhnich V.P. Management and decision-making under the control of the operational strength of the vessel on the basis of the modern theory of catastrophes // Marine intelligent technologies, №1(19), 2013, p. 7 – 14.
3. Iwao Watanabe, Hidecimi Ontsubo. Analysis of the accident of the MV Nakhodka. Part 1. Estimation of wave loads // Marine Science and technology. Vol.3, No 4, Springer, 1998, p.171 – 180.
4. Keuning J.A., Pinkster J., F. van Walree. Further Investigation into the Hydrodynamic Performance of the AXE Bow Concept, HSMV 2002, Naples, Italy, 2002.
5. Khudyakov L.Yu. The research design of ships. – Leningrad: Shipbuilding, 1980.
6. Kuchling H. Handbook of physics: Trans. from German. 2nd ed. – Moscow: Mir, 1985.
7. Nechaev Yu. I. Catastrophe theory: a modern approach to decision-making. St. Petersburg: Art-Express, 2012.
8. Russian Maritime Register of Shipping. Rules for classification and construction of ships. Vol.1 / RMRS. – St. Petersburg: RMRS, 2013.
9. Tetsuya Yao, Yoichi Sumi, Hiroyasu Atsushi Kumano, Hidetoshi Sueoka, Hideami Ontsubo. Analysis of the accident of the MV Nakhodka. Part 2. Estimation of the structural strength // Marine Science and technology. Vol.3, No 4, Springer, 1998, p.181 – 193.
10. The bow: application №2013121521/11 Russian Federation: Int. Cl. B 63 B 1/06 / Burakovskiy P.E .; applicant "KSTU"; appl. 07.05.2013.
11. Voytkunskiy Ya.I., Faddeev Yu.I., Fedyaevskiy K.K. Hydromechanics. – Leningrad: Shipbuilding, 1982.

Safety of Dynamic Positioning

R. Gabruk & M. Tsymbal
Odesa National Maritime Academy, Odesa, Ukraine

ABSTRACT: This paper presents an innovation methodology for quantifying the safety of dynamic positioning in locally confined area of technological work under nonlinear dynamic influences of environmental factors. The methodology is based on integrated paradigm of prediction coordinated interactions of components which form dynamic positioning system. Proposed methodology implementation effect is based on the analysis of complex models that form the knowledge base. The comprehensive reserve of controllable thruster's reactions was adopted as quantitative characteristic of dynamic positioning safety.

1 INTRODUCTION

The intensive growth of mobile water transport objects (MWTO) fleet with dynamic positioning systems (DPS) displays a complex problem dedicated to safety of navigation. Effective implementation of high precision navigation processes interfere with environmental factors (EF): wind, current, waves (Fig. 1).

Figure 1. Environmental factors disturbances.

Wind, waves and current produce nonlinear external disturbances, which try to move MWTO outside of the locally confined area permitted boundaries. Environmental forces represent potential hazard to safety of navigation and technological work performance.

During dynamic positioning (DP), MWTO experiences motion in 6 degrees of freedom (DOF). The motion in the horizontal plane referred to as surge (longitudinal motion, usually superimposed on the steady propulsive motion) and sway (sideways motion). Heading or yaw (rotation about the vertical

axis z). The remaining three DOFs are roll (rotation about the longitudinal axis x), pitch (rotation about the transverse axis y), and heave (vertical motion). DPS implies stabilization of the surge, sway and yaw modes.

DPS may be defined as a system that automatically controls MWTO to maintain position and heading exclusively by means of active thrust. DPS also provides a manual joystick control that may be used for joystick control alone or in conjunction with a position measuring system for combined manual or auto control. Without a position measuring system, DPS can provide automatic stabilization and control of the MWTO heading using the gyrocompass as the heading reference. In addition to the standard operational modes and functions, various tailored functions are available to optimize MWTO operation for a wide range of technological works (offshore loading, trenching, dredging, drilling, pipe and cable laying, etc.).

DP-capable MWTO must have a combination of power, maneuverability, navigational ability and computer control in order to provide reliable positioning ability. This forms an integrated system, which consists of different elements.

Nonlinear forces of external disturbances play a major role. This forces (R_d) are generated by wind, current and waves and trying to move MWTO from holding position. DPS compensates these disturbances by means of controls that produce the necessary control forces R_c. Forces of controlled reactions should be equal or greater than forces of environmental disturbances.

$$\begin{cases} R_c \geq R_d, \\ M_c \geq M_d. \end{cases} \Rightarrow \begin{cases} \sum_{p=1}^{n} R_{Xcp} \geq \sum_{j=1}^{3} R_{Xdj}, \\ \sum_{p=1}^{n} R_{Ycp} \geq \sum_{j=1}^{3} R_{Ydj}, \\ \sum_{p=1}^{n} M_{XOYcp} \geq \sum_{j=1}^{3} M_{XOYdj}, \end{cases}$$

where R_c - force of controlled reaction; M_c - moment of controlled reaction; R_d - nonlinear force of EF disturbances (wind, current and waves forces); M_d – nonlinear moment of EF disturbance; R_{Xc}, R_{Yc}, R_{Xd}, R_{Yd} - projections of forces on the corresponding axis of MWTO coordinate system; p - number of MWTO thrusters; j – number of EF.

A comprehensive reserve of controlled thrusters reactions ΔR_c ($\Delta R_c = R_{c\ max} - R_c$) considered to be a quantitative characteristic of DP safety. The complex dynamic system of threshold type MWTO – EF functioning normally when the process of nonlinear disturbance and typical parameters of the system don`t extend beyond all established limits.

Presently safety of DP operations assessed by dynamic positioning operator (DPO) on the basis of Company Safety Management System (SMS) requirements, Failure Modes & Effects Analyses (FMEA) and using capability plot diagrams.

Capability plot diagrams (Fig. 2) commonly used for environmental forces affect assessment.

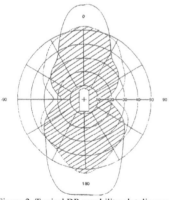

Figure 2. Typical DP capability plot diagram shape.

Capability plot diagrams provide DPO maximum safe acceptable wind speed with considered rotatable current in MWTO coordinate system. Almost all diagrams for self-propelled MWTO have large uninformative protrusions in sectors around 0 and 180 degrees (mostly more than 80 knots). Capability plot diagrams do not provide information about active thrust forces distribution under real environmental forces disturbances. All this lacks do not allow to form a clear picture for DPO.

The following methodology is proposed for guaranteeing safety of navigation during DP operations and DP process optimization, as well the verification of the decision validity at real risks of commercial MWTO exploitation.

2 DP PROCESS DECOMPOSITION

The main element of MWTO system is a Hull that has mass, hydrodynamic and aerodynamic characteristics (Fig. 3). It contains all other subsystems. The EF influence on the Hull subsystem. Therefore, the Hull is the central subsystem.

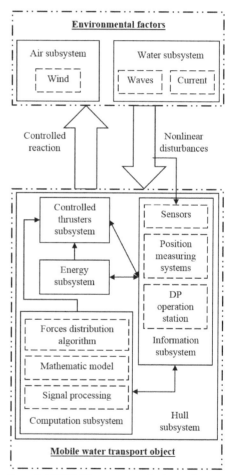

Figure 3. MWTO-EF systems complex interaction.

Energy subsystem acts as a resource of safety for the MWTO. Input and output data from different sensors, position measuring systems and MWTO subsystems form data flows. Information subsystem distributes data flows circulation during the interaction of DPS functional elements. As well, it provides user interface and operation station inputs.

Computation subsystem on the ground of processed data, which Information subsystem provides, computes signals that should compensate forces of external influences. Mathematic model is designed to compute the difference between the set point values and offset values of heading and position. Mathematic model issues forces to counteract offsetting forces to return MWTO to the set point heading and position. This special calculating module continually calculates MWTO response to EF nonlinear disturbances. Forces distribution algorithm calculates force for each thruster. Computation subsystem should be considered as an appropriate open system that connects data from the various sources and produces control signals for the Controlled thrusters subsystem.

Controlled thrusters subsystem produces appropriate control force vectors, which are necessary for the MWTO reaction to EF nonlinear disturbances. At the next decomposition level, each of these subsystems could be represented as composition of interacted subsystems.

EF system EF consists of Water and Air subsystems. On the fluidity property basis, it is possible to describe these subsystems as various models of fluid. Models of fluid motion reflect following subsystems: Wind, Waves and Current.

Mathematical models of interacting subsystems represented by equations of a rigid body motion in a fluid, equations of hydrodynamics and aerodynamics, equations of electric drives electrodynamics, equations of thruster's mechanics, equations, that describe processes in DP control systems.

3 METHODOLOGY

The proposed methodology is based on a new integrated paradigm of coordinated element actions to support decision-making process regarding DP safety in locally confined area of technological work (Fig. 4).

Data, that describe MWTO physical characteristics, enter at the beginning. These data could be entered manually or could be selected from the library of existing objects. Mathematical description of complex MWTO system on these data.

Step 1. MWTO characteristics determination. The determination of hydrodynamic and aerodynamic characteristics, added masses and moments of inertia are carried out by well-known methods for various types of MWTO standard hulls and superstructures.

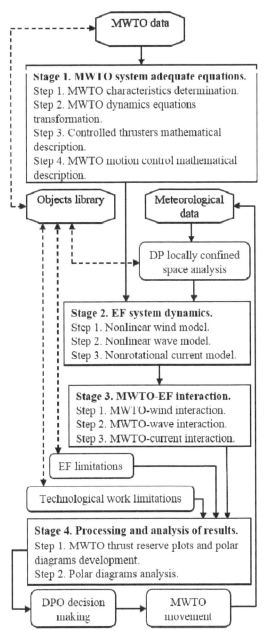

Figure 4. Information and logical model of methodology.

Step 2. MWTO dynamic equations transformation. It is possible to represent MWTO movement mathematical dynamics model in Cauchy form by one nonlinear differential general equation:

$$\frac{dv}{dt} = f(d,c,t),$$

where v- MWTO state vector; d - vector of disturbances; c - vector of controlled reactions; f - non-linear function; $t_0 \leq t \leq t_i$- simulation step interval.

The MWTO state vector general form includes 12 variables:

$$v_i = [x_{ig}, y_{ig}, z_{ig}, V_{ix}, V_{iy}, V_{iz}, \theta_i, \varphi_i, \psi_i, \omega_{ix}, \omega_{iy}, \omega_{iz}].$$

When analyzing MWTO motion it is convenient to define the following well-known coordinate systems.

The North-East-Down (NED) coordinate system is defined as relative to the Earth's reference ellipsoid (World Geodetic System 1984). It is defined as the tangent plane on the surface of the Earth moving with the MWTO, but with axes pointing in different directions in comparison with the body-fixed axes of the MWTO. For this system, the x-axis points towards true North, the y-axis points towards East while the z-axis points downwards normally to the Earth's surface. During MWTO DP operations in a local confined area, it is possible to consider that longitude and latitude are constant. Therefore, it is convenient to use for navigation an Earth-fixed tangent plane on the surface. This is usually referred to as flat Earth navigation and it will for simplicity be denoted as the NED frame. For flat Earth navigation, it is possible to assume that the NED frame is inertial such that Newton's laws still apply.

MWTO coordinate system (Fig. 1) is a moving coordinate frame, which is fixed to the MWTO. The MWTO orientation uniquely determined by the appropriate angles between axis of coordinate systems.

In MWTO complex mass center flat movement consideration, some state variables were excluded, because they are not controlled by DPS. These state variables are: complex mass center movement along the vertical axis of the MWTO coordinate system, projection of the linear speed on a vertical axis of the NED coordinate system, angular movements around the longitudinal and transverse axes of the MWTO coordinate system, angular speeds around the longitudinal and transverse axes of the NED coordinate system. Thus, MWTO state vector includes six variables:

$$v_i = [x_{ig}, y_{ig}, V_{ix}, V_{iy}, \varphi_i, \omega_{iz}].$$

DPS controls the movement of MWTO in the plane of the horizon and doesn`t control the vertical movement, rolling and pitching. This significantly simplifies the model without harming desired scientific results. A Vector of controlled reactions represents the projection of forces and moments generated by controlled thrusters. A vector of external disturbances represents the projection of EF forces and moments.

Step 3. Controlled thrusters mathematical description. During mathematical description of controlled thrusters subsystem following should be the considered: thrusters and hull interaction, forbidden zones formation for each thruster, influence of MWTO dynamic on the thrust effect, all possible DP thrusters allocation modes. Power resources, which are represented by Energy subsystem, should be simulated according to MWTO classification society approved Power Management Plan.

Step 4. MWTO motion control mathematical description. During this step, hierarchical performance of Computation and Information subsystems are described. Computation subsystem mathematical description involves development of station keeping and low speed maneuvering algorithms on MWTO mathematical model basis. Information subsystem provides data from different sources for these algorithms. This data include MWTO position, kinematics and dynamics, disturbance and reaction forces in different coordinate systems, power reserve, thruster allocation and set point, etc. Forces distribution algorithm computes necessary forces for each available thruster to perform DP operations.

Stage 2. EF system dynamics mathematical description is carried out on the basis of meteorological data and DP locally confined area analysis. At this stage, depending on the used method and accuracy requirements, the following should be determined: EF characteristics, character of hydrological features (Froude number, tide heights, etc.). The library of water areas is created on this stage. Systematic EF water area data include the following: general weather conditions in locally confined area of DP operations, characteristics of wind speed and its direction, characteristics of sea state (wave heights, swell direction), characteristics of current speed and its direction (peculiarities of changes).

Step 1. Nonlinear wind model. Wind is defined as the movement of air relative to the surface of the Earth. Mathematical models of wind forces and moments improve performance and robustness of the system in extreme conditions. The nature of the nonlinear component depends on the water area of MWTO operation. The appropriate spectral characteristics may also be used.

Step 2. Nonlinear wave model. The process of wave generation due to wind starts with small wavelets appearing on the water area surface. This increases the drag force, which in turn allows short waves to grow. Short waves continue to grow until they finally break and their energy is dissipated. A developing sea, or storm, starts with high frequencies creating a spectrum with a peak at a relative high frequency. A storm, which has lasted for a long time, creates a fully developed sea. After the wind has stopped, a low frequency decaying sea or swell is formed. These long waves form a wave spectrum with a low peak frequency. If the swell from one storm interacts with the waves from another storm, a wave spectrum with two peak frequencies may be observed. In addition, tidal waves will generate a peak at a low frequency. Hence, the resulting wave spectrum might be quite complicated in water areas, where the weather changes rapidly.

Step 3. Nonrotational current model. Current is defined as horizontal motion of water systems with a constant average speed. Vertical movement of water particles from one layer to another is not considered.

Stage 3. MWTO-EF interaction. Step 1. MWTO-wind interaction. Assumption of wind flow homogeneity and quasi-stationary properties plays a major role in the mathematical description of the MWTO reaction to wind disturbance. Calculation of wind forces and moments acting on MWTO results from relative wind speed and angle is done.

Step 2. MWTO-wave interaction. The interaction of MWTO hull with waves is a complex physical process. The mathematical description of the MWTO reaction on the wave disturbance requires consideration on regular and irregular waves. In the first case, the model will have deterministic nature, and in the second - a stochastic nature. Wave-induced forces and moment on MWTO are calculated using force transfer function. Wave height is a governing factor of this function. Research of the stochastic model is more complicated and time consuming, but results more accurately describe the reaction to the wave disturbance.

Step 3. MWTO-current interaction. The nature of the MWTO reaction depends on the current speed and direction. All hull protractions hydrodynamic effect is calculated on this step as well. Consideration of current speed changes, caused by hydrological features of the water area, improves the accuracy of results.

In order to receive parameters of the state vector to evaluate MWTO safety of DP operations, which undergo nonlinear EF disturbances, it is necessary to make description of the MWTO subsystems interaction, which are coordinated to solve the following tasks:

– Decomposition of complex systems and tasks into more simple (typical or standard).
– Relationship determination between selected components in the logic algorithms form.
– Distributed processing by Computation subsystem of various primary data from Information subsystem. During this process, navigation regarding MWTO dynamic positioning in locally confined space are formed and represented to DPO.
– Distributed secondary processing of aggregation or group data, which together reflect MWTO

state vector characteristics and EF disturbance main vector characteristics (course over ground, heading, speed, direction and force of disturbances).
– Determination of threat directions that form extreme situation concerning DP safety.
– Parameters identification (for components of Controlled thrusters and Energy subsystems), which are required for safe and effective DP process realization in locally confined area.
– Realization of identified parameters by DP process control laws.

Stage 4. Processing and analysis of results. Step 1. MWTO thrust reserve (load) plots and polar diagrams development. EF disturbance lead to DPS reaction, which through the forces distribution algorithm specifies load set points to controlled thrusters. The comprehensive reserve of controllable thruster's reactions represents a quantitative characteristic of MWTO DP safety. The curves of thrust reserve received for specific weather conditions, together with the assessment of the MWTO state vector provide an opportunity to assess the DP safety and execution of technological work expediency.

However, these curves describe only one possible MWTO heading. The picture of thrust reserve will be different if MWTO heading will change during DP. This happened due to changes in MWTO-EF interaction parameters. A similar situation observed when parameters of EF changed.

For the complex situation assessment, it is necessary to know how thrust reserve distributes on all possible MWTO headings. For this purpose, it is convenient to use polar diagrams that characterize spatial distributions of thrust reserve.

Step 2. Polar diagrams analysis. Polar diagrams represent thrust reserve (load) on MWTO possible headings range (360 degrees) and expand proposed methodology opportunities. Polar diagrams allow DPO to ensure MWTO safety of navigation during DP, establish safe abortion route in case of any emergency, optimize DP process and reduce costs of MWTO commercial exploitation. Identified limitations, imposed by the character of technological work and EF, have a great importance and should be reflected in the diagram.

Library formation of existing MWTO objects, limitations due to EF disturbance and technological work take place during methodology performance.

Decision making by the DPO is a crucial part of the methodology. To make proper decision DPO should consider results of solving the following tasks during limited time horizon:

– Assurance of DP process safety and optimization in locally confined area of technological work execution.
– Assessment of the current situation, which can rapidly change because of external disturbances

29

or because of MWTO factors (failure of DPS components, etc.).
- Situational decision-making, especially in the critical and extreme conditions of DP operations.
- Implementation of decisions taken according to identified resources and limitations of active control. That provided adaptation of the MWTO operative control synthesized laws to the practice of achieving targeted results.

4 IMPLEMENTATION

First author conducted implementation of proposed methodology during his work on board DP-1 vessel. This vessel is typical UT 733-2 project vessel, which is equipped with two bow thrusters (800 BHP each) and two azimuth stern thrusters (3600 BHP each). Vessel performed DP in locally confined area (Fig. 5) during supply operations of "Sea Island" offshore facility (Persian Gulf).

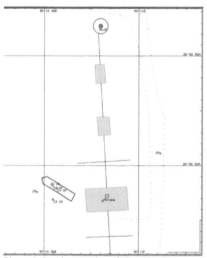

Figure 5. Supply operation of "Sea Island".

The following weather conditions were observed in the locally confined area of technological work: wind direction 25°, speed 11 knots; current direction 50°, speed 0.9 knots; wave direction 300° degrees, observed significant wave height 1 m.

Company SMS requirement: 80% maximum thruster load. Following polar diagram of thruster loads was calculated (Fig. 6). On the diagram, which referred to NED coordinate system, loads of the bow thrusters and azimuth stern thrusters were represented as MWTO DPS reactions on concrete EF disturbances. Reference to NED coordinate system allows DPO to plan easier MWTO movements and abortion route. This diagram could be attached to navigation chart or put near ECDIS screen for quick reference.

Loads greater than 80% form "EF limitation" sectors. These limitations correspond to currently observed weather conditions. If the EF values change, the polar diagram will also change.

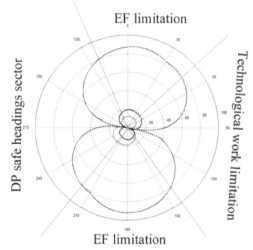

Figure 6. Polar diagram.

MWTO's cargo and cargo hoses connections are located on the stern of the vessel. It means that during supply operations the vessel should be parallel or stern to the offshore facility. This condition superimposes with the "EF limitation" sectors and forms "Technological work limitation" sector.

The remaining sector forms "DP safe headings sector". All headings within this sector are safe and allow performing supply operations. 300° heading was selected as final practical heading to conduct DP.

Safety of DP process is guarantee according to the proposed methodology by using researched analytical regularities of interdependence of parallel processes of discrete navigation and continuous DPS control of MWTO state vector under nonlinear EF disturbances.

5 CONCLUSIONS

The proposed structuration of ergatic interaction provides adaptation processes of high performance and operational problems solving, which arise during DP operations in extreme situations where the time factor is crucial.

The formed knowledge base allows DPO to ensure MWTO DP safety in various conditions. Introduced polar diagrams referred to NED coordinate system, have an adaptive and dynamic nature and favorably differ from known capability plot diagrams, which are static and referred to MWTO coordinate system.

REFERENCES

Bray, D. J. 1999. *Dynamic Positioning Operator Training. The official guide to The Nautical Institute training standards.* 2nd edition. London: The Nautical Institute.

Fossen, T. I. 2002. *Marine Control Systems.* Trondheim: Norwegian University of Science and Technology.

IMCA M 140. 2000. *Specification for DP Capability Plots.* London: IMCA.

IMCA M 178. 2005. *FMEA Management Guide.* London: IMCA.

IMCA M 117. 2006. *The Training and experience of Key DP Personnel.* London: IMCA.

IMCA M 182. 2006. *International Guidelines for the Safe Operation of Dynamically Positioned Offshore Supply Vessels.* London: IMCA.

IMCA M 103. 2007. *Guidelines for the Design and Operation of Dynamically Positioned Vessels.* London: IMCA.

IMCA M 190. 2011. *Guidance for Developing and Conducting Annual DP Trials Programmes for DP Vessels.* London: IMCA.

IMO MSC Circular 645. 1994. *Guidelines for vessels with dynamic positioning systems.* London: IMO.

Morgan, Dr. M. J. 1978. *Dynamic Positioning of Offshore Vessels.* The Petroleum Publishing Company. Tulsa, Oklahoma.

Sizov, V.G. 2003. *Ship theory.* Odessa: Fenix.

UK Offshore Operators Association/Chamber of Shipping. 2002. *Safe Management and Operation of Offshore Support Vessel.* London.

New Layout of Mrzeżyno Port Entrance Design - Results by Real Time Simulation Study

A. Bąk & L. Gucma

Maritime University of Szczecin, Szczecin, Poland

ABSTRACT: Construction of new ports requires applying the entire range of methods and the close cooperation between designers and specialists of different field including safety of navigation or oceanography. At present there is no possible to conduct of such investments without carrying out simulation examinations. Such studies most often take place as multistage where the next stage is a result previous one. The paper presents complex method of water areas optimization with consideration of navigational safety. The real time simulation method was implemented in presented study. Methods of the simulation of the real time leaning against manoeuvre simulators were applied. Two alternatives differing between oneself were examined. The best one under safety of navigation consideration was chosen as final. The results were used as design guidelines for small Polish sea Mrzeżyno Port development.

1 INTRODUCTION

The Marine Traffic Engineering (MTE) research team of Maritime University of Szczecin (MUS) since 70-ties is engaged in research works concerned with evaluation of navigation safety for port design and optimization of water areas [Gucma 2012, 2009]. The researches described in this paper are focused on Mrzeżyno Port modernisation which is one of the most important research studies of MTE team deled with complex sea port planning and design [Computer simulation 2012]. The main objective of researches was concerned with realisation of following tasks [PIANC 2014]:

1 Determination of optimal parameters of:
   - new sea port in Mrzeżyno area with respect to shape, width and depth,
   - outer and inner port breakwaters with respect to its shape taking into consideration wave height in port;
   - turning places with respect to its shape and optimal depth;
   - the berthing places in inner port in respect to its shape, length, depth, maximal energy of ships contact, maximal speed of ships propeller and bowthruster streams on the bottom.
2 Determination of safety conditions of port operation in respect to:
   - admissible meteorological conditions for given kind of ships and manoeuvres;
   - other navigational conditions and limitations like presence of other ships on berths, use of position fixing systems on approach, navigational markings, vessel traffic service.
3 Determination of manoeuvring procedures during berthing and unberthing for different kind of ships and propulsion systems.
4 Determination of underkeel clearance by Monte Carlo method.
5 Determination of usage of main engine during entrance.
6 Determination of researches ships distances to the most dangerous objects.
7 Carrying out most typical emergency runs (typical failures on entrance) and describe necessary emergency action for the captains.

Mrzeżyno Port is former military port, now mostly concentrated on fishing and recreational aspects with some ambitions of cargo and passenger port at west Polish (Fig. 1).

As a characteristic ship (m/f *Design*) for new Mrzeżyno Port design in presented study according to investors economic analysis and their needs the typical Baltic Sea small passenger ship of length L=60m (fig. 2) has been chosen. The essential parameters of design ship are presented in the Table 1.

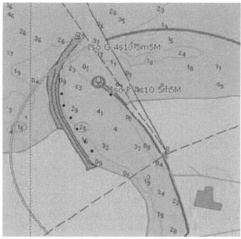

Figure 1. Mrzeżyno Port. [S-57 navigational chart]

Table 1. Main parameters of design passenger ship (L= 60m) operated on the Baltic Sea area [Computer simulation (2012)]

Parameter	m/f Design
Length - LOA	60 m
Breadth	12 m
Draft	2.5 m
Machinery	total 2x1.200kW at 900rpm.
Propeller	1 variable pitch propeller
Propeller	diameter=1.83m at rpm 360
Speed	approx. 18 kn. (at 90%)
Rudder	1x35 deg. / area=2.7m^2 /conventional
Bowthruster	150kW
Lateral wind area	approx. 500m2

Figure 2. General arrangement of m/f Design (based on plans of m/f Adler Dania) [www.adler-schiffe.de]

2 THE DESIGN OF BREAKWATERS MRZEŻYNO PORT ASSUMPTIONS

The most important aim of Mrzeżyno Port design is to provide access to the port by passenger ships up to 60m length and enable future port development in respect to cargo and fishing activities [Computer Simulation 2012]. Nowadays due to wrong breakwater layout the silting process is very rapid and bottom reveals considerable changes which is dangerous for navigation. The main design restrictions are as follows (Fig. 3):
1 to design additional pair of outer breakwaters (for better wave dumping) well sheltered from extreme weather conditions,
2 to design turning place inside the port,
3 to test and validate at minimum 2 alternative port designs made by two independent design teams.

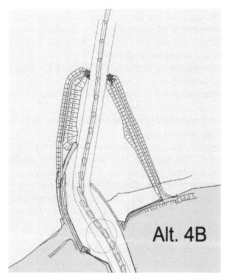
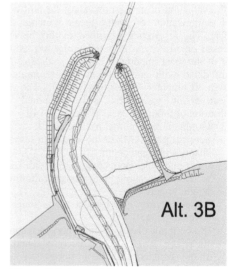

Figure 3. Two design alternatives at Mrzeżyno Port (Var. no 4B – left and Var. no 3 design – right) [Computer simulation (2012)]

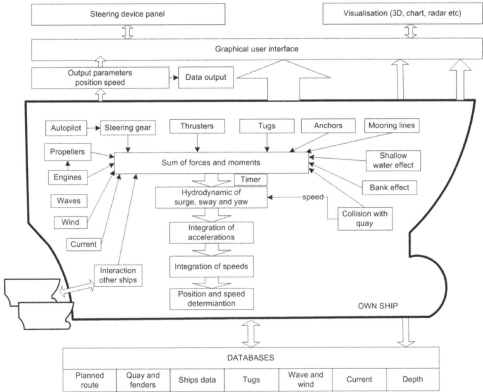

Figure 4. The main functional diagram of simulation model [Gucma L. 2005]

3 METHODS APPLIED

The real time simulation interactive method with captains and pilots engaged in ships manoeuvring trials was applied. This method is assumed as most reliable and suitable in this kind of research studies [Gucma L. 2005]. MTE research team possess several kinds of manoeuvring simulators: form limited task with 2D display to modern full mission simulator with 3D and real control systems.

3.1 Real time simulation method – limited task simulator

Two classes of hydrodynamic models in MTE team own limited tasks simulators are utilized. First class of models are used when only limited parameters are known (usually when non existing ships or general class of ships are modelled) the second class models are used when exact and detailed characteristics of hulls, propellers and steering devices are known. Additionally real manoeuvring characteristics are used for validation of models. The model of m/f *Design* used in researches is based on modular methodology where all influences like hull hydrodynamic forces, propeller drag and steering equipment forces and given external influences are

modelled as separate forces and at the end summed as perpendicular, parallel and rotational ones.

The model is operating in the loop where the input variables are calculated instantly (settings and disturbances) as the forces and moments acting on the hull and momentary accelerations are evaluated and speeds of movement surge, sway and yaw. The most important forces acting on the model are:

1. thrust of propellers;
2. side force of propellers;
3. sway and resistant force of propellers;
4. bow and stern thrusters forces;
5. current;
6. wind;
7. ice effects (neglected);
8. moment and force of bank effect (neglected);
9. shallow water forces;
10. mooring and anchor forces (neglected);
11. reaction of the fenders and friction between fender and ships hull;
12. tugs forces (neglected);
13. other depending of special characteristics of power and steering ships equipment.

The functional idea of the ship manoeuvring simulation model is presented in Figure 4.

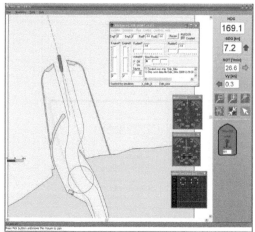

Figure 5. Interface of simulation model m/f *Design* turning at Mrzeżyno port (limited task simulator, alternative 2)

Interface of model is typical 2D nautical chart – like interface (Fig. 5). The interface covers information of ships state (position, course speed, yaw etc), quay and shore line location, navigational markings, soundings, external conditions, tug and line control and control elements of the model. The model is implemented in Object Pascal with use of Delphi™ environment and Visual C™ with use of C++ language.

Limiting to the usual 3DOFs (the horizontal planar motion), the ship movement over the ground (thus the so-called dynamic effect of the water current is introduced) is given by [Artyszuk 2005]:

$$\begin{cases}(m+m_{11})\dfrac{dv_x^g}{dt}=(m+c_m m_{22})v_y^g\omega_z+(m_{11}-c_m m_{22})v_y^c\omega_z+F_x\\[2mm](m+m_{22})\dfrac{dv_y^g}{dt}=-(m+m_{11})v_x^g\omega_z+(m_{11}-m_{22})v_x^c\omega_z+F_y\\[2mm](J_z+m_{66})\dfrac{d\omega_z}{dt}=-(m_{22}-m_{11})(v_x^g-v_x^c)(v_y^g-v_y^c)+M_x\end{cases} \quad (1)$$

$$\dfrac{dx_0}{dt}=v_{NS}^g,\quad \dfrac{dy_0}{dt}=v_{EW}^g,\quad \dfrac{d\psi}{dt}=\omega_z \quad (2)$$

$$\begin{bmatrix}v_{NS}^g\\v_{EW}^g\end{bmatrix}=\begin{bmatrix}\cos\psi & -\sin\psi\\\sin\psi & \cos\psi\end{bmatrix}\cdot\begin{bmatrix}v_x^g\\v_y^g\end{bmatrix} \quad (3)$$

where:

v_x^g, v_y^g, ω_z = ship surge, sway and yaw velocity over the ground, x_0, y_0, ψ = position Cartesian coordinates and heading, m = ship mass, m_{11}, m_{22}, m_{66} = added masses, c_m = empirical factor, F_x, F_y, M_z = external excitations (resultant/total surge, sway force and yaw moment), generally consisting of the following items (denoted by additional subscripts) and being generally the functions of ship speed through the water ('v_w'):

$$\begin{cases}F_x=F_x\left(v_x^w,v_y^w,\omega_z\right)\\[1mm]F_y=F_y\left(v_x^w,v_y^w,\omega_z\right)\\[1mm]M_z=M_z\left(v_x^w,v_y^w,\omega_z\right)\end{cases} \quad (4)$$

$$v_x^w=v_x^g-v_x^c,\quad v_y^w=v_y^g-v_y^c \quad (5)$$

$$\begin{bmatrix}v_x^c\\v_y^c\end{bmatrix}=\begin{bmatrix}\cos\psi & \sin\psi\\-\sin\psi & \cos\psi\end{bmatrix}\cdot\begin{bmatrix}|\vec{v}^c|\cos\gamma_c\\|\vec{v}^c|\sin\gamma_c\end{bmatrix} \quad (6)$$

where: $|\vec{v}^c|$ and γ_c represent the velocity and geographical direction of the water current (a uniform current by default).

4 STATISTICAL METHODS OF DATA PROCESSING

Ship simulators are very widely used today. The hydrodynamic models are becoming more and more reliable. Without efficient statistical data processing it is not possible however to draw conclusions from the conducted experiments. Usually different kind of data processing analysis is applied in case when horizontal and vertical ships movement is considered.

4.1 *Safe manoeuvring areas – method of simulation result data processing*

The most important factor is safety horizontal area needed for navigators for performing manoeuvres [Gucma 2005, 2012, Irribaren 1999]. The assumption of such model is that the ship moves along predefined route x (Figure 6.a) with following probability of accident:

$$P_{AW}=P_{SA/A}P(Y\geq y_{MAX})=P_{SA/A}\int_{y_{MAX}}^{+\infty}f(y)dy$$

where: $P_{SA/A}$ = conditional probability of serious accident, $f(y)$ = the distribution of ships position, y_{MAX} = distance from to the centre of the waterway (route) to the waterway border.

Probability of serious accident $P_{SA/A}$ could be defined by the Heinrich ratio or more detailed consequence analysis. One of the most important stages of accident probability evaluation is statistical analysis of the results. The probabilistic concept of safety manoeuvring area is presented in Figure 6.a. The distributions are strongly dependant of waterway area arrangement and could be evaluated in simulations and validated in real experimentations.

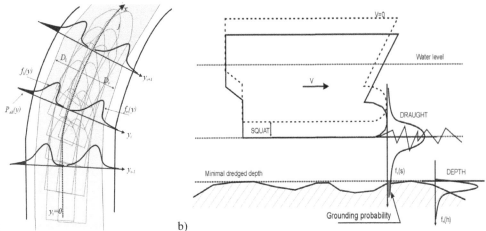

a)   b)

Figure 6. Probabilistic concept of safe manoeuvring area determination on the waterway (a) and underkeel clearance of ships determination (b)

Table 2. The detailed plan of simulations

No.	Breakwater design alternative	Manoeuvre	Wind [m/s (B)]	Wave [m]	Current [kn]
1	4B	Entrance and turn	0	0	0
2			NW 11m/s (6B)	0,9m	out 1w
3			NW 17m/s (8B)	1,6m	aa.
4			NE 11m/s	0,9m	aa.
5			NE 17m/s	1,6m	aa.
6	3	Entrance and turn	0	0	0
7			NW 11m/s (6B)	0,9m	out 1w
8			NW 17m/s (8B)	1,6m	aa.
9			NE 11m/s	0,9m	a.
10			NE 17m/s	1,6m	aa.

5 RESEARCH PLAN

The following conditions for manoeuvring have been considered:
- zero conditions – for validation and comparing of manoeuvring areas;
- wind 11m/s (lower limit of 6°B);
- wind 17m/s (lower limit of 8°B).
  Two directions of wind have been considered:
- NW as navigationally most difficult for port entrance.
- NE as problematic wind form stern directions that could course the problems with enter due to moment of inertia on ships in stern area.
  Height of significant wave is according to wind and typical wind-wave conditions on the Baltic are presented in Table 2.

5.1 Simulation series

There have been conducted 10 simulation series each for 14 to 16 ship passages (entrances only). Detailed plan of simulation passages is presented in Table 2.

6 RESULTS OF RESEARCHES

All the simulation trials have been conducted by skilled captains and pilots having experience in this kind of ships and manoeuvres. The simulation data have been recorded and analyzed. Analysis of simulation results was made in basis of following criterions:
1 ship manoeuvring lanes widths (horizontal safe manoeuvring area dimension),
2 under keel clearance (Monte Carlo method),
3 energy induced in contact point with berth structures,
4 velocities of propeller bottom stream,
5 engine and rudder settings,
6 probabilities of collision with given points,
7 time of manoeuvre,
8 emergency manoeuvres.
  In this paper only the results dealing with the first criterion are presented. The safe manoeuvring areas on 95% level of confidence are widely used in analysis [Gucma, 2005]. The results from collective series (all simulation trials in all conditions) are presented in Figure 7. Additionally maximum area

as the area of all ferry passages in given simulation series is presented.

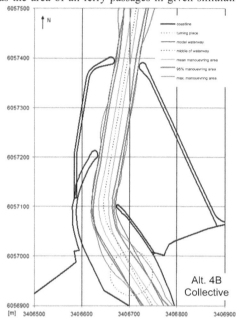

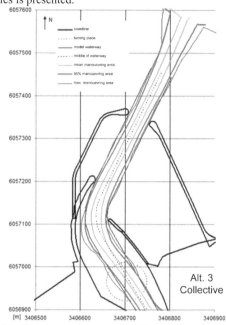

Figure 7. Manoeuvring areas of m/f *Design* ferry during entrance with for all ship passages – comparing 4B an 3 alternative. The 95% confidence level presented in red (mean and maximal area are presented blue and magenta).

The wave analysis have been made for all alternative solutions. The results for significant wave height of 2m from direction NNE are shown in Figure 8. It show well damping factor of waves by breakwater structure and outer port.

wind with wave height equals 1.8m outside the port.

7 CONCLUSIONS

Presented complex study could be used for guidelines for port design and operational weather limits estimation of port and breakwaters designs in navigational aspects. The design study is multistage process considering

Conclusions related to conditions of safe manoeuvring passenger ship m/f *Design* in Mrzeżyno Port might be stated as follows:

1 The alternative 4B is much better under safety of navigation consideration in compare to alternative 3 so the alternative 4B should be taken into consideration for further development.

2 Some minor corrections (up to 5m shape changes of breakwater) are necessary to provide in alternative 4B (Fig. 9).

3 The alternative B could used as a the final stage of port design of Mrzeżynono Port in scope of navigational safety with taking into account wave height inside the inner port.

4 The acceptable conditions of wind for safe entrance of L=60m length ferry is NNE 17m/s

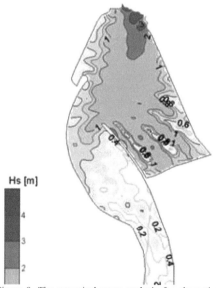

Figure 8. The numerical wave analysis for alternative no. 3 (waves system in port significant wave height outside the breakwaters NNE 2.9m and period of 8.2s) [Study 2011]

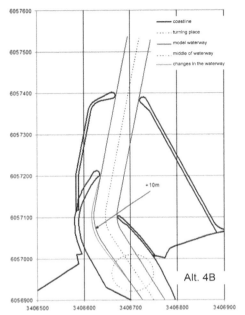

Figure 9 Proposed minor changes after simulation researches in alternative 4B

REFERENCES

Artyszuk J. 2005. Towards a Scaled Manoeuvring Mathematical Model for a Ship of Arbitrary Size. Scientific Bulletin, Szczecin: Maritime University of Szczecin.

Computer simulation for new port breakwater layout in Mrzeżyno 2012. Unpublished report. Maritime University of Szczecin.

Gucma L. 2005. Risk Modelling of Ship Collisions Factors with Fixed Port and Offshore Structures (in Polish) Szczecin: Maritime University of Szczecin.

Gucma L. 2009. Wytyczne do zarządzania ryzykiem morskim (Guidelines for maritime risk assessment). Szczecin: Wydawnictwo Akademii Morskiej w Szczecinie.

Gucma L. 2012. Zarządzanie ryzykiem w rejonie mostów usytuowanych nad drogami wodnymi w aspekcie zderzenia z jednostkami pływającymi (Risk assessment in bridge area in respect to ship–bridge collision). Szczecin: Wydawnictwo Akademii Morskiej w Szczecinie.

Iribarren J.R. 1999. Determining the horizontal dimensions of ship manoeuvring areas. PIANC Bulletin No 100, Bruxelles.

PIANC 2014. Harbour Approach Channels Design Guidelines. PIANC Report PIANC Secretariat General. Bruxelles.

Study 2012. Wave analysis study in Mrzeżyno Port. Bimor. Szczecin.

Safety of Offshore Supply Operations

J. Pawelski

Gdynia Maritime University, Gdynia, Poland

ABSTRACT: Offshore marine operations are conducted in all climatic zones, including stormy and polar areas. All units engaged in such operations require continuous supply of fuel, construction, drilling equipment and provision. Supply operations must be carried out even in very difficult conditions but safety of personnel and environment is given the highest priority. It can be only achieved by setting appropriate safety rules.

1 INTRODUCTION

Offshore petroleum industry at each field life cycle requires vast amount of cargo being delivered on regular basis. Initially, main volume of transported goods are: drilling equipment, construction elements and large quantities of gas oil as a fuel. Production phase of the field is less demanding on supplies and mainly consist of productions chemicals, spare parts, provision and waste skips. Waste is taken by supply vessels on their trip back to shore base. According to current legislation, offshore installations cannot dump any garbage into water except food, commuted to 25 mm and only if they are further than 12 Nm from shore. At this stage natural gas is chiefly used as a fuel at most of the installations. Abandoning field involves simultaneous work of drilling rig and several other vessels. Again much higher volume of cargoes is needed from shore to the field and back. Supply vessels have to make more frequent trips with materials, fuel and equipment being sent back. Most of the local regulations call for complete removal of field infrastructure, including subsea installations. Picture below shows oil containment operation involving simultaneous work of several offshore units:

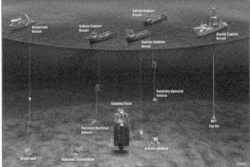

Figure 1. Oil containment operation [4].

There are two kinds of vessels used for offshore supply operations. Both of them have length limited to 70 – 80 meters for good manoeuvrability and are equipped with dynamic positioning system (DP) IMO class 2. First group consist of multipurpose anchor handling, tow and supply vessels (AHTS), but their versatility means smaller cargo capacity and they are expensive to run. Real workhorse of offshore operations is purpose designed platform supply vessel (PSV). Her construction is optimized mainly for carriage of bulk and packaged cargoes at reasonable fuel and hiring costs. Typical modern PSV is pictured below:

Figure 2. Typical modern PSV [4]

Figure 3. Excerpt from offshore installation data card [1].

Their construction is under continuous development including: wider use of liquefied natural gas as fuel for economical and environmental purposes, better propulsion systems and other improvements. There is also class of supply vessel specifically designed for polar operations. Very often offshore logistic operations are carried concurrently out with other operations in the field and they are included in simultaneous operations document (SIMOPS) for safety purposes. Today offshore operations are conducted in all climatic zones including tropics and ice covered polar waters.

2 VOYAGE PLANNING

Voyage planning of supply vessel comprises several consecutive steps, critical for safe and efficient operations. Voyage plan is always being agreed upon before commencement of loading at shore base. Planning starts with vessel routeing which consists of sequence of calls at installations, communications lines and reporting, advice on estimated time of arrival (ETA), estimated time of departure (ETD) and required sailing speed (economic or best speed). Logistic or operating company provides latest weather forecast for intended destinations. Important part of information available for master is installation and port data card. It provides very detailed information regarding berthing places, contact details, dangers to navigation around offshore installation, which are unavailable on official charts. Additional information given there are cargo procedures specific for particular installation and their communication details. Excerpt from installation data card is shown below:

Voyage planning also takes under consideration local weather conditions so planned route avoids heavy weather areas. Offshore installations are closely monitoring local weather and sea conditions and vessel is well advised regarding local weather. Unfavourable weather conditions for cargo operations may delay PSV arrival which in such case usually slows down or takes a longer route.

3 ENTERING SAFETY ZONE AND SAFE WORK CONDITIONS

Offshore installations are protected by establishment of 500 meters safety zone around fixed structure, floating unit or vessel. Prior to entry vessel is required to fill out approved checklist and obtain permission from offshore installation manager (OIM) or control room operator. Before entry into 500 meters zone, vessel dynamic positioning system (DP) needs to be fully operational and checked. Most of offshore procedures call for vessel approach at a tangent at safety zone as shown on picture below:

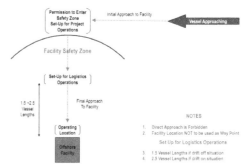
Figure 4. Approach to Facility [2]

At certain distance from destination point vessel is required to set up on final course and verify ability to hold position in conditions near installation. For logistic operations supply vessel should set up in distance not less 1.5 ship's length for drift-off situation and not less than 2.5 ship's length for drift-on circumstances. Usually time 10-15 minutes is given before moving into working position. This time allows DP operator to asses fully environmental conditions and to ensure that all station keeping arrangements are stable. Final approach is made with speed over ground not exceeding 0.5 knots. All operations are documented by filling out pre-entry and pre-approach check lists.

4 ADVERSE WEATHER WORKING

Offshore industry established certain criteria, called trigger points, when certain precautions should betaken under consideration. Typical trigger points are shown below in Table 1:

Table 1. Trigger points [3]

Trigger	Precaution
Thruster and Propulsion Utilization	Typically, where a vessel is required to take up and maintain station close to a facility the continuous power utilisation of any manoeuvring thruster (including main propulsion) must not exceed 45% of the available power. Where a vessel has been fitted with a consequence analyser, the DP Operator must follow warnings from this system.
Unfavourable Wind Direction 20 knots mean wind speed at 10m level. 25 knots mean wind speed at 10m level.	No installation overboard venting or discharges whilst working supply vessels, unless previously agreed with vessel Master. Master may cease operations if Safe operation within the Safety Zone is compromised due to overboard venting or discharges. Secure loose items and advise greater caution to prevent injury to personnel and damage to equipment. Consideration must be given to ceasing operations. Master, OIM and Crane Operator should evaluate the weather conditions and forecast. If necessary, a risk assessment should be carried out before commencing / continuing the operation. Consider vessel motion, possible injury to crew and potential cargo
Sea State 3m – 4m Significant Wave Height	Master, OIM and Crane Operator should asses situation on positioning and cargo handling before arrival within safety zone. Account for vessel motion, hose work, any awkward lifts, potential cargo damage due to heave and potential effect on hose work.
Tidal Streams Strong Currents or Tides	Consider delaying cargo operations, especially hose work, until slack tides if vessel cannot hold station satisfactorily against tide.
Visibility On Approach to Installation	Remain outside safety zone to avoid collision with installation or another vessel. Maintain radar watch.
Visibility <250m During Operations Poor Visibility	Cease cargo operations if crane operator is unable to see vessel deck crew clearly.

Vessel and equipment Vessel moving violently. Forecast for adverse weather	Master may cease operations if vessel movement starts to affect vessel station keeping or crew safety. Consider making for sheltered waters or port to avoid risk to personnel to equipment or to cargo. Such considerations must take into account the time taken to reach sheltered waters or port.

Vessel takes working position on lee side of installation whenever it is possible. If vessel is working on weather side of installation both parties should provide written risk assessment covering listed requirements.

5 CONTEINERIZED CARGOES

Most of goods are shipped to offshore installations in 10 ft. and 20 ft. offshore containers. Their construction is very similar to typical cargo containers but they are more sturdy so can withstand more severe impacts. Due to smaller size there is no need to use lifting frame and 4-legged certified slings are used for handling of such containers in offshore supply operations. All containers are certified by classification society for offshore use and are periodically inspected and re-certified. Container with expired certification date cannot be loaded and must be returned empty to shore base. Example of offshore container is shown below:

Figure 5. 10 ft. offshore container [5]

Containers can only be carried of deck in single layer only. For storage on shore two layers are allowed. Drilling equipment and loose items are sent offshore in 20 ft. long steel baskets. Vessel cargo plan contains information regarding weight of each item loaded on deck. For loads up to 4 ton offshore crane can use single fast line at maximum range. Heavier containers require usage of crane main tackle. I such situation cargo operations are more difficult due to slowly moving cargo hook and limited safe operational range of crane. Containers are loaded on deck of supply vessel in groups, occupying designated bays. Each group is destined for separate installation. Vessel is trying to keep about 10% of her deck free to allow first some back loading from installation to free up some storage space. Anchor handling tow and supply vessels (AHTS) have steel cladding at part of he deck. To prevent uncontrolled movement of stored containers it is recommended to use bedding ropes underneath deck cargo. Very serious safety issue regarding deck cargo operations is so called "Cherry Picking". It can be defined as being "selective discharge of cargo from within the stow".

The term "Cherry Picking" includes:
– Cargo lifting arrangements not being directly accessible from deck level.
– Breaking stow from an open location with no clear and secure access / escape routes to adjacent safe havens.
– Any requirement for personnel to use unsecured ladders or to climb on top of other cargo or ship's structure and to enter any container to connect lifting arrangements is prohibited at all times.

Supply vessel master is fully authorized to refuse any such request and to stop cargo operations until safe handling routines are re-established.

Common threats for crew working on deck of supply vessel are objects falling from containers. Objects which constitute such risk mostly include:
– Left over loose tools for servicing equipment.
– Foreign objects in container fork lift pockets.
– Ice formed when water entrained in a cargo item freezes.

Where is practical and safe to do so deck cargo should thoroughly inspected. During cargo lifting all crew should stay well clear in safe havens. Cargo items with such objects identified need to be quarantined until unwanted items are removed or full risk assessment is provided if they cannot be removed.

6 TUBULAR CARGOES

This type of cargo is frequently carried on deck of offshore supply vessels to and from installations. Drilling and field development need large quantities of marine risers, drill pipes and well casings. Marine pipe laying operations solely rely on continuous supply of pipe sections. Smaller diameter tubes with length up to 6 meters are handled in offshore baskets for faster and safer operations. Larger equipment is carried on deck, sometimes as full deck cargo in case of large lots. Properly stowed pipes, as pictured below, are stable during sea passage and can be easily offloaded with proper lifting equipment at destination point:

Figure 6. Marine riser full deck cargo [3]

More difficult is carriage of only few round items on deck. Traditional way of securing included wooden wedges nailed to the deck and lashings made of wire ropes and bulldog clips. Another way of securing were pipe stanchions. Today, some modern PSV are fitted with Automatic Sea Fastening Arrangement (ASFA) as picture below:

Figure 7. PSV with Automatic Sea Fastening Arrangement [3].

Introduction of automatic system improved both crew safety and time efficiency of cargo handling operations. When cargo lot contains piping with different diameters it is advised to load larger piping first on deck to create stable base. Smaller pipes can be loaded on top. Pre-slung pipe bundles must have lifting gear extended and laid across the pipes to avoid become wedged between piping. During sea passage in heavy weather water may enter piping through protective caps and reduce vessel stability.

7 BULK CARGOES OPERATIONS

Offshore supply vessels are designed for carriage relatively large quantities of bulk cargoes either in liquid or dry powder conditions. Liquid cargoes include light gas oil, potable water, water or oil based drilling muds, base oils, brines and chemicals. Dry bulk cargo consists of cement, barite and bentonite. All bulk cargoes are transferred by hoses.

Liquids are pumped but bulk powders are discharged with compressed air. Some of bulk cargoes are rated as flammable liquids, toxic or corrosive substances and marine pollutants. All drilling muds and brines permitted to carried in bulk are to be included in vessel International Noxious Liquid Substances Certificate (INLS). They can be carried only in designated tanks listed in INLS and not used for other purposes. Even best maintained cargo handling equipment may unexpectedly spring a leak. Such risk is addressed in cargo handling procedures based on risk assessment and in "Shipboard Marine Pollution Emergency Plan" (SMPEP). During cargo transfer all pollution prevention equipment is placed, as per SMPEP and fire fighting is ready immediate for deployment. When dealing with corrosive and toxic liquids crew is provided with proper personal protective equipment (PPE). Transfer operation of such substances requires proper risk analysis and issuance of permit to work (PTW). Before beginning of transfer vessel and receiving facility must agree on sequence of transfer and complete required checklists for pre-commencement preparations. All hazardous cargoes must be accompanied with material data sheets (MSDS) and crew familiarized with associated hazards. Hoses used for cargo transfer should be sufficient length and slung in such way to enable hose to be landed on deck and crane pendant slackened before hose is secured and pendant disconnected. All the time hoses must remain afloat with aid of suitable flotation devices. It is recommended to provide hose string with self-sealing weak ling couplings. The function of these items is to avoid over-tensioning or even breaking the hose and therefore having to discard the complete length of hose string. Spills from this type of assembly are avoided by the self-sealing action of the coupling.

7.1 *Liquid bulk cargoes – common liquids*

Offshore industry uses term common liquids for fuel and potable water. Fuel supplied to offshore units is light gas oil for Diesel engines. Very few installations, like some FPSO's, require heavy fuel oil for their boilers. Offshore supply vessels are not designed to carry such kind of fuel in heated tanks. Heavy fuel oil is available only from bunker barges. Transfer of gas oil from PSV to installation is a routine operation like ordinary bunkering. During transfer samples are taken as required by Convention MARPOL Annex VI for fuel quality assurance. Watchman on deck should always have unobstructed at fuel hose. Receiving installation cannot shut valve against working pump. After completion of bunkering hose is drained dry by gravity with help of installation crane. Blowing hose with compressed air is not recommended since an increased risk of explosion will result. Hydrocarbon transfer operations must be stopped during electric storms.

Water supply is the safest bulk supply operation but it must comply with certain hygienic standards. Potable water tanks must be solely used for this purpose, maintained under vessel's planned maintenance plan and audited under International Safety Management (ISM). Pumps and pipework should be flushed with super chlorinated water and than with fresh potable water prior refilling tanks. Hoses used for water transfer should be dedicated for this purpose and cleaned with chlorinated water on weekly basis. Each year hose need to be replaced. Water in potable water tanks of PSV typically may be stored for 7 days. After this time water is transferred to drill water tanks. Specific rules regarding water quality management usually are given in national regulations.

7.2 *Dry bulk cargoes*

Bulk dry cargoes handling is potentially dangerous operation due to the way this kind of materials are being handled. This must be done in controlled manner to minimize risk of serious accident. Dry powder is moved with help of compressed air and whole system during operation remains pressurized. Air supply for cargo system comes only from pressure regulated source. Failure of large volume tank at elevated pressure may have catastrophic effect. Before setting up system for cargo handling all components must be bled to atmospheric pressure before any joint is open. Handling of dry bulk materials involves systems containing large volume of compressed air. The stored energy in such system is sufficient for serious personal injury. Before operation all piping and hoses are blown with air to make sure they not clogged. When staring operations special care should be taken to sequences and manner in which valves are opened to avoid the risk of inadvertently over-pressurizing any element of the system.

7.3 *Bulk transfer of particular concern*

Special precautions should be exercised when transferring special products which include but not limited to zinc bromide aqueous solution and methanol.

Zinc bromide brine is heavy liquid used for displacement of drilling mud during well completion. It is highly corrosive and environmentally contaminating substance. There is no risk of fire as it is non-flammable liquid but subjected to high temperature may form harmful compounds. Carriage in bulk requires INLS certificate. Due to highly corrosive properties it is essential to protect crew against injury from exposure. All personnel handling hoses should be provided with rubber boots, nitrile heavy duty gloves, chemical suit and face protection. Tank washings are considered special waste. Tanks cleaning and washings discharge must be monitored by approved surveyor.

Methanol is highly flammable liquid with flashpoint at 11°C. Offshore installation are using it as natural gas hydrates inhibitor Flame of burning methanol is invisible in daylight and is very difficult to detect. Vapors are heavier than air. Small fires can be put out with dry chemical or CO_2 portable extinguishers. Large fires are extinguished with alcohol resistant foam. Fine spray or water fog are useful for dispersing vapors or washing out of spilled liquid but 25% solution may still burn. Very strict rules regarding ignition sources must be adhered to during transfer operations. During electrical storms operation should be terminated. Methanol is considered water pollutant and water used for firefighting should be contained for disposal. Indigested is highly toxic for humans. In case inhalation, eye exposure and short time skin exposure acts as irritant. Long time skin exposures allow absorption of large dose of methanol.

8 CONCLUSIONS

Supply operations are conducted on large scale even in adverse weather conditions to provide wide variety of cargos necessary for uninterrupted operation of offshore installations. To ensure safety of supply operations, offshore industry developed wide range of safety rules specific for this kind of operations. These rules are not replacement for international regulations which are fully applicable for offshore petroleum industry.

REFERENCES:

[1] Norwegian Oil and Gas Association: 'NWEA Guidelines for Safe Management of Offshore Supply and Rig Move Operations', 2009
[2] Norwegian Shipowners Association, Norwegian Oil and Gas Association, Netherlands Oil and Gas Production Association, Danish Shipowners Association, Oil & Gas UK, United Kingdom Chamber of Shipping : 'Guidelines for Marine Operations', 2013
[3] Oil & Gas UK, United Kingdom Chamber of Shipping : 'The Guidelines for Offshore Marine Operations UKCS Supplement', 2014
[4] http://www.offshore-technology.com
[5] http://www.shippingcontainersforsaleuk.com

Risk Evaluation Model for Management of Navigation Safety in an Entire Ship Route Area

S. Hwang, E. Kobayashi & N. Wakabayashi
Graduate School of Maritime Sciences, Kobe University, Japan

N. Im
Division of Marine Transportation System, Mokpo Maritime University, Korea

ABSTRACT: This paper introduces a new model for risk evaluation of navigation safety in an entire ship route area. The model takes into consideration the perceptions of navigational officers by means of a risk evaluation algorithm developed using a questionnaire. The results obtained with the questionnaire reveal the types of factors that may affect navigation officers' perceptions. The results were quantified and incorporated into the new model. A numerical simulation was carried out to validate the proposed model, using Osaka Bay as an example. A new approach was employed to monitor the level of navigation safety throughout a ship's route area. The ship's route is divided into small sections for which the level of navigation safety can be quantified in terms of a safety index based on the ship's navigation data within a specified distance range. Using the numerical results for each section along the route, a hazard map for the route can be developed in real time. This model can be used by vessel traffic centers and port safety authorities to evaluate the navigation safety level of ship routes in real time.

1 INTRODUCTION

Because marine accidents can result in the loss of ships and material damage to marine infrastructure, safety is a significant issue in shipping and ship navigation in industrial fields. Therefore, many safety regulations pertaining to ship design and equipment have been developed to promote ship safety. However, significant marine accidents continue to occur, and it has come to light that one of the main causes of marine accidents is human error. Human factors were adopted among the causes to be investigated in marine accidents in Resolution A.884 (21) (IMO, 2000). A large proportion of marine accidents are ship collisions caused by human error. As one of the major sources of human error, navigators play an important role in navigating ships. Previous studies have addressed the need for navigational safety evaluation methods for use in assessing a ship's navigation situation and enhancing ship navigation safety. Various navigational safety evaluation methods have been proposed.

Fujii (1971) and Macduff (1974) indicated that a safe navigational zone can be determined from the probability of ship collisions or groundings. In these studies, the level of navigational safety in the observation zone was calculated using statistical analysis of marine accident and traffic data. Hence, the safety level obtained indicates the potential risk of a ship collision or grounding. This type of safety evaluation method can be helpful in distinguishing between navigational safety and hazard zones. On the other hand, the safety level determined using this type of method is calculated using historical data. It is not possible, using this type of method, to evaluate the safety level of a ship route using navigational situations in real time. In addition, this approach does not consider human factors which are the main causes of marine accident.

The concept of a ship domain has been introduced in research on navigational safety evaluation method. A certain area around a ship that should remain clear of other ships is identified on the basis of the safe distance between ships. This area is referred to as the ship domain. Based on the shape of the ship domain, navigators can maintain a safe distance from other ships. The safe distance can be calculated on the basis of statistical analysis of marine traffic data (Goodwin, 1975; Fujii, 1971), fuzzy logic (Pietrzykowski, 2008, Wang, 2010), or questionnaire results and fuzzy logic (Pietrzykowski, 2009; Wang, 2010). The shape and size of a ship's domain is determined by the calculated safe distance. Two-dimensional domains such as circular, rectangular, elliptical, and polygonal shapes have been proposed. Several methods for determining the safe distance have been proposed for use in this approach to safety evaluation. As a result, this approach has been helpful in supporting navigators' decisions. However, this approach has limitations in terms of its ability to consider navigational situations in the safety evaluation.

It has been proposed that risk assessment for navigational areas should consider the decision-making processes of navigators in avoiding ship collisions. The risks associated with navigation situations can be quantified on the basis of navigators' knowledge and competence. Hasegawa (1997) and Im (2004) have proposed risk quantification methods that consider the DCPA (distance to the closest point of approach) and the TCPA (time to the closest point of approach). A risk quantification method that considers factors such as the distance between ships, the rates of change of the ships' directions, and the approaching speed was developed by Hara (1995). These previous studies have shown that the risk associated with a navigational situation can be determined using a risk quantification method based on reasoning rules derived from fuzzy membership functions. The value of the reasoning rules was analyzed using a ship handling simulator with navigators serving as experts. Inoue (1996, 1997, 2000) proposed an ES (environment stress) model to evaluate the difficulty of ship handling for navigation purposes. In this model, risk is quantified by measuring a navigator's physical stress when operating a ship handling simulator and by employing a questionnaire to obtain input from navigators. This risk quantification model considers factors such as the distance between a ship and another ship or an obstacle, the rate of change of the relative directions, and the approaching speed.

The purpose of these previous studies was to evaluate the risk associated with a navigational situation using risk quantification. As a result, this type of safety evaluation method is useful in evaluating marine traffic situations on the basis of factors that influence a navigator's perspective. On the other hand, this type of approach does not consider a variety of factors that can affect a navigator's perspective while navigating. In addition, this type of approach can only be used to evaluate the navigational safety of the surroundings of an individual ship.

The common point of the various methods of navigational safety evaluation that have been mentioned is that they involve various methods for risk quantification. These methods have been shown to be useful tools for evaluating the risk associated with ship navigation. However, safety evaluation methods that use risk quantification based on a navigator's knowledge are limited to being applicable only to evaluating the ship navigation

situation of an individual ship. Therefore, while these methods help navigators on board ships to avoid collisions, they cannot be used to determine the navigational safety zone anywhere along an entire ship route at a specific time, which would be useful to port safety authorities and vessel traffic service centers in managing ship navigation safety. Other safety evaluation methods using risk probability assessment can be used to determine the navigational safety zone, but the risk is quantified on the basis of analysis of historical data. As a result, this approach does not reflect real-time navigational situations and navigators' perspectives as factors in the occurrence of marine accidents.

This paper introduces a new model for evaluation of the navigation safety zone throughout an entire ship route for use by a port safety authority or vessel traffic service center. In evaluating the risks associated with a navigation situation, this model considers a variety of factors that affect a navigation officer's perceptions while navigating. A risk quantification method reflecting the knowledge of navigators was incorporated in this model, and a new algorithm was developed for evaluating safety in an entire ship route.

To verify the effectiveness of the proposed model, a simulation was carried out for the Osaka Bay area. The proposed model was found to be effective in quantifying navigation safety throughout an entire ship route in Osaka Bay. This model can be helpful to vessel traffic centers and port safety authorities in ship navigation safety management.

2 A NEW SAFETY EVALUATIN MODEL

A new safety evaluation model is presented in this section for use in assessing risk throughout a ship's entire route area. The approach used for risk quantification for evaluation of navigation safety takes into consideration a variety of factors that affect a navigator's perceptions while navigating a ship. The factors identified as affecting a navigator's situational awareness are shown in figure 1.

A navigator recognizes other ships around his own ship and then tries to assess the risks presented by these other ships to determine their navigation actions. The factors involved in estimating a ship's maneuverability are considered in the navigator's decision-making. In addition, the model considers these factors in calculating the proper time for taking action—either standing on or giving way. For this reason, the model incorporates various factors that affect a navigator's perceptions during ship navigation, as shown in table 1.

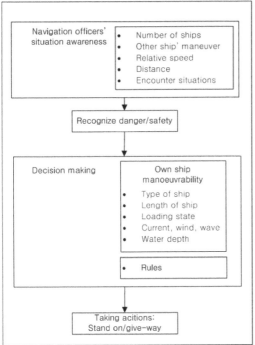

Figure 1. The process of a navigator's decision making in ship navigation

Table 1. Factors in safety index model

Safety index factors	
Ship information	Ship type
	Length of ship
Relationship between ships	Relative speed
	Distance between ships
	Encounter situations
	- head to head
	- crossing (starboard/port)
	- overtaking
Environmental situation	Time
	Day

2.1 Risk quantification based on navigator's perception

The elements that should be considered in evaluating the safety level were introduced in the previous section. The elements must be quantified to assess the risk associated with a navigation area based on navigators' perceptions. Hence, a questionnaire was used to obtain input from navigators. The design of the questionnaire used to explore navigators' perceptions is shown in table 2.

Table 2. The design of a questionnaire for risk quantification

Items	Details	
Type of ship	Container ship, LNG, VLCC, Ferry, Passenger ship, Bulk carrier, Fisher, LPG, PCC, Reefer ship, Tug boat	
Length of ship	Under 100 m, 101–150 m, 151–200 m, 201–250 m, 251–300 m, over 301 m	
Relative speed	0–1.0 k't, 1.1–2.0 k't, 2.1–3.0 k't, 3.1–4.0 k't, over 4.1 k't–	
Distance (L, length of ship)	Under 5L, 6–10L, 11–15L, 16–20L, 21–30L, over 31L	
Encounter situations	Head-on (give-way)	On the centerline of the ship situations from right ahead of 30 degrees abaft the beam of either side of ship
	Crossing on starboard (give-way)	On the starboard side showing from right ahead of 30 degrees to 112.5 degrees
	Crossing on port (stand-on)	On the port side showing from port ahead of 30degrees to 247.5 degrees
	Overtaking (stand-on)	At the stern showing 67.5 degrees from right aft on each side of ship,
Time (LT, local time)	1st officer's	04:00–08:00, 16:00–20:00
	2nd officer's	00:00–04:00, 12:00–16:00
	3rd officer's	08:00–12:00, 20:00–24:00
Day	Mon., Tue., Wed., ThuR., Fri., Sat., Sun.	

In the questionnaire, navigators were asked how much each element affects their perception, using a nine-level evaluation scale (level 1: no influence; level 9: significant influence). The results reflect the navigators' opinions in a quantitative manner that can be incorporated into the safety evaluation model. In this model, each element in question is quantified using equation (1):

$$I_{ij} = \sum_{1}^{N} R_{ij} \times \frac{1}{N} \qquad (1)$$

where:
I_{ii} : average of numerical values for j^{th} element of ith item
R_{ii} : answer value for j^{th} element of ith item (R_{ii} =1-7)
N : number of respondents
i : item number of questionnaire (i =1-8)
j : element number of each item
The results of the quantification of each element obtained using the questionnaire is shown in table 3.

Table 3. Risk quantification of each element determined using questionnaire

Items	Score		
Type of ship	5.3-8.1		
Length of ship	4.5-8.1		
Loading state	Ballast	4.58	
	Full	5.45	
Relative speed	5.1-7.5		
Distance (L, length of ship)	3.8-7.8		
Encounter situations	Head-on (give-way)	Passing 7.9	Meeting 4.0
	Crossing on starboard (give-way)	7.9	3.2
	Crossing on port (stand-on)	7.3	2.2
	Overtaking (stand-on)	7.4	2.4
Time (LT, Local time)	4.38-5.50		
Day	4.91-5.08		

2.2 Risk assessment for each section

This section describes an algorithm for assessing the risks throughout a ship route area using risk quantification based on the results obtained using the questionnaire. The stepwise process of the algorithm is presented in figure 2. Using the proposed algorithm, the safety level reflecting a navigator's perceptions can be calculated for an entire navigational area.

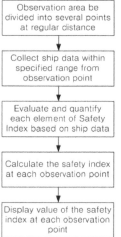

Figure 2. The process of evaluating safety of an entire ship route area

To evaluate the safety throughout an entire ship route area, the area is divided into small sections, and an observation point is established in each section. After establishing these points, ship data are collected from within a specified range of each observation point. Each element based on ship data is evaluated and quantified with respect to the navigator's perception. In the calculation, each element influences the risk quantification. The safety level of each section is calculated by summing the quantified risks associated with the elements in the model. As a result, in this study, the safety level of each section is expressed by a representative value called a safety index. The safety index is calculated using equation (2):

$$SI = \sum_{1}^{n}\sum_{1}^{i} I_{ij} \tag{2}$$

where:
SI : Safety index for each section
n : Number of ships within specified range of observation point
I_{ij} : Risk quantification of each element in question
i : item number of questionnaire (i =1-8)
j : element number of each item

3 SIMULATION RESUTLSS

In section 2, a new model was described that can be used to evaluate the safety of an entire ship route area on the basis of navigators' perceptions. This section presents an assessment of the suitability of the proposed model for use as a safety evaluation method in assessing risk for an entire ship route area. A simulation was conducted for Osaka Bay using this model. Osaka Bay is located at the eastern end of the Seto Inland Sea, which is Japan's largest semi-enclosed sea. This bay has two entrances for the Osaka/Kobe port areas, which are the Akashi Strait and the Tomogahima Channel. According to the Port Authority of Japan (2010), all ships are requested to maintain their heading when entering or leaving the port areas in Osaka Bay, as shown in figure 3.

Figure 3. Traffic rules in Osaka Bay (From port authority in Japan, 2010)

The whole navigational area was divided into small sections, and the safety level of each was calculated. Figure 4 shows that 53 observation points were established in Osaka Bay. Ship data were collected from within a range of 3 miles of each observation point for use in evaluating the navigation situation, using AIS (automatic identification system) data. The safety index for each section of Osaka Bay was calculated by summing the quantified risks calculated using ship navigation situation data such as the ship type, length, speed, distance, and so on. The whole navigational area was divided into small sections, and the safety level of each was calculated. Figure 4 shows that 53 observation points were established in Osaka Bay. Ship data were collected from within a range of 3 miles of each observation point for use in evaluating the navigation situation, using AIS (automatic identification system) data. The safety index for each section of Osaka Bay was calculated by summing the quantified risks calculated using ship navigation situation data such as the ship type, length, speed, distance, and so on.

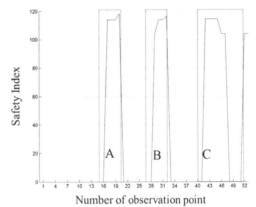

Figure 5. Safety index of each section in Osaka Bay at 15:10 on February 3, 2013

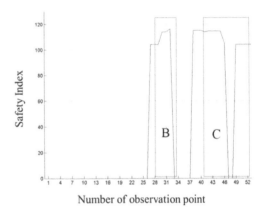

Figure 6. Safety index of each section in Osaka Bay at 15:15 on February 3, 2013

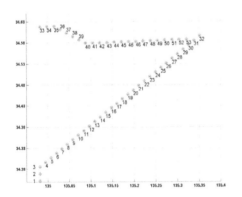

Figure 4. Observation point for each section in Osaka Bay for safety evaluation of whole ship route area

The simulation was carried out to determine the navigational safety zone along the whole ship route area at specific times, based on the calculated risk associated with each point determined using the proposed model. The safety index values calculated for Osaka Bay at 15:10 on February 3, 2013 are shown in figure 5. The results of the navigational safety evaluation after five minutes are shown in figure 6. The safety index varies between 100 and 120, depending on the navigation situation, in figures 5 and 6. A comparison of figures 5 and 6 shows that the safety index changes over time in section A, while of the safety index for sections B and C remain unchanged.

The safety index level at a specific point is calculated for a period of time to determine the navigational safety zone. At observation point 25, ships are in a passing or overtaking situation. As figure 7 shows, the safety index level changes over time at observation point 25. The safety index level varies around approximately 100. Various encounter situations, such as head-on, passing, and crossing starboard/port were considered for observation point 32. At this point, there are many ships arriving and leaving the Osaka/Kobe Port. The results of the evaluation of the safety of the zone are shown in figure 8. The safety index level remains consistent during the time period of the risk evaluation. The safety index level tended to be higher at point 32 than at point 25, depending on the encounter situation.

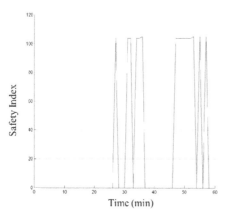

Figure 7. Safety index of observation point No. 25

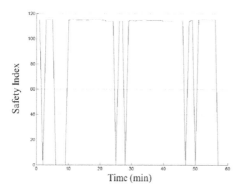

Figure 8. Safety index of observation point No. 32

The risk calculation results were plotted in color to illustrate the navigational safety zone easily and quickly. The whole ship route in Osaka Bay is plotted with respect to the level of the safety index in figures 9 and 10. The safety index level in figure 9 is slightly higher than in figure 10. In addition, changes in the navigational safety zone depending on ship movements in real time are well illustrated.

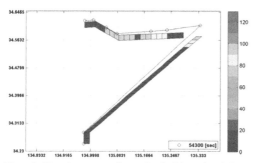

Figure 9. Hazard map according to level of the safety index at 54300 sec

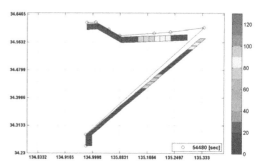

Figure 10. Hazard map according to level of the safety index at 54480 sec

4 CONCLUSIONS

This paper describes a new safety evaluation model that can be used to support a port safety authority or vessel traffic service center. This model takes into consideration the perceptions of navigational officers in the risk evaluation algorithm, which was developed using a questionnaire to obtain input from navigators. A numerical simulation was conducted to verify the usefulness of the proposed model, using Osaka Bay. A new approach was employed to monitor the level of navigation safety along an entire ship route. Using the numerical results, the safety index along the ship route was plotted to illustrate the level of risk in each area along the route. This type of plot is called a hazard map.

This model is expected to be able to serve as a new tool for determining the level of navigation risk throughout a ship's entire route area in real time, more quickly and more easily than is possible using other navigation safety evaluation methods. The procedure developed in this study can be used by vessel traffic service centers and port safety authorities to evaluate the navigation safety level of a ship route in real time.

REFERENCES

IMO (2000), Resolution A.884 (21), Amendments to the code for the investigation of marine causalities and incidents
Fujii Y., Shiobara R. 1971, The analysis of traffic accidents, The Journal of Navigation, 24(4): 537-543.
Macduff T. 1974, The probability of vessel collisions, Ocean Industry: 144-148.
Goodwin E.M. 1975, A statistical study of ship domains, The Journal of Navigation, 28: 329-341.
Fujii Y., Tanaka K. 1971, Traffic capacity, The Journal of Navigation, 24: 543-552.
Pietrzykowski Z. 2008, Ship's fuzzy domain - a criterion for navigational safety in narrow fairways, The Journal of Navigation, 61: 499-514.
Pietrzykowski Z., Uriasz J. 2009, The ship domain – a criterion of navigational safety assessment in an open sea area, The Journal of Navigation, 62: 93-108.

Wang N. 2010, An interlligent spatial collision risk based on the quaternion ship domain, The Journal of Navigation, 63: 733-749.

Hasegawa K., et al. 1997, Reconfiguration of ship auto-navigation fuzzy expert system(SAFES), The Kansai Society of Naval Architects of Japan, 8: 191-196.

Im N. 2004, Automatic control for ship collision avoidance support-II, The Journal of navigation and port research, 28: 9-16.

Hara K., Nakamura S. 1995, A comprehensive assessment system for the maritime traffic environment, Safety Science, 19: 203-215.

Inoue K, Hara K, Kankeko M., Masuda K. 1996, Assessment of the correlation of safety between ship handling and environment, The Kansai Society of Naval Architects of Japan, 94: 147-153.

Inoue K, Kubono M., Miyasaka M., Hara D. 1997, Modeling of mariners' perception of safety when being faced with imminent danger, The Kansai Society of Naval Architects of Japan, 97: 235-245.

Inoue K. 2000, Evaluation method of ship handling difficulty for navigation in restricted and congested waterways, The Journal of Royal Institute of Navigation, 53: 167-180.

The Osaka harbor information center for security of ship navigation 2010, Port of Osaka Entrance and Departure Manual.

Safety of Navigation
Activities in Navigation – Marine Navigation and Safety of Sea Transportation – A. Weintrit (ed.)

The Identification of typical Hazards and Limitations to the Commercial Shipping Safety, Created by Offshore Activity and Crew Transfer High Speed Crafts, Operating in the Vicinity of the Intensive Traffic Flow Areas

G. Szyca
Humber Estuary Service, United Kingdom

ABSTRACT: The authors do the effort to approach the problematic aspects of main identified hazards caused by intensive offshore activity development as well as increased high speed crafts traffic impeding commercial ships' flow in the restricted areas such as: pilot boarding points, Traffic Schemes, Precautionary Areas and narrow channels. The crucial challenges and perception limitations to the VTS staff will be emphasized accordingly. The Harbour Master Humber annual report will be recalled in order to near miss and potential threats to the shipping safety. The recent proposal as well as the new concepts under the discussion stage will be evoked accordingly.

1 INTRODUCTION

In the recent years huge demanding of the renewable source of green energy enforced to establish new wind farms locations mostly in the coastal waters at open sea. Such location is imposed by suitable distance from shore, close vicinity of the port infrastructure, well as acceptable bathymetric factors and flat seabed profile. The Humber river estuary can be treated as a benchmark fulfilling all of these criteria. Within the last few years the Humber estuary region has rapidly developed into an area of national focus and excellence for the renewables industry, in particular offshore wind farming. To date, there are three operational wind farms to the south of the estuary and two to the north - due for completion by mid-2015. Several 'Round 2 and 3' projects to the east of the region have also been consented for development, posing a greater challenge to constructors and operators due to their distance from a suitable home port. In the area of the Humber estuary at present there are two wind farms: Donna Nook wind farm (fully operational) of Lincolnshire in south of Humber traffic separation scheme TSS and Northern Gateway wind farm (under construction) in north of the TSS.

The impact which these larger projects will have upon the Humber and its existing commercial shipping lanes is something of conjecture at this time, as strategies for their development must surely evolve beyond that of an onshore base. The locations of these wind farms from shore (in excess

of 40nm) suggest that construction and operation & maintenance (O&M) bases must be developed and managed from mother ships, 'float hotels' or offshore platforms due to the costs associated in running crew transfer vessels out to these distances. O&M strategies will also undoubtedly incorporate helicopter support in the future, which will further negate the usage of crew transfer vessels (CTVs) from an onshore base.

There are around 40 daily movements of high speed crafts recorded engaged with wind farms sector in conjunction of average 80 commercial vessels in the TSS area and Shore Traffic Zones posing the significant challenges both to the Vessel Traffic Operators, ships' navigators and local pilots. The wind farms mentioned above have their operational bases located at the Port of Grimsby which is situated seven miles inside the Humber Harbour Area. Operators use the port as a base for both the construction and O&M phases of their projects, which generally have a 25 year 'in service' life span. These projects involve the transportation of large numbers of technicians and workers.

2 IDENTIFIED HAZARDS TO THE SHIPPING SAFETY IN RELATION TO OPERATING HIGH SPEED CRAFTS (HSC).

Developers look to maximise technician's 'time on turbine' in good weather, by seeking the shortest and most direct routes to their farms at speeds approaching 30 knots. Where their home ports are located within estuaries such as the Humber, the management of CTVs and their interaction with larger commercial vessels within channels and traffic separation schemes (TSSs) may well be coordinated and safeguarded on a daily basis by a Vessel Traffic Service (VTS). Not all ports and harbours are served by a VTS. This is a matter for the Statutory Authority to address through a process of formal risk assessment and cost benefit analysis. Through this process, a VTS may be deemed necessary and afforded in order to mitigate the risks involved with conducting maritime activities within the port or harbour. However, they do not all offer the same level of service. Some provide an 'information' service only whereas on the Humber, a 'traffic organisation' service exists which is fully capable of interacting with and managing vessel movements within the area of jurisdiction.

At the height of the construction phases of the two northern wind farms in summer 2014, a maximum of 40 CTVs operated from Grimsby on a daily basis. Many of these vessels would cross the inner traffic separation scheme of the Humber where commercial traffic is funnelled into the river and pilot boarding and landing operations take place – often in challenging weather conditions. Movements during a construction phase will be concentrated between specific times of the day owing to craft being tidally bound to their home port or reliant upon daylight working hours. It is worth noting that during the O&M phase of a project, CTV numbers reduce dramatically and are not required on location every day, as they may be during construction.

After numerous risk assessments done for the Humber Traffic Separation Schemes and adjacent areas the following hazard been identified:
1 Excessive speed of the HSC seriously limiting perception of the VTS operator.
2 Intensive traffic flow of other commercial vessels reducing time reaction of the VTS operator and ship's Master/Pilot.

3 The wide variety of duties (e.g. pilot boarding arrangements, directing the vessels for the specific ETA, communicating with dock masters/berth operators, etc.) to the VTS officer seriously reducing proactivity in order to shipping safety.
4 Adverse weather conditions, strong tidal streams imposing extra "navigational vigilance" causing fatigue,
5 Seasonal increased traffic of the pleasure crafts and fishing boats navigating in close vicinity of commercial traffic flows and crossing traffic lanes.

3 MEASURES ALREADY UNDERTAKEN TO MAINTAIN DESIRED SAFETY LEVEL

Measures already undertaken to maintain desired safety level:
1 Delineation of recommended routes for windfarm transfer vessels to minimize impact to commercial shipping.
2 Speed restrictions to 20 knots in the Humber Harbour Area.
3 Further speed limitation to 10 knots in condition of restricted visibility less than 0.5 nautical miles.
4 Publication and updating of "Recommended Route for Wind Farm Transfer Vessels" guide.

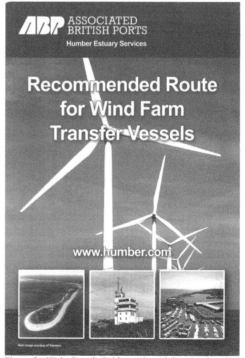

Figure 2. High Speed Guidance issued by Humber Estuary Service [9]

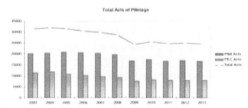

Figure 1. The Pilotage Acts occurred on Humber River in recent years [8]

5 Continuous monitoring of effectiveness of implemented measures

6 Periodical liaison committee with involves parties (Humber Estuary Services HES, Dong Energy, Siemens Energy, etc.).

7 Re-establishing new recommended routes after wide consultations and after several near miss situation analysis:

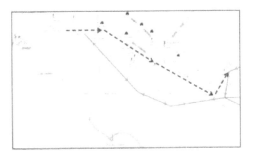

CAPT. A. FIRMAN
HARBOUR MASTER, HUMBER

30 January 2015

THE INTERNET

In order to widen communication between ABP Humber Estuary Services and those with an interest in the estuary, you are invited to visit our website which carries a wide range of information, including current live weather and Buoy positions, charts, tidal information and copies of this and other Notice to Mariners.

www.humber.com

ASSOCIATED BRITISH PORTS

P.O. Box No. 1, Port House, Northern Gateway, HULL HU9 5PQ

NOTICE TO MARINERS

No. H. 10/2015

H U M B E R APPROACHES

RECOMMENDED ROUTES FOR WIND FARM TRANSFER VESSELS – OUTWARD ROUTE TO THE NORTH

NOTICE IS HEREBY GIVEN that commencing on the 1st February 2015 a change to the current recommended route for Wind Farm Transfer Vessels outward from Grimsby to wind farms North of the Humber approaches will be implemented.

The new recommended outward route instigated prior to and in conjunction with the Tetney Mono Buoy Subsea Pipeline Replacement Project will follow the main channel from Grimsby through the Inner Traffic Separation Scheme to the No 2B Buoy.

Windfarm Transfer Vessels should call VTS Humber on Channel 14 for permission to cross the channel to the North in the precautionary area at the No 3 Chequer Buoy on approaching the No 2B Buoy.

If required to wait before crossing, Wind Farm Transfer Vessels should clear the traffic separation scheme at the No 2B Buoy to the South and await permission from VTS Humber to cross to the North when clear to do so.

MARINERS ARE ADVISED that the recommended route for Wind Farm Transfer Vessels bound to and from wind farms to the **South** of the Humber approaches will remain in use.

Figure 3. The amendment of the recommended route for wind farm transfer vessels issued by Harbour Master Humber [9]

4 PLANNED CONCEPTS FOR UPGRADING SHIPPING SAFETY

The Humber district Authority, ship Operators/Owners, Pilots and Captains are aware that present solutions maintaining set up safety shipping safety targets are insufficient, however presently wide range consultations are being held within all engaged parties to achieve intended objectives. More of them are still at the discussion stage but some of are under the trials with eventual further implementation. Ones amongst of the most essential are:

1 Utilizing of the new software to more efficient evaluation of the potential dangerous radar targets.

2 Work out new anti-collision domains and implementation them into the existing systems.

3 Concept of creating new Traffic Centre focusing on offshore High Speed crafts in relation to impeding commercial traffic within the district.

4 Launching modern, highly sophisticated identification and anti-collision devices simultaneously with new methods of utilization the existing ones.

5 Seeking new concepts and ideas within consultation groups, Safety Navigation Review Committee and liaison teams.

6 Providing new mode of the simulator courses including high speed crafts traffic.

5 THE CHALLENGES FOR VESSEL TRAFFIC SERVICES (VTS) OPERATORS

The extent of management required by the VTS to safely integrate such high numbers of CTV movements through busy shipping channels without having a negative effect on the safety or efficiency of commercial vessel traffic, has increased significantly. The methods available to a VTS and practices utilised to influence the movements of CTVs will vary from location to location but will, without exception, be based upon a harbour authorities regulatory powers.

Powers of general direction vary widely between each individual port or harbour. However, where a designated VTS exists, the 'Merchant Shipping Vessel Traffic Monitoring and Reporting Regs 2004' will apply. Hence the Master of any vessel shall comply with the rules, regulations and reporting requirements of that VTS. This is the basis of control that can be applied to CTV movements within a VTS area of jurisdiction, together with the overarching COLREGS.

The rapid expansion of this industry, and the numbers of CTVs required in support of these wind farms, clearly creates a significant challenge for a

VTS to manage on several levels. Primary factors which influence the extent of this challenge range from the powers and regulating directions in force (if any), existing levels of traffic density, geographical and hydrographical limitations, the complexity of navigable waterways and, of course, the number of CTVs operating within their jurisdiction. Secondary factors would include the scope of possible segregation of CTV traffic from commercial shipping lanes or reducing the frequency of potentially conflicting interaction between the two at crossing points.

The effective control of multiple CTV movements requires prior knowledge of their departure from home port and the routes followed to their respective wind farms. This involves a large number of craft reporting to the VTS to confirm their departure, the number of persons on board and their intended track - adding to the operator's radio workload and associated administration duties.

An operator's attention to the traffic image can quickly be drawn to multiple CTV movements within their area owing to their rate of progression and rapidly changing vectors which can impinge quickly upon other river users. The inherent speed and manoeuvrability of these vessels may result in them becoming involved with other commercial traffic within the turn of a radar scanner. An operator has to be attuned to the tempo at which such craft operate and be more vigilant and proactive in their identification of traffic conflicts. This has an obvious impact upon the operator's span of attention and may lead to overload at peak times, or draw the focus away from other vessels and activities such as pilot boarding and landing operations.

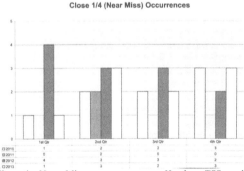

Close 1/4 (Near Miss) Occurrences

Figure 4. Near Miss occurrences on Humber TSS and Precautionary areas in the years 2010-2013 [7][8]

At locations en route, where the interaction between CTVs and commercial traffic is unavoidable, (such as at TSS crossing points/precautionary areas etc.) the application and use of any directions or general agreements in place, the COLREGS, or byelaws becomes particularly

important. This can be particularly onerous upon the operator who is continually identifying traffic conflicts and resolutions to such. The amount of interaction between an operator and vessels within his/her area will depend upon the level of service offered. An organisation service is obliged to confirm the intent of such vessels and, where necessary, intervene by issuing 'information' or 'instructions' - to prevent a close quarters situation developing. Constant repetition of these exchanges and 'clearances' to cross TSSs, issued to multiple vessels on the VHF, loads the designated channel and adds again to the operators workload.

6 SUMMARY

The establishment of recommended CTV routes through consultation, risk assessment and mutual agreement, between a Harbour Authority and wind farm developers has become industry best practice under the Port Marine Safety Code (PMSC). Having agreed the routing of such vessels, it removes any doubt as to their intended tracks, and proactive promulgation of this information raises other mariner's awareness of the operations and the areas in which they can expect to encounter CTVs. The additional application of agreed speed limits, where considered necessary within controlled waters, can also add to operational safety in particularly hazardous areas or during times of restricted visibility.

The movements of CTVs can be influenced by tides, weather, daylight hours and the type of work being undertaken within the farm. Operational experience on the Humber has shown that at some point in the weekly tidal cycle, other tidally restricted vessels, or vessels on scheduled routes will come into contact with a large number of CTVs. It is at these times when the VTS operator has to be particularly vigilant and proactive in identifying potential close quarter's situations and may be required to deny clearance requests to cross shipping channels at agreed reporting points. These decisions can be challenging to make when dealing with a large numbers of vessels moving in confined waters, with largely differing characteristics by way of speed and manoeuvrability.

One of quarter situations will occur between CTVs and other vessels from time to time, and upon the Humber there have been documented incidents of such. Unfortunately, there has also been a documented incident of a CTV colliding with a floating mark on the edge of the VTS area. We are not alone in this regard, as there have been numerous incidents involving CTVs at other ports in recent years. Perhaps following such incidents, and with the safety spotlight of the MCA and MAIB firmly fixed upon this developing industry, we should address the somewhat controversial subject of CTV skipper competence and certification.

The required level of certification required by an operator to skipper a CTV can vary greatly. The MCA have issued guidance on competency standards (The Workboat Code), but in a rapidly growing industry where demand for such craft and skippers is presently so high, the marine industry's normal tendency to seek higher levels of competency than recommended is perhaps ignored. This issue may contribute to incidents such as above and, without doubt, add to the workload of a VTS operator who is trying to ensure such craft move throughout the district in accordance with the COLREGS, local rules and regulations. Often, skippers new to the industry are unfamiliar with navigating at such relatively high speeds, in somewhat unconventional craft, in congested waters - or lack the skill set required to do so with confidence. Several developers should be commended for their acceptance of these issues. Keen to operate to high standards of safety, they have taken it upon themselves to bridge these skill gaps and provided further training for their skippers in high speed navigation and fast ARPA usage.

A resolution to this issue cannot be laid at the door of a single organisation or operator and will not be found in the short term. A Harbour Authority committed to the PMSC cannot ignore or stand idle to such an issue, if they become aware of it manifesting itself into a threat to the safe management of their harbour. Powers to positively influence competency levels may not be an option for a Harbour Authority, but education under the PMSC Guide to Good Practice can be just as effective and more positively received by developers than a host of regulatory restrictions. It is therefore imperative that a good relationship exists between a Harbour Authority and operators/developers of wind farms in order to promote and progress such education and training to their CTV skippers. In turn this will raise navigational awareness of the port/harbour within which they operate and will yield reductions in the numbers of incidents occurring between other river users and CTVs.

The importance of stakeholder liaison and the development of good working relationships with developers/operators prior to the construction phase and throughout the O&M phase of their wind farms

cannot be stressed enough in relation to navigational safety and the appreciation of an open port duty.

The offshore wind farm is relatively young sector and still difficult to define all existing potential hazards to shipping safety. Every day we learn something new and facing new challenges but permanent and continuous monitoring methods can allow in the near future to work out the efficient models to maintain and upgrade required safety level. The essential is to mutual understanding the specification of operational issues of the offshore industry, applying all available modern technologies, participating in appropriate model courses (as Bridge Resource Management course), creating the new ones and developing the existing ones. The key role of the International Maritime Organization IMO to establish the identical standards for the same sectors seems crucial.

ABBREVIATIONS USED IN BELOW PAPER

1 VTS – Vessel Traffic Service
2 TSS – Traffic Separation Scheme
3 HSC – High Speed Craft
4 CTVs – Crew Transport Vessels
5 O&M – Operation and maintenance
6 ETA – estimated time of arrival
7 IMO – International Maritime Organization
8 MCA – Maritime and Coastguard Agency
9 MAIB – Marine Accident Investigation Branch

REFERENCES

[1] IALA VTS Manual, London 2011.
[2] Annual Maritime and Coastguard Agency Report, London 2014.
[3] Role of Human Element – Assessment of the impact and effectiveness of implementation of the ISM Code, IMO MSC 81/17.
[4] IALA Recommendation V-103 for the Training and Certification of VTS Personnel.
[5] MSC.351(92) Amendments to International High Speed Crafts, London, 21.06.2013.
[6] MSC.352(92) Amendments to International High Speed Crafts, IMO, London, 21.06.2013.
[7] Harbour Master Humber Annual Report, Hull 2014.
[8] Harbour Master Humber Annual Report, Hull 2013.
[9] Recommended Route for Wind Farm Transfer Vessels, Humber Estuary Service, 2012.

Safety of Navigation
Activities in Navigation – Marine Navigation and Safety of Sea Transportation – A. Weintrit (ed.)

Physical Characteristics of Virtual Aids to Navigation

R.G. Wright
GMATEK, Inc., Annapolis, MD USA
World Maritime University, Malmö, Sweden

M. Baldauf
World Maritime University, Malmö, Sweden
Institute of Innovative Simulation and Maritime Systems, Rostock, Germany

ABSTRACT: This paper considers issues necessary to be addressed to ensure the reliable and verifiable integration of virtual aids to navigation into the present system of short range aids to navigation. Such virtual aids are intended to replicate many of the functions and capabilities of physical aids to navigation in terms of assisting navigators in determining their position, determining a safe course, warning of dangers and obstructions, and promoting the safe and economic movement of commercial and military vessel traffic and cargo of strategic military importance. Specifications for the establishment of virtual aids to navigation as well as verification of their correct deployment to ensure they are watching properly must necessarily follow a new and more comprehensive path for development and implementation based upon physical attributes, performance characteristics and also their data representations within electronic databases.

1 INTRODUCTION

Physical aids to navigation (AtoN) have been used for thousands of years to guide vessels along their routes and provide assurance of safe passage using known landmarks and structures to indicate safe waters. In the modern era technology has provided us with buoys, lighthouses, light ranges, day marks and other devices to accomplish this same purpose. Complemented with radar, depth sounders and precision timing devices, the capability to broaden situational awareness in identifying physical environmental features and to track vessel progress while underway has never been greater.

New to the scene are virtual or electronic aids to navigation (eAtoN) defined by the International Association of Lighthouse Authorities as something that "does not physically exist but is a digital information object promulgated by an authorized service provider that can be presented on navigational systems" (IALA O143). This concept is revolutionary to vessel navigation in much the same manner as was the introduction of radar – with many of the same problems likely to be encountered in terms of training and operation. The potential to instill new vessel navigational capabilities that cannot be achieved using traditional, physical AtoN is real. However, the probability of encountering many limitations and fragilities unique to eAtoN is high, and it is necessary to anticipate and adequately

prepare for such eventualities to ensure safety of navigation is maintained.

An attempt is made to address eAtoN needs critical to ensure the proper installation and verification of performance by authorized service providers as well as their safe and reliable use by mariners. An expanded range of eAtoN physical, performance, environmental and computational limitations are considered in this analysis. Strategies are also provided to overcome some of the potential vulnerabilities of such devices at various points in the eAtoN lifecycle to avert threats by opportunists to render such devices themselves useless or even hazardous to navigation.

2 THE CASE FOR VIRTUAL ATON

Virtual aids to navigation are intended to supplement and not replace existing AtoN in areas where the timely marking of hazards to navigation can be performed faster and more effectively than placing physical AtoN. This may be on a temporary basis until physical AtoN can be installed such as in marking new wrecks or where previously uncharted hazards to navigation are detected. Virtual aids to navigation can also be installed on a permanent basis where the use of physical AtoN is problematic or not possible. This includes coral reefs where sinkers cannot be placed due to their adverse environmental effects, in the Arctic where ice movement can carry away physical AtoN, and along rivers and tributaries where water levels and channel locations are subject to frequent change. Another possibility is that eAtoN functionality may provide flexibility in terms of purpose and positioning that may be tailored to the unique requirements of individual vessels for determining adequate widths of channels, placement locations and other capabilities such as aid to vessels having lost their track and need position assistance.

3 EATON IMPLEMENTATION TECHNOLOGIES

The IALA definition of a virtual aid to navigation cited earlier provides no direction as to the implementation technologies through which such a capability may be achieved. However, earlier guidance recognized that Automatic Identification Systems (AIS) can be applied to AtoN to further improve and enhance services to mariners and assist AtoN authorities to ensure the safe provisioning of such aids to navigation as the volume of traffic justifies and the degree of risk requires, as stated in SOLAS, Chapter V, Reg13 'Establishment and operation of aids to navigation' (IALA G1081).

The use of AIS to effect eAtoN implementations must rely on physical infrastructure to accomplish their objectives. This in itself is not problematic in areas where ready access is available and adequate financial resources exist to install and maintain such physical infrastructure. However, this is not the case over vast portions of the planet where eAtoN capabilities are needed most – the Arctic and in sensitive tropical regions. Indeed, all regions that are without adequate financial resources and those affected by war can benefit from the ability to rapidly install eAtoN without physical infrastructure that fulfill the IALA definition, "does not physically exist but is a digital information object". Such an approach requiring no physical infrastructure has recently been presented that can overcome other limitations such as the lack of hydrographic survey, sporadic and low-bandwidth communications, and an absence of government support (Wright & Baldauf 2014). Both such implementations require eAtoN presentation on navigational systems, with the primary system for navigation being the Electronic Chart Display and Information System (ECDIS) (MSC 82). AIS signals are also presented on radar and other appropriate displays.

As of January 2015 the US Coast Guard has 145 AIS eAtoN in 7 of the 9 districts, with 15 more (approximately) being delivered in March 2015. Deployment of these AtoN types is being accomplished in an effort to best determine their use and application for future waterway guideline development (USCG 2015). Examples have also been provided of Local Notice to Mariners chart corrections and illustrated IHO and IEC-compliant eAtoN portrayals on paper charts, electronic nautical charts (ENCs) and radar for: (USCG 2014; see also IALA G1062).

– Physical AIS eAtoN: AIS signal broadcasts originate from a physical AtoN,
– Synthetic AIS eAtoN: AIS signals originate from a remote AIS base station and are broadcast to a location where a physical AtoN exists,
– Virtual AIS AtoN: AIS signals originate from a remote AIS base station and are broadcast to a location where no physical AtoN exists but are displayed on ENCs and ECDIS.

The National Oceanographic and Atmospheric Administration (NOAA) announced an expanded set of symbols used to portray AIS eAtoN on ECDIS and that NOAA charts would be updated to add AIS eAtoN locations (OCS 2014). These symbols include a magenta radio ring surrounding the eAtonN reflecting the transmitted nature of the signal, which does not necessarily apply to non-AIS virtual eAtoN.

4 CHARACTERISTICS

The term "characteristics" when used in relation to AtoN have generally referred to their physical and performance aspects as can be readily seen and measured to determine whether they are watching properly. However, with the introduction of eAtoN that exist as a digital information object this concept has become somewhat muddled. Even virtual eAtoN have a physical presence on navigation display devices such as ECDIS and radar.

The following paragraphs attempt to clarify these issues by introducing their digital representations within the context of 'characteristics' by which an assessment may be determined of whether they are watching properly. A discrepancy is defined as any failure of an AtoN to display its characteristics as described in the Light List or to be on its assigned position. When a discrepancy is reported, a response level for its correction is determined based upon severity and availability of assets (USCG 2005).

The lines of demarcation as to whom discrepancies are to be reported must also be redrawn. The US Coast Guard is the cognizant organization for reporting AtoN discrepancies, while NOAA is cognizant for charting discrepancies that include ENCs and ECDIS. If the eAtoN is portrayed incorrectly on a chart this would be reported to the US Coast Guard.

4.1 *AtoN Characteristics*

For traditional AtoN, two main aspects of their design encompass various characteristics that must

be verified to determine they are watching properly. These include:
– Physical, and
– Performance.

Physical AtoN characteristics consist of nominal operating and discrepant conditions and include type (buoy, daymark, range, lighthouse, racon) color, shape, numbering, light features (red/green/yellow, flashing/steady/occulting), sound features (bell/gong /horn/whistle), position (lat/long, on station, off station, adrift, missing, not marking best water) as well as condition (sinking, stranded, capsized, excessive rust). Performance aspects include light and sound intensity, racon (operational, not operational, operating improperly), rhythms and rates of installed devices, and visibility (dayboards are faded, lights/numbers are obscured, etc.) (USCG 2010).

AtoN are documented and described in databases (ATONIS/USAIMS, ENC) and data products (e.g., Light List, Notice to Mariners, Coast Pilot). However, these data objects and representations are secondary to their physical manifestation in terms of performance. Indeed, physical AtoN have existed and stood watch properly for centuries with little more representation as "data objects" than a written note on a hand-made chart.

4.2 eAtoN Characteristics

Both physical and synthetic AIS eAtoN share the characteristics cited in the previous paragraph with their associated physical AtoN that must be considered during verification and when reporting discrepancies. However, this does not necessarily apply to AIS or non-AIS virtual eAtoN since neither are associated with a physical AtoN other than for type, position and operational status.

Although the IALA definition of virtual AtoN describes them as data objects, they actually do exist in the physical sense when they are depicted on a navigational display to be observed and acted upon by a watchstander. In this sense the physical characteristics of AIS eAtoN include the symbols for physical, synthetic and virtual; and non-AIS virtual (USCG 2014).

ENC depiction of physical characteristics for both AIS and non-AIS virtual eAtoN on ECDIS is more comprehensive and includes symbols for cardinal marks (N/E/S/W), lateral marks (IALA A/B port and starboard), isolated danger, safe water, special purpose and emergency wreck marking (OCS 2014).

The performance characteristics of eAtoN are determined in part by the specifications for each specific device. From a practical perspective for operational verification they either work as specified (operational) or don't work (not operational) in much the same fashion as a racon installed on AtoN.

The bulk of eAtoN characteristics exist in the form of data object representations in the domain of the authorized service provider. In the United States this is the US Coast Guard and NOAA. These characteristics include the Light List number, type of aid, name, position, class, inspection dates and other information.

5 LIMITATIONS AND VULERABILITIES

Potential limitations and vulnerabilities associated with the implementation of eAtoN technology exist, some of which are described below. With careful planning and diligent design and implementation practices these limitations may be managed and overcome to ensure their reliable and verifiable operation is achieved.

5.1 AIS Broadcast Range

The range of AIS broadcasts in the VHF Frequency spectrum is limited to line of sight based primarily upon the height of the base station transmitting and vessel receiving antennae. The range of VHF signals is estimated at nominally 20 miles at sea (USCG 2015a). This limits the placement of AIS eAtoN to achieve reliable performance at other than remote locations to a distance less than 20 miles, especially inland where terrain and ground-based structures can interfere with signal propagation (Baldauf, 2008).

AIS is also subject to the effects of Tropospheric ducting that can propagate VHF signals hundreds of miles from their origin (Biancomano 1998). Such effects can introduce interference sources to signals from AIS stations within the nominal AIS reception range and can result in performance reduction of AIS both ashore and on vessels (ITU 2007).

5.2 AIS Spoofing and Jamming

The ability to spoof and jam AIS broadcasts has particular significance where AIS eAtoN signals are used for vessel navigation. A lack of security controls can facilitate a ship being diverted off course by placing eAtoN in undesirable or even dangerous locations inadvertently, for hijacking or for other nefarious purposes (Simonite 2013).

The vulnerabilities of AIS have also resulted in its use by criminals to an attempt to evade law enforcement (Middleton 2014). Another report found that AIS data is being increasingly manipulated by ships that seek to conceal their identity, location or destination for economic gain or to sail under the security radar, and concludes that this is a fast growing, global trend undermining decision makers who rely, unknowingly and unwittingly, on inaccurate and increasingly manipulated data (Windward 2014).

5.3 GNSS Spoofing and Jamming

Similar to AIS, Global Navigation Satellite System (GNSS) signals can also be spoofed and jammed causing unreliable and even deceptive navigation signals to be received by vessels (Forssell B. (2009). A recent example is an experiment by a group of University of Texas at Austin researchers where a yacht was driven well off course and essentially hijacked using spoofing techniques (Zaragoza 2013). This phenomena was also the subject of a recent article in the US Coast Guard Proceedings acknowledging this as being of concern beyond the maritime industry to include the transportation sector as a whole (Thompson 2014). Jamming can have the same effects as an outage, as was demonstrated in 2010 when numerous, low power personal privacy jammers were detected as interfering with GPS involving airport operations at Newark, NJ (Grabowski 2012).

5.4 GNSS Outages

The worldwide GNSS is comprised of the United States GPS, Russian GLONASS, European Galileo and Chinese BeiDou systems which are at various stages of completion. These multiple systems imply that backup capabilities exist if one or more of these systems were to go out of service, either temporarily or on a permanent basis. This was demonstrated during the ten-hour GLONASS outage that occurred on 1 April 2014 where a Broadcom 47531 receiver performing the simultaneous tracking of GPS, GLONASS, QZSS and BeiDou signals was able to successfully identify and remove the bad GLONASS satellite positions (Gibbons 2014). Multiple GNSS receivers are only beginning to come into the commercial marketplace. Under normal conditions the performance of these systems is likely to equal or exceed existing, single technology systems.

All GNSS regardless of technology used are subject to the same atmospheric and signal propagation limitations, multipath interference, orbit errors, satellite geometry and orbital debris. One or a combination of such factors may degrade GNSS signals to reduce their accuracy or make their signals unreliable or unusable. With the discontinuance of Loran and no commitment to establish any backup system to GNSS using a fundamentally different positioning technology such as that used by eLoran, there is presently no alternative available for navigation other than that provided by traditional aids to navigation.

5.5 Database Hacking

One of the greatest dangers of eAtoN is their primary existence as data objects in cyberspace, without having a traditional physical presence to provide backup in the event of their electronic corruption or disappearance. This property makes them susceptible to hacking and denial of service attacks that can render them useless or even detrimental and hazardous to navigation.

Widespread corruption can occur at the source databases within which eAtoN objects reside at the authorized service provider. In the United States this responsibility is shared between the Coast Guard for AtoN and the Light List, and NOAA for ENC that form the nation's navigation charts. Corruption can also occur at the local level, where individual or groups of eAtoN in the same geographical area may be corrupted.

Initiatives exist at both the Coast Guard and NOAA aimed at defending their computer networks from attacks (Radgowski 2014; NOAA 2014). Both initiatives acknowledge the threats involved and are steps in the correct direction to manage and even overcome the adverse effects on national security imposed by these threats. Issues that pertain to eAtoN design, development and implementation cross agency lines, barriers and firewalls; making the solution to these problems even more difficult.

6 METHODS FOR VERIFICATION

There are three levels at which verification of eAtoN must be considered. The first level focuses on where they are represented in electronic form as data objects. Numerous vulnerabilities can exist ranging from simple data entry errors to the intentional hacking, manipulation or destruction of the data content. Compounding the severity of the problem is that eAtoN data is represented in multiple data systems across Government agencies that may be altered or modified from their original content, making the ENC a product of collaborative datasets.

The second level of verification is the actual technical performance of the eAtoN device and mechanisms themselves. The third level involves verification of the physical eAtoN characteristics as they are manifested at the deployed location.

6.1 Data Object

A data object is defined as an item or group of items, regardless of type or format that a computer can address or manipulate as a single object that will inform the user as to the characteristic of a virtual AtoN (IALA G1081).This definition indicates the characteristics contained within the database are highly correlated with the unique eAtoN object it represents. The corruption of these data can fundamentally alter the behavior, functionality and/or performance of the eAtoN. Such corruption can occur throughout the lifecycle of the object from the characterization of data as requirements, design of the structure in which these data reside, initial entry of the data into the data structure, process of extracting the data, its fusion with other data to effect a process or outcome using a navigational display, and the final representation of the data in its intended use for navigation.

The process flow depicted in figure 1 provides a simplified example of a generic verification process that could be used on a continuous basis by cognizant service provider(s) to examine the contents of multiple databases to detect errors and inconsistencies caused by database hacking as well as errors inherent to database operations.

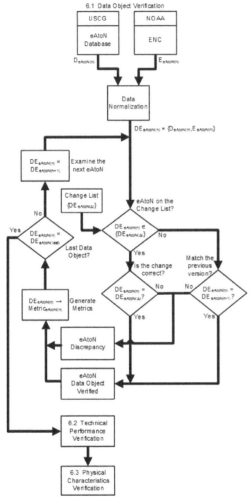

Figure 1. eAtoN Data Object Verification Process.

This approach is conceptually aligned with the Department of Homeland Security Continuous Diagnostics and Mitigation (CDM) Program designed to protect government networks and their data.

6.1.1 Database Structures

Multiple databases and data structures distributed geographically across different government agencies host the data required to create, implement and support eAtoN operations. These include legacy systems already supporting AtoN characteristics modified to support eAtoN and the equivalent data requirements within the ENC, with legacy processes used to integrate these data and create their final products. The implementation and hosting of eAtoN data representations exists on different data platforms and host software, with diverse formats and timing of system updating and maintenance. How these data and relevant metadata are shared, the flow of these data managed, and the processes and frequency through which this occurs is the focus of the Committee on the Marine Transportation System (CMTS) in the United States. This should be accomplished in a manner that coincides with the update and revision cycles of the contents of the data structures independent of the development of products derived from the data contents. This also requires proper filtering and assurance that the destination system and associated processes be sufficiently robust so as not to be overwhelmed by the volume of data received.

6.1.2 Data Normalization

Prior to initiating data object verification it is necessary that the contents of the data streams be processed and normalized from their native formats contained within the source databases to a common format ensuring proper comparisons of the data may be accomplished. This includes data for both AtoN and eAtoN since they are integrated together into the same legacy systems, ensuring data objects for both forms are verifiable and can be verified using the same process. This also requires that inputs of metadata, human interface and guidance, a priori data, and other machine data necessary to perform verification are prioritized and properly associated with the data for subsequent processing.

The product of the data normalization stage represents the totality of the data from all sources necessary to accomplish verification:

$$DLE_{eAtoN(n)} = \{D_{eAtoN(n)}, E_{eAtoN(n)}\} \qquad (1)$$

where D=AtoN/eAtoN; and E=ENC data objects. Note that the process flow of figure 1 has been simplified to show eAtoN data, however both AtoN and eAtoN data objects from the same data sources can be verified using this same technique. These data are provided at a rate sufficient to process changes in synchronization with the data at their source databases.

6.1.3 Verification Process

Verification of normalized data is accomplished with the knowledge of changes that are supposed to have occurred and the implication that any other changes that may be detected are therefore discrepancies. Each individual normalized AtoN/eAtoN data object is compared to the change list to determine whether it is contained within the set of changes expected for that specific individual process:

$$DE_{eAtoN(n)} \in \{DE_{eAtoN(\Delta)}\} \qquad (2)$$

If the data object is part of the set of changes then the characteristics of the normalized data object are compared to those on the Change List to ensure their proper implementation:

$$DE_{eAtoN(n)} = DE_{eAtoN(n\Delta)} \qquad (3)$$

where an affirmative result causes a determination of the AtoN/eAtoN as a verified data object and a negative result causes a determination of a data object discrepancy.

If the data object is not part of the set of changes then the characteristics of the normalized data object are compared to those of the previous revision (n') of the data object:

$$DE_{eAtoN(n)} = DE_{eAtoN(n')} \qquad (4)$$

where an affirmative result causes a determination of the AtoN/eAtoN as a verified data object and a negative result causes a determination of a data object discrepancy.

Completion of individual data object verification is achieved with a determination of verified or discrepancy, wherein metrics are generated followed by the examination of the next data object:

$$DE_{eAtoN(n)} \rightarrow Metric_{eAtoN(n)} \qquad (5)$$

$$DE_{eAtoN(n)} = DE_{eAtoN(n+1)} \qquad (6)$$

Upon detection of the last data object, data object verification for this process run is completed and initiation of Technical Performance is then followed by Physical Characteristics verification.

6.1.4 *Metrics*

Data object examination is complete when steps necessary to determine verification or discrepancy have been achieved. Metrics to measure verification progress and resultant products must be established to indicate process completion and performance scores are created to indicate product quality and deficiency levels. Such metrics must also ensure feature and capability traceability to product specification and design, stability of software configuration, adequacy of depth and breadth of testing, and overall product maturity. Configuration controls and trouble reporting procedures need to be established to track the rate, type and severity of discrepancies as well as required changes to software, processes, design, and requirements resulting from discrepancies found and corrected.

6.2 *Technical Performance*

Verification of the technical performance of AIS eAtoN lies primarily in determining that the system is operational and performs the required functions. General guidance on this subject may be found in the appropriate IALA guidelines (IALA G1028). Guidance on verification of AIS equipment should be found in the technical specifications, acceptance test procedures and maintenance test procedures appropriate for specific equipment configurations.

6.3 *Physical Characteristics*

The existence of eAtoN as data objects without having a traditional physical presence does not necessarily preclude their verification using many of the same physical parameters as AtoN. This may provide an ideal opportunity to demonstrate the use of technology to resolve doubts and concerns regarding navigation strictly by electronic means rather than traditional methods by using live environmental sensor data to obtain fixes to known landmarks, structures, bottom terrain features and buoys.

Many of the physical characteristics of physical and synthetic AIS eAtoN are shared with their associated AtoN. Characteristics unique to physical, synthetic and virtual eAtoN include type, position and operational status as well as the presentation of these characteristics on navigational displays, e.g., radar and ENC/ECDIS. The highest priority is depiction of position, which is closely followed by the other characteristics.

6.3.1 *Position*

The easiest and most risky means of verifying the eAtoN characteristic associated with position is though the use of GNSS to compare the measured position with the charted position. In the case of physical and synthetic eAtoN there is a physical AtoN present at the location as well as an AIS/ECDIS representation to corroborate the GNSS fix, assuming that verification of AtoN position has already been accomplished. Prudence would dictate that bearings to physical landmarks and features be also made to further confirm the reliability of the fix.

For AIS and non-AIS virtual eAtoN, the problem becomes more complicated since there is no physical AtoN presence at all. A fix developed based upon bearings taken to physical landmarks and features would be a suitable method for verifying location only in the case where such features were visible and not obscured or out of visual or radar range. However, there is another means to take such a fix through reference to ground. This may be accomplished, again using modern technology, by ground referencing to obtain position verification using known surface landmarks (radar bearings to known landmarks, etc.). This can include bottom features obtained through wireform and/or point cloud ENC models compared to live radar and/or echosounder measurements made over time intervals using running averages and first derivative trend information. ENC information is already on board vessels in the charting equipment (e.g., ECDIS) and requires the proper resolution and correlation to determine bearings, produce the necessary fixes and generate warnings.

Fix and bearing information to known physical environmental features for each eAtoN can be taken during initial installation and encoded as part of its characteristics. These characteristics can then be used anytime thereafter to verify position accuracy during normal use and subsequent verification. Data encryption of position characteristics can also be used to ensure their security and validity.

Such methods can also be used to detect the effects of AIS and GNSS jamming and spoofing since the presumed location based upon GNSS would not coincide with environmental features. Used with inertial backup, it would also be possible to verify position in the event of GNSS outage.

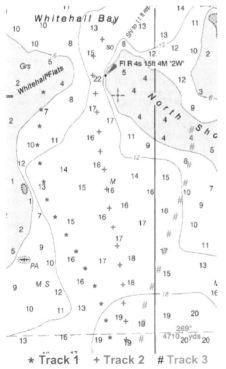

★ Track 1　　+ Track 2　　# Track 3

Figure 2. Divergent Ground Tracks 1 and 3 Compared to the intended Ground Track 2.

Data obtained from echosounder measurements using this technique on ground tracks 1, 2 and 3 may appear similar to that shown in table 1.

Table 1. Difference and Rates of Change of Error Track 1 and Error Track 3 from Intended Track 2 shown in Figure 2.

		t=	1	2	3	4	5	6	7	8	9	10	11	12	13	14	15
* Track 1		Depth	19	17	15	16	16	16	15	15	14	13	13	10	8	7	3
		dD/t	0	2	4	2	2	1	2	1	2	3	4	9	14	8	11
		dD/dt		2	2	-2	0	1	1	-1	1	1	1	5	5	-6	5
+ Track 2		Depth	19	19	19	18	18	17	17	16	16	16	17	19	22	15	14
		dD/dt		0	0	1	0	1	0	1	0	0	-1	-2	-3	7	1
# Track 3		Depth	19	19	19	18	15	17	17	16	12	10	8	4	4		
		dD/t	0	0	0	3	0	0	0	4	6	9	15	18			
		dD/dt		0	0	0	3	-3	0	0	4	2	3	6	3		

time (t) track from bottom to top of chart.

Transit of the intended track shown in figure 2 (Track 2) is dependent upon accurate GNSS position correlation with chart location data. Should either AIS or GNSS spoofing or jamming occur resulting in inaccurate positioning, differences in both the depth and/or the rate of change of depth profiles between the intended and actual transited courses would be detectable. For example, should spoofing occur where the vessel believes itself to be on the intended Track 2 based upon AIS/GNSS sensor readings yet follows either of error Tracks 1 or 3, deviation from the proper course will be detected through comparison of derived bottom feature and contour data illustrated in table 1 with independently obtained echosounder readings even though AIS and/or ECDIS is falsely displaying the intended course. Should GNSS jamming or outage be encountered, these same bottom reference data can be used with inertial system backup to continue navigation and accurately update the vessels position.

6.3.2 Other Characteristics

Once verification of accurate eAtoN positioning has been accomplished, verification of additional characteristics that include eAtoN type, name, etc., would be performed by examining the contents of the AIS/ECDIS information on the navigation display. For example, the type indication should correlate with the proper valid symbol for cardinal marks (N/E/S/W), lateral marks (IALA A/B port and starboard), isolated danger, safe water, special purpose or emergency wreck marking as published for that location. eAtoN name and other characteristic verification would be accomplished using the same method.

6.3.3 Metrics

Physical Characteristic examination is complete when all steps necessary to determine verification or discrepancy have been accomplished. Metric results provide traceability of verification and identify areas where further product and process maturation is needed. Data collection for many metrics can also be automated, ensuring measurable progress in completing verification and generating performance scores to aid in their understanding.

7 PRESENT STAGE OF MATURITY

eAtoN technology is very much in an early stage of development with only a handful of AIS eAtoN deployed in experimental evaluation program locations worldwide. Non-AIS eAtoN configurations are even less mature as theoretical concepts and implementations have yet to materialize outside the laboratory. Participation of the maritime community is being actively solicited by cognizant authorities and authorized service providers to ensure progress is constructive and meeting user needs. This is evidenced by U.S. Army Corps of Engineers (USACE), NOAA, and Coast Guard invitations to maritime stakeholders to participate in Future of Navigation Public Listening Sessions and Navigation Information Days throughout the country to collect comments and feedback regarding requirements for navigational information and service delivery system needs (Smith 2014; NOAA 2014a). Additional eAtoN installations are being deployed throughout the United States in an attempt to fulfill a wider set of maritime needs.

8 CONCLUSIONS

Installation of AIS-based eAtoN system facilities will continue as operational experience and results showing their utility are documented. A critical need exists for non-AIS eAtoN technology for use in remote and sensitive environments. Further research and development should be encouraged in this area.

Significant limitations and vulnerabilities exist in the AIS and GNSS technologies that support eAtoN operations. Spoofing and denial of service attacks will accelerate due to the lack of security in both of these areas as states and criminal organizations gain experience in using and misusing these technologies.

Opportunities exist using currently available data fusion and sensor technology to mitigate these problems and reduce the severity of and even eliminate their effects, increasing marine safety in general and specifically the safety of navigation.

The proposed techniques can also be applied to improve existing AtoN verification practices.

REFERENCES

Baldauf 2008. Baldauf, M.; Benedict, K.; Motz, F.; Wilske,E.; Grundevik, P. (2008). Aspects of Technical Reliability of Integrated Navigation Systems and Human Element in Case of Collision Avoidance. In: Proceedings of the Navigation Conference–Navigation and Location: Here we are! Church House, Westminster, London (UK), 28-30 October 2008

Biancomano 1998. Vincent J. Biancomano and Scott W. Leyshon, Practical Tropospheric Model for VHF/UHF Propagation, *National Weather Digest*, http://www.nwas.org/digest/papers/1998/Vol22No2/Pg15-Biancomano.pdf

Forssell B. 2009. The dangers of GPS/GNSS. *Coordinates*, February 2009, pp. 6-8

Gibbons 2014. Glen Gibbons, GLONASS Suffers Temporary System wide Outage; Multi-GNSS Receiver Overcomes Problem, Inside GNSS, April 2, 2014, http://www.insidegnss.com/node /3979

Grabowski 2012. Joseph C. Grabowski, Personal Privacy Jammers: Locating Jersey PPDs Jamming GBAS Safety-of-Life Signals, GPS World, April 1, 2012, http://gpsworld.com/ personal-privacy-jammers-12837/

IALA G1028. Guideline No. 1028, *Automatic Identification System (AIS)*, pg. 8, Volume 1, Part I (Operational Issues)

IALA G1062. Guideline No. 1062-*Establishment of AIS as an Aid to Navigation*, Edition 1, pg. 6, December 2008.

IALA G1081. Guideline No. 1081-*Virtual Aids to Navigation*, Edition 1, pg. 7, March 2010.

IALA O143. O-143, *IALA Recommendation on Virtual Aids to Navigation*, pg. 5, March 2010.

ITU 2007. REPORT ITU-R M.2123, *Long range detection of automatic identification system (AIS) messages under various tropospheric propagation conditions*, (2007), pg. 32, Appendix 5 AIS propagation observations – Ducting.

Middleton 2014. Hide and Seek, Managing Automatic Identification System vulnerabilities, *Coast Guard Proceedings*, Winter 2014-2015, pg. 50.

MSC 82. MSC 82/24/Add.2. Annex 24, Res. MSC.232(82) (adopted on 5 December 2006) *Adoption of the Revised Performance Standards for Electronic Chart Display and Information Systems (ECDIS)*.

NOAA 2014. Office of the Chief Information Officer, IT Security Office, 15 Sept. 2014, https://www.csp. noaa.gov/

NOAA 2014a. NOAA Office of Coast Survey, Navigation Industry Day, 10 October 2014, Annapolis, MD

OCS 2014. NOAA Office of Coast Survey, Nautical Charts and Pubs, *Symbols for AIS ATON*. http://www.nauticalcharts.noaa.gov/mcd/updates/ais_aton.html

Radgowski 2014. Capts. J. Rogowski and K. Tiongson, Cyberspace-the Imminent Operational Domain, *Coast Guard Proceedings*, Winter 2014-2015, pg. 18.

Simonite 2013. Tom Simonite, Ship Tracking Hack Makes Tankers Vanish from View, *MIT Technology Review*, October 18, 2013.

Smith 2014. CAPT Scott J. Smith, 21st Century Waterway, Future of Navigation, *eNavigation Conference*, 12-13 Nov.2014, Seattle, Washington.

Thompson 2014. Brittany M. Thompson, GPS Spoofing and Jamming-A global concern for all vessels, *Coast Guard Proceedings*, Winter 2014-2015, pg. 50.

USCG 2005. US Coast Guard, *Aids to Navigation Manual, Administration*, COMDTINST M16500.7A, 2 MAR 2005, pgs 9-1, 9-3.

USCG 2010. US Coast Guard *Aids to Navigation Manual, Technical*, COMDTINST M16500.3A.

USCG 2014, News Release, *Coast Guard Testing Nation's First "Virtual" Aids to Navigation in San Francisco Bay in Effort to Enhance Safety*, August 4, 2014, 11th District Southwest Public Affairs.

USCG 2015. USCG Navigation Systems Div., R.D. Lewald.

USCG 2015a. US Coast Guard Navigation Center, Automatic Identification System, *How AIS Works*, http://www.navcen.uscg.gov/?pageName=AISworks

Windward 2014. AIS Data on the High Seas: An Analysis of the Magnitude and Implications of Growing Data Manipulation at Sea, http://www.windward.eu

Wright & Baldauf 2014. Collaborative Navigation through the Establishment and Distribution of Electronic Aids to Navigation in Real Time. *Joint Navigation Conference*, Institute of Navigation (ION); June 2014, Orlando, FL.

Zaragoza, S 2013. University of Texas at Austin, S. Zaragoza, Spoofing a Superyacht at Sea, 30 July 2013, http://www.utexas.edu/know/2013/ 07/30/spoofing-a-superyacht-at-sea/

Analysis and Decision-making for Control of Extreme Situation of Fishing Vessels on the Base Dynamic Model of Catastrophe

E.P. Burakovskiy, Yu.I. Nechaev, P.E. Burakovskiy & V.P. Prokhnich
Kaliningrad State Technical University, Kaliningrad, Russia

ABSTRACT: The problems of management and decision making while operative controlling the emergency situations in the field of operational durability of fishing vessels are discussed. Implementation of management and decision making is executed in on-board intelligent systems of new generations functioning in the real-time mode. Algorithms of controlling the emergency situations are made on the basis of the catastrophes dynamic model in the multiprocessor computing environment. Practical applications of the developed computing technology are considered in tasks of operational durability at the different level of acting indignation in a complex dynamic environment.

1 INTRODUCTION

The modern catastrophe theory application while controlling the emergency situations is connected with formalization of some important aspects of vessel's operational durability problems on the basis of achievements of intelligent technologies and high-performance computations [1, 6, 7]. The formal apparatus of controlling the emergency situations is formulated in the form of theorems and axioms of the catastrophes dynamic model [12]. The theoretical basis of this model determines integration of knowledge about vessel's behaviour in a complex dynamic environment in virtual polygons (VP) and intelligent systems (IS) of new generations [12, 13]. Herewith, the structure of management and decision making is introduced on the basis of axiomatics of set theory, achievements of operational durability and modern computer mathematics. Presentation of materials is accompanied by considering important practical applications in the field of dynamically complex systems.

The peculiarity of the proposed research is that integration of computing technologies of naval structural mechanics with methods and models of neuro dynamic systems (ND-systems). These systems connect neuro illegible and neuro evolutionary modeling. Realization of the structural and functional configuration of ND-systems on the basis of the modern catastrophe theory is executed with the help of fractal geometry and formal conceptual analysis [1-5]. Graphic interpretation of physical regularities is presented by a fractural structure. With the help of such structure processes of developing emergency situations within the frame of paradigm of information processing in the multiprocessor computing environment [13]. The conception of fractal structure is introduced as a result of abstraction process; empirical testing of theoretical results and statements accuracy is checked on the basis of the possibility principle. This principle is fundamental while designing new theories in different fields of scientific-practical applications, including operational durability problems.

Emergency situations beginning, developing management mechanisms and supporting decision making while interpretation of a vessel and external environment interaction is described in the integrated software-hardware environment of onboard computing system. Examples of application of considered problems in the context of the most conceptual decisions –virtual polygons (VP) and onboard intelligent systems (IS) of new generations ensuring the safety of vessels' seafaring and technological tools of ocean development. Functioning of VP and IS is executed in the real-time mode while intensive impact of non-stationary wind flow and wave systems. The dynamics of interaction is realized within the frame of synergetic theory of management with the help of ontology of flow of events [12, 15].

2 THE PROBLEM OF EMERGENCY SITUATIONS CONTROLLING ON THE BASIS OF THE MODERN CATASTROPHE THEORY

Principal difference of the theoretical basis of emergency situations control consists of the fact that the problem of the dynamics of a vessel and external environment interaction is realized on the basis of the modern catastrophe theory [12]. The catastrophes dynamic model integrates the interactive component of the knowledge formal system describing the vessel's behaviour on the basis of fractal geometry and entropy analysis [9]. Herewith, fundamental results of the classical theory are explicated in new interpretation of vessels' operational durability in conditions of continuous changes in the object's dynamics and external environment. The main aim of the article is to make a reader interested in the circle of new ideas and arouse interest in new research field, including general aspects of the dynamics of complex systems.

The conceptual model of information processing while interpretation of functional area of the control system in a complex dynamic environment simulates:

$$S(U) = \langle F(Com) : \{T(t,\tau) \times X(KB) \times Q(V,W)\} \to Y(R) \rangle \quad (1)$$

where $S(U)$ determines multitude of management strategies, and its components represent the following structures: $F(Com)$ – multitude of elements realizing the competition principle; $T(t, \tau)$ – multitude of time moments determining the model of developing controlled situations; $X(KB)$ – multitude of elements of data operative base; $Q(V, W)$ – multitude of meaning of the vector of input signals (conditions of no-stationary external environment); $\{T(t, \tau) \times X(KB) \times Q(V, W)\}$ – multitude of data regularities; $Y(R)$ – multitude of the rules of information generalization, τ – the time interval of realization.

The algorithm feedback in the conceptual model (1) is used for modeling formation of control actions. While supporting decision making procuring interaction is realized on the basis of criteria of maximum efficiency. The principle of the algorithm feedback is the residual of the system's entropy before and after obtaining information which reduces the uncertainty in evaluation and analysis of the current state of the system and lack of uniqueness in choice of ways of formation of control actions depending on criticalness of arising situations.

Interpretation of the system's topological dynamics is performed within the frame of the modern catastrophe theory [12]. Configuration of the computing complex is presented with a fractal graph

$$G(F_R) = \langle V(E,U), A(E,C) \rangle, \quad (2)$$

formalizing events $(V(E,U)$ describing actions in the system; and conditions $A(E,C)$ – in logical description of its condition. As follows from this statement, the complex control module ensures processing information flow connected with the system's conditions and description conditions while the process of evolution in the non-stationary dynamic environment.

The computing complex formal model of the emergency situations control system functioning in the multiprocessor computing environment is presented in Figure 1.

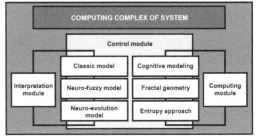

Figure 1. The conceptual model of the emergency situations control system within the frame of the integrated computing complex

The functional block of the system's dynamics contains interpreting and computing modules ensuring neuro illegible and neuro evolutionary modeling with the use of fractal geometry and entropy analysis. Integration of the mentioned components is realized within the frame of the catastrophes dynamic model and allows to formalize processes of information processing on the basis of the data of the physical experiment, achievements of modern intelligent technologies and high-performance computations [13].

The task of controlling the system's dynamics is in investigating its behaviour while different level of acting disturbance. Within the frame of such interpretation measurements and information processing are performed to determine the system's dynamic characteristics.

The vector of independent parameters $\Lambda(\pi)$ characterizes the system's movement modes within the limits of the permitted area. Equilibrium of the system's state in the range of variables from $\Lambda(\pi)_0$ to $\Lambda(\pi)_k$ is determined on the basis of formulations:

$$\Lambda(\pi) = (\pi_1,...,\pi_p) \in \Omega \subset R^p;$$
$$\Omega : \forall q \in 1,..., p, \pi_q \in (\pi_q^{\min}, \pi_q^{\max}); \quad (3)$$

$$\forall q \in 1,..., p, \pi_q \in [\pi_q^0, \pi_q^k] \subset \Omega;$$
$$\pi_q^0 \in \Lambda(\pi)_0, \pi_q^k \in \Lambda(\pi)_k, \quad (4)$$

where $y_0(\Lambda(\pi)_0)$, $y_k(\Lambda(\pi)_k)$ are original and finite values of the system's equilibrium.

The control action changing the system's state in accordance with the desired law is described with the help of correlations:

$$y(\Lambda(\pi)) = y^*(\Lambda(\pi));$$
$$\pi_q = \lim_{t \to \infty} v_q(t) = v_q = const;$$
(5)

$$y^* = y_0(\Lambda(\pi)_0), \quad y^*(\Lambda(\pi)_k) = y_k(\Lambda(\pi)_k),$$
(6)

where $y^*(\Lambda(\pi))$ is the vector characterizing the diversity of the system's equilibriums determined in finite parametric areas; $\Lambda(\pi)$ is the vector of parameters $(\pi_1,...\pi_p)$, determining the system's characteristics and disturbing actions for the desired evolution (it is stated according to the system's operation modes within the limits of the permitted area Ω "input-output").

Thus, the problem of the system's behaviour control on the basis of formulations (3) – (6) is in considering the stochastic discrete system with r-dimensional area of inputs. Outputs of the system in the time moment t present themselves vector-columns of parameters $(\pi_1, \ldots ,\pi_p)$. converting this information is connected with solving difficult problems of analysis and interpretation of physical measurements data presented by the information vector $J(\pi_1, \ldots ,\pi_p)$.

Evolution of the system is developing according to the peculiarities of interaction and is formalized as applied to the control of emergency situations [1-5, 8-13]. Methods of the system's management are oriented to ensuring functioning of the computing complex in the conditions of continuous changes in the object's dynamics and external environment. Adaptation of the system is provided with the help of mechanisms of fuzzy control determining configuration "input - output". Herewith, feedback may be negative while stabilization of the system in the attractor's basin, or positive leading to bifurcation, instability and chaos.

Formation the structure of the system's data while modeling dynamic situations is realized in the direction of the organization conversion in the multiprocessor computing environment [13]. The process of conversion is accompanied by the phase transition as a result of which the quality jump in changes of the system's elements, at that the sequence of conversion is performed as a chain of information change:

$$F(SA) \to Fluct \to Bifurc \to Reorg \to F(NSA),$$
(7)

where $F(SA)$ is functioning the system in the field of attraction of the original structure-attractor; *Fluct* – fluctuations; *Bifurc* – bifurcation; *Reorg* – reorganization (the phase transition); *F(NSA)* – functioning in the field of the new structure-attractor.

Within the limits of the presented conceptual model processes of structural and functional IS configuration on the basis of the apparatus of the logical structure knowledge organized on the basis of the synergetic management theory [15] and cognitive paradigm [12] with the use of the complication principle [14] and the concept of the minimum description length [8] are formalized.

The principal difference of the considered approach to information processing in the system's functional area is in the use of integrated models of analysis and interpretation of dynamic situations. Realization of the models is determined by a complex character of mathematical description of discontinuity of investigated phenomena. The advantages of the offered approach are in the possibility of not only geometrical but analytical interpretation of the system's dynamics; and also in the form of simple geometrical images of fractal displays of dynamic situations.

3 THE INTEGRATED DYNAMIC ENVIRONMENT OF IS EMERGENCY SITUATIONS CONTROL

The integrated dynamic environment of interpretation of emergency situations allows to execute the synthesis of IS while continuous control of a vessel's behaviour and different conditions of external environment. Functioning of the programming complex is performed on the basis of computing procedures with the high level of parallelism (Fig. 2). The basis of plotting the coherent theory explaining and forecasting the features of the control system is the information approach as namely information is a universal category for describing systems and phenomena [8].

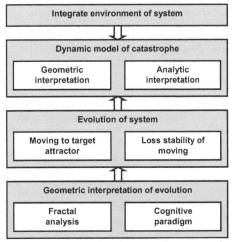

Figure 2. The integrated environment of emergency situations control

Let's formulate the description of the modern catastrophe theory in the form of interpretation of the dynamic system of emergency situations control. Plotting the theoretical basis of the system within the frame of this theory is defined by original axioms. All the rest statements of the theory are logical conclusions of axioms. In the basis of the axiomatic synthesis of models of catastrophes lies the method of interpretation. Each definition and relation of the system's axiomatic theory is correspondent to some formalized description.

Definition 1. Summation of focused views is called the field of system's interpretation:

$$P(Int) = \langle Q_1(CSE), ..., Q_m(CSE) \rangle, \tag{8}$$

where $< Q_1(CSE), ... , Q_m(CSE) >$ – mathematical descriptions of models of identification, control and forecast of emergency situations determining a general structure of functional modules of the computing complex of the catastrophes dynamic theory.

Some outgivings about elements of the system's interpretation field are put into correspondence with all statements of the axiomatic theory. The interpretation field ad its features are the object of study of the basic theory - the catastrophes dynamic theory which also may be presented as the axiomatic theory. This method allows to prove consistency of the system's theory depending on the axiomatic catastrophes theory.

4 THE CATASTROPHES DYNAMIC MODEL WHILE CONTROLLING EMERGENCY SITUATIONS IN SEA CONDITIONS

Methodological foundations and formal apparatus of used methods and models of IS determine solving the problems of diagnostics and modeling in difficult dynamic environment. The developed methodology allows to realize the problem of emergency situations control on the basis of the modern catastrophe theory.

The catastrophes dynamic model (Fig. 3) determines the interpretation function in the form of bifurcation multitude $B(\theta,t)$ in the area of control variables; multitude $GZ(\theta,t)$ defining the control dynamic environment integrating impact on external disturbances and peculiarities of the object's dynamics; multitude $F(W(t),D(t))$, determining external environment and structural changes in the vessel's behaviour.

While changing control parameters the system can evolve to the condition of equilibrium correspondent to the field of stabilization while the system's movement to the target attractor.

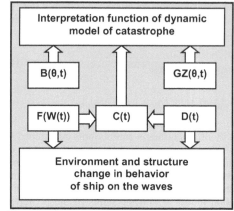

Figure 3. The catastrophes dynamic model of emergency situations control

Statement 3. The theoretical basis of the catastrophes dynamic model supposes the topological analysis of the system's behaviour in the following marginal (ultimate) conditions:

$$\Phi(Int): \Omega(W) \xrightarrow{U_1(t)} ... \xrightarrow{U_m(t)} \Omega(Stab) \tag{9}$$

$$\Omega(W) \xrightarrow{U_1(t)} ... \xrightarrow{U_n(t)} \Omega(Cat) \tag{10}$$

Here $\Omega(W)$ – the fields characterizing the system's conditions adjusted on the basis of fractal geometry; $\Omega(Stab)$ - the attraction field determining the system's movement to the target attractor; $\Omega(Cat)$ - the stability loss field (the catastrophe's beginning), $[U_1(t),...,U_m(t)]$ and $[U_1(t),...,U_n(t)]$ – control impacts for events of the system's evolution.

The target attractor in displays (9), (10) is considered as an asymptotic limit ($\tau \to \infty$) of decisions, to which the system's original conditions don't have straight influence, at that, if the laws of saving permit several equilibriums (decisions) than the movement condition is realized to which the minimum entropy growth is correspondent.

Definition 2. The dynamics of the system's emergency situations control in the area of state is characterized by the interpretation function determining the attraction field and the control actions in the system's evolution process under the influence of external disturbances.

Interpretation of the system's behavior on the basis of the competition principle allows to ensure continuous control of its dynamics. Within the frame of this conception the system's evolution is determined by external conditions and is realized via self organization and adaptation of the system to external disturbances. In the process of the system's development its state and complexity are changing which are determined by the stored information.

5 THE AXIOMATIC BASIS OF THE EMERGENCY SITUATIONS CONTROL SYSTEM

Main operations executed in IS within the frame of information processing paradigm [13] is performed on the foundation of the axiomatic basis [12], allowing to describe the system's topology at the level of structural and functional configuration (Fig. 4).

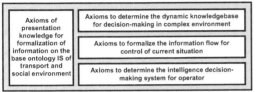

Figure 4. The axiomatic basis of the emergency situations control system

The primary virtue of such technology is presenting the system's evolution at the interval of realization [t_0, t_k] in the form of fractural structures, and interpretation of the interaction dynamics with the help of entropy analysis [9]. As a result simplicity and visual expression of displaying the development process of the current situation in complex dynamic environment.

In the considered field the axiomatic presentation of knowledge while the system's interpretation is determined by formalization of information flow on the basis complex ontology [12], in the form of the following statements.

Axioms 1. The appointment of identification axioms is description of all types of variables and relations determining the system's structure formalized field:

$$Axiom(Ident) = \langle Var, Rel \rangle \in \Omega(Str), \qquad (11)$$

where Var, Rel – variables of the problem and relations between them; $\Omega(Str)$ -the structure.

Axioms 2. The planning axioms set the rules (the order and conditions of correctness) of computations. In the basis of topological objects computations in the phase area lies the feature of interval arithmetics permitting to execute convergence of the interactive procedure of computations of interval meanings of parameters to some localizing intervals containing demanded decisions while interpreting the system:

$$Axiom(Plan) = \langle Evol(PH), Int(Cal) \rangle, \qquad (12)$$

where $Evol(Ph)$ – the evolution field of the phase area; $Int(Cal)$ – procedures of interval arithmetics.

Axioms 3. The computation axioms set the rules of computing relations in the phase area of the NN-system. The computation axioms also include optimization axioms.

$$Axiom(Calcul) = \langle Rule(Cal), Axiom(Opt) \rangle, \qquad (13)$$

where $Rule(Cal)$ –rules of calculations; $Axiom(Opt)$ – optimization axioms ensuring solving the problem of optimal decisions search, the structure of which is presented in the form of Logical Axioms Tables.

Thus, knowledge axiomatic presentation allows to execute support of modeling procedures while interpretation of the system's evolution in complex dynamic environment.

6 THE AXIOMATIC BASIS OF FRACTAL GEOMETRY

Interaction dynamics interpretation in the system is realized on the base of fractal structures. These structures ensure modeling of the situations development process and mechanisms of managing changes of fractal displays in the frame of the catastrophes dynamic model [12].

Axioms 4. The system's organization is described by the graph-structure in the form of events and conditions (states of the system):

$$g(v(E), a(E)) \in G(V(E), A(E)), \qquad (14)$$

determining the sequence of configurations $g_1(F_R)$, ... ,$g_N(F_R)$ of the system's movement while its equilibrium (movement to the target attractor) and while loss of stability (a catastrophe beginning).

On the current configuration of fractal families original $g_0(F_R)$ and finite $g_N(F_R)$ conditions of the system are fixed depending on interpretation peculiarities.

Axioms 5. The system is developing at the time interval [t_0, t_k] in the form of the sequence of discrete states:

$$S(t) \in S[t_0, t_k], \qquad (15)$$

formalized within the frame of the quasi stationarity [12, 13].

The area of the fractal system's states $S(t)$ is determined by ordered collections of fractal displays. These displays present themselves self similar structures developing in the process of the system's evolution. In the general form the fractal ensemble is imagined in the form of compact multitude presented on the basis of interpretation of the system's dynamics. The sequence of tick marks of evolution turns the fractal system $S(t)$ from the current configuration determining the beginning of its functioning into subsequent states setting its dynamics at the investigated time interval [t_0, t_k].

Axioms 6. The stages of the system's stable development $S(t)$ determine the system's movement to the target attractor. In this situation the system stays in the state close to equilibrium, and its organization doesn't undergo considerable changes.

While the stable state of the system external disturbances are described by flow load changing the fractal structure Fr depending on peculiarities of the situation. Movement to the target attractor *S(G, Attr)* while this state of the system is determined by configuration:

$$S(G, Attr) \rightarrow Stab(F_R). \qquad (16)$$

Axioms 7. The stage of the critical condition (a catastrophe's beginning) is characterized by the output of the system's investigated parameters from the given range and beginning of bifurcations – the system's stability loss, which is connected with formation of alternative variants of its organization. Tick marks of the system's functioning determine the process of self organization *S(G,SO)* of the fractal structure F_R in conditions of continuous system's dynamic changes. The system's configuration while loss of stability (the phase transition and forming a catastrophe) is characterized by the condition:

$$S(G, SO) \rightarrow Cap(F_R). \qquad (17)$$

Thus, conditions and their realization within the frames of the catastrophe dynamic model allow to formulate the axiomatic basis and present ontology of interpreting current situations. Developed on the basis of this formalization ontology models are considered as constituents of the common problem of the systems' ontology and are defined by the truth criterion *C_R(True)* based on fullness demand *Dem(Full)* and consistency *Dem(Non-Contr)* of formulated axioms and rules of conclusion:

$$Ont(SAU) = \langle C_R(True) [Dem(Full), Dem(Non-Contr)] \rangle. \qquad (18)$$

7 THE TOPOLOGICAL ENTROPY OF THE EMERGENCY SITUATIONS CONTROL SYSTEM

In problems of IS dynamics the concept of topological entropy is used. In the theory of dynamic systems this entropy is defines in accordance with the following formulation [12]:

$$h(T) = \lim_{\varepsilon \to \infty} S_\varepsilon(T), 0 \le h(T) < \infty, \qquad (19)$$

and points out exponential asymptotics of trajectories number different form the given one but arbitrarily with high precision.

The axiomatic basis of topological entropy while system's investigation in complex dynamic environment is formalized on the basis of the following definitions.

Axioms 8. The system's dynamic interpretation in the evolution process in uncertainty conditions is determined on universal multitude – topological entropy. The size of entropy of the investigated parameter is calculated within the frame of this multitude [9].

Axioms 9. The topological entropy realization while controlling the system's behaviour creates the multitude core, containing reorganization, permitting to present the entropy size *H(y)*, entropy potential *Δэ* and its adhesion *δΔэ* on the basis of the catastrophe dynamic model (movement to the target attractor and the stability loss):

$$H(y) = -\int_{-\infty}^{\infty} p(y) \ln p(y) dy = -\int_{-\Delta}^{\Delta} \frac{1}{2\Delta} \ln \frac{1}{2\Delta} dy = \ln 2\Delta \qquad (20)$$

$$\Delta_э = K_э \sigma ; \ \Delta_э \rightarrow \min ; \ \delta\Delta_э = \Delta_{э1} - \Delta_{э2} \qquad (21)$$

where *p(y)* – the law of distributing the investigated parameter; *$K_э$* – entropy coefficient; σ - the size of standard deviation characterizing the degree of dispersion of the system's parameter.

Axioms 10. The system of axes *(σ, Kэ)* is considered as a special situation of the phase area of the system (entropy plane). The display point *(σ, Kэ)* defines the rectangle with sides σ and *Kэ* on correspondent axes. While *Δэ = const* multitude of points on the plane *(σ, Kэ)* form the isentropic line – the system's condition "isotropy".

Entropy potential of the system's condition uncertainty for every point of the isotropic line will be constant and equal Δэ. The system's change from one condition to the other one on one isotropic line doesn't change its entropy potential. The summation of entropy trajectories for different situations of the system's interaction creates the entropy portrait.

8 THE STRATEGIES OF INFORMATION PROCESSING WHILE CONTROLLING THE EMERGENCY SITUATIONS DYNAMIC CHARACTERISTICS

The system's evolution is displayed on the basis of synergetic paradigm [15]. Within the frame of such formalization the approach is used realized with the use of summarized dissipation principle formulated by N.N. Moiseev [10]. As applied to emergency situations investigation this principle has the following interpretation: if not the only system's state is allowed but a whole set of states correspond to the existing laws and relations, inflicted to the system, then that state is realized which is met by the minimum entropy growth. This principle has a universal character and is used while investigating behaviour of complex dynamic systems.

The evolution process is in the system's transition from one state to another one at the expense of control disturbances changing. As a result the system begins movement itself to the side of new state which it has during some time interval. In synergetics this approach is based on the principle of risks bifurcation management. In this case the system's dynamics development is executed with the use of the following axioms formulated within the frame of the generalized dissipation principle [10]:

Axioms 11. The bifurcation diagram displaying the system's movement to the target attractor presents itself a cascade of bifurcations. Herewith, this movement along regular sites of evolution is interchanged with bifurcation points characterizing the situation of "the order parameter" in the self organizing system on the basis of the synergetic paradigm.

Axioms 12. Movement along the bifurcation cascade in case of the system's stability loss is characterized by the movement trajectory displaying increasing the danger of a critical situation (a catastrophe) beginning.

Axioms 13. The principle of enclosure of the dynamic system's attractors defines basic definitions of the synergetic paradigm which can be interpreted as a decomposition of areas of states in the self organization process and segregation of the order parameters to which other system's parameters tune in the evolution process.

Realization of axioms 11-13 allows to segregate "hidden" effects and regularities in the system. Owing to decomposition of states' areas certain set of self organization processes forms and target attractors are created having their attraction fields depending on peculiarities of multitudes in the structure of the catastrophe dynamic system [12]. This feature of nonlinear dissipative systems is explained by self organization effect and opens new possibilities for risks' analysis in the systems while interpreting emergency situations [1, 12, 13].

The common approach to interpretation of decisions is based on using the complexity principle [14], within the frames of which the problems of mathematical description of modes of external environment and interaction dynamics are considered. The formal model of choice is realized on the basis of the criterion function formed depending on the solving problem of the system's behaviour control:

$$F(CR) = \Phi_1(Ident), \Phi_2(Appr), \Phi_3(Forecast), \qquad (22)$$

where $\Phi(Ident, Appr, Forecast)$ – the function of decision choice while the system's dynamics control, at that, Φ_1 - realizes the decision choice while identification of the current situation (external disturbances and object's dynamics evaluation), Φ_2 - the decision choice in approximation problems, and Φ_3 - adaptive forecast while the system's evolution control in accordance with the conception of the catastrophes dynamic theory.

While practical realization of the criterion function $F(CR)$ in models of the system's behaviour

the preference is always given to mathematical description built on the base of physical modeling data as the most efficient means of achieving a priori information about the object's behaviour. If in the process of the system's behaviour control appear an emergency situation then it is considered as a new problem and while its interpretation the class of possession is defined or a new class is formed on the base of the adaptive resonance principle [12, 13].

9 IDENTIFICATION OF THE EMERGENCY SITUATION ON THE BASE OF ALTERNATIVE DISPLAYING OF THE SYSTEM'S DYNAMICS

The system's control at the realization interval $[t_0, t_k]$ of the catastrophes dynamic theory within the frames of "the running window" of information processing is realized on the basis of the situation's identification from multitude of alternative variants:

$$\{S_i\} \quad (i = 1,...,z < \infty), \qquad (23)$$

each of which according to chosen criterion R is correspondent to one of alternative algorithms:

$$\{A_i\} \quad (i = 1,...,q). \qquad (24)$$

The correspondence between the investigated situation S_i and the optimal structure of the algorithm $S_i \rightarrow A_i$, is realized by the function

$$i = \phi(j), \quad (i = 1,...,q; j = 1,...,z). \qquad (25)$$

The function definition (25) solves the problem of the adaptive structure with the use of the algorithm's work results on previous stages of investigation and also special tests defining possession of the controlled situation to one of (23).

In the case using continual multitude each situation is code by the vector $S = (S_1, ... , S_u)$ in the u-dimensional area. Each point S_i of this area fits the best algorithm A_i, determined as

$$R(A_i, S_i) = \min_{j=1,...,q} R(A_j, S). \qquad (26)$$

Thus, we have a q-class problem of images distinguishing on the basis of the finite teaching haul $<S_j, A_j>$ $(j = 1, ... , N)$, elements of which are achieved by solving the optimization problem

$$R(A, S_i) \rightarrow \min_{A_1,...,A_q} \Rightarrow A_j, \qquad (27)$$

where S_j $(j = 1, ... , N)$ – situations presenting different parts of the area $\{S\}$.

While information processing in the real-time mode short realizations of investigated processes are used. The pointed strategy of information processing finds reflection in problems of designing algorithms of parameters' interpretation characterizing the vessel's behaviour in the emergency situation [12].

As an illustrative example let's consider the real-time mode system of evaluating the system's dynamics parameters under the influence of external disturbances (Fig. 5).

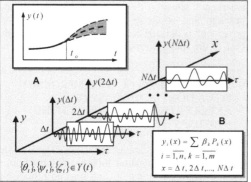

Figure 5. Modeling the system's dynamics: A-the measurements data (full curve) and forecasted meaning of the output characteristics (dotted line); B- the adaptive model structure; x = Δt, 2Δt,..., NΔt the current time of the situation development; τ - the realization interval; y(Δt), ... , y(NΔt)- the output characteristic.

Here the graphic interpretation of the work of the algorithm control is presented in the conditions of continuous change of the object's dynamics and external environment. The adaptive system realizes evaluation and forecast of the current situation development on the basis of the physical experiment data.

The sequence of operations of the algorithm functioning provides procedures of converting measurement information aimed at forming "the running window" and evaluation of dynamic characteristics depending on peculiarities of interaction regimes.

10 INTERPRETATION OF THE EMERGENCY SITUATIONS IN THE REAL-TIME MODE

As a result of information processing about the system's topological dynamics considering formulated above axioms at the interval [t_0, t_k] interpretation operations are executed. While executing the first event the rule P_1 is realized which fits the algorithm A_1 and the original fractal structure G_0 and so on. This sequence is determined by fixed system's position which are characterized by fractal displays G_1, ... , G_N in the current situation:

$$\langle P_1,...,P_N ; A_1,...,A_N \rangle . \qquad (28)$$

Thus, multitude of the system's dynamics realizations is characterized by ordered rules P_1, ... , P_N and algorithms of information processing A_1, ... , A_N with the help of which operative control of the

system's evolution is ensured in complex dynamic environment. The systematic approach while interpretation of the current situations by methods of the catastrophes theory allows to formalize the modeling process of the system's dynamics and present a network of topological areas in the form of the conditions tree of complex interaction (Fig. 6).

At this figure the conditions tree is displayed in a compact form showing the system's generalized topological structure on the basis of the catastrophes dynamic model. Here at the levels of the system, the subsystem and the element components are accentuated ensuring solving the problem of the current situation distinction and its analytical and geometrical interpretation with evaluation of the achieved result. The system integrates methods of classical mathematics, fuzzy logic, neuron networks and evolutionary modeling in the aims of solving the problems of decision making in complex dynamic environment.

Forming control impacts is realized depending on the emergency of appearing situations with the use of relative entropy [9] while interpretation of the system's dynamics in the event of movement to the target attractor [12]. As an original condition is adjusted the permitted system's entropy $\Im_{per}=1$, and limits are in the following form:

$$R(S)=(1-\Im_E/\Im_{per}) \leq R(P), \qquad (29)$$

where $R(S)$ – the degree of the system's declination from equilibrium; $\Im_E$ and $\Im_{per}$ – current and permitted system's entropy; $R(P)$ – the set degree of the system's declination from equilibrium which is set by parametric installation.

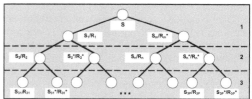

Figure 6. The conditions tree of the system's dynamics: S - the system's components; R –the result; 1- the system; 2 – the subsystem; 3 - the element.

Using conversion [12], we obtain a common decision for a number of necessary information in the following form

$$F_n(T_{CR})=(T_{CR})^n \exp(-T_{CR}), \qquad (30)$$

where T_{CR} –the time of information updating.

11 CONCLUSIONS

Thus, a new paradigm of calculation technology for dynamics of complex objects, realized in IS of ship strength control brings about the following advantages:

Thus, the paradigm of information processing on the basis of topological structures of the catastrophes dynamic model promotes the effective dialogue of an operator while management and decision making in the emergency situations. Modeling the system's dynamics within the frames of the paradigm ensures:

1 The choice of rational models and algorithms of IS functioning in the set parametric areas;

2 Forming the service of cognitive images and control impacts in the process of the system's evolution in complex dynamic environment;

3 Teaching neuro dynamic systems on tests examples from the library of cognitive images and management rules.

Thus, using the paradigm of information processing permits to increase the management and decision making quality owing to realization of correct algorithms and cognitive image phenomenon possessing high dynamics and information while the system's topological synthesis.

The developed conception and approach to analysis of operational durability problems on the basis of the modern catastrophe theory is executed in onboard IS of new generations. The knowledge structure allows to formalize information flow via different relations between input and output parameters of investigated nonstandard (non-nominal and emergency) situations. The developed software of IS functioning permit to use the advantages of different formalisms – compactness, simplicity and fullness of knowledge content in models realized in the multiprocessor computing environment [13, 15].

REFERENCES

1. Burakovskiy E.P., Burakovskiy P.E., Nechaev Yu. I., Prokhnich V.P. Problems of forecasting and disaster monitoring methods of the modern theory of catastrophes // Marine intelligent technologies, №2(16), 2012, p. 50 – 60.

2. Burakovskiy E.P., Burakovskiy P.E., Nechaev Yu. I., Prokhnich V.P. Operational strength of the hull of fishing vessels: a tutorial. – St. Petersburg, 2012. – 392 p.

3. Burakovskiy P.E., Nechaev Yu. I. Fractal structure of the operational strength of vessels fishing fleet on the basis of the modern theory of catastrophes // Marine intelligent technologies, №1(23), 2014, p. 9 – 14.

4. Burakovskiy P.E., Nechaev Yu. I. Practical implementation of graph-analytical control system of the overall strength of the fishing vessels on the basis of the modern theory of catastrophes // Marine intelligent technologies, №2(24), 2014, p. 9 – 13.

5. Burakovskiy P.E., Nechaev Yu. I. Practical implementation of graph-analytical control system of strength of fishing vessels at groundings, collisions between vessels and the entrance to the ice field on the basis of the modern theory of catastrophes // Marine intelligent technologies, №3(25), 2014, p. 99 – 104.

6. Burakovskiy P.E., Burakovskiy E.P., Nechaev Yu.I., Prokhnich V.P. Catastrophe theory in intellectual control system of vessel operational strength // The 10th Jubilee International Conference on Marine Navigation and Safety of Sea Transportation.– 19-21 June 2013, Gdynia, Poland, p. 26

7. Burakovskiy P.E., Burakovskiy E.P., Nechaev Yu.I., Prokhnich V.P. Catastrophe theory in intellectual control system of vessel operational strength // Marine Navigation and Safety of Sea Transportation: Navigational Problems.– UK, Croydon: CRC Press / Balkema, 2013, p. 29–34.

8. Kolmogorov A.N. On the representation of continuous functions of several variables as a superposition of continuous functions of one variable and addition // Reports of the USSR Academy of Sciences. 1957. Vol. 114, Issue 5, p. 953-956.

9. Lazarev V.L. The theory of entropy potentials. – St. Petersburg: Publishing House of the Polytechnic University, 2012.

10. Moiseev N.N. Selected Works. Moscow: Tayreks Co, 2003.

11. Nechaev Yu. I. Artificial intelligence: concept and application. – St. Petersburg, 2002

12. Nechaev Yu. I. Catastrophe theory: a modern approach to decision-making. St. Petersburg: Art-Express, 2012.

13. Onboard intelligence systems. Part 2. Ship Systems. – Moscow: Radiotechnika, 2006.

14. Solodovnikov V.V., Tumarkin V.I. Complexity theory and design of control systems. – Moscow: Science, 1990.

15. Synergetic Paradigm. The variety of searches and approaches. – Moscow: Progress – Tradition, 2000.

A Case Study in Flawed Accident Investigation

E. Doyle
Cork, Ireland

ABSTRACT: The Irish motor fishing vessel TIT BONHOMME, with a crew of six persons on board, was returning to Union Hall, her home base, during early morning darkness on 15 January 2012. In the final approach to the unlit harbour entrance the vessel stranded on the seaward side of Adam's Island, a well-known and charted hazard at the entrance to Glandore Harbour. In the adverse weather and sea conditions prevailing, the vessel quickly broke up and sank with the loss of five crew.

The report of the subsequent accident investigation found "the single overriding causal factor" was considered to be crew fatigue, giving rise to a situation where the vessel simply "steamed into and stranded on Adam's Island".

This paper examines a probable 'causal chain' and identifies factors not given due consideration in the MCIB Report; if, due to the shortcomings of an investigation, the error chain is only partially revealed the proper lessons from a serious casualty can never be adequately learned.

1 INTRODUCTION

1.1 *Incident Overview*

In the early morning darkness of Sunday, 15 January 2012, a local trawler TIT BONHOMME was returning to her home port of Union Hall, located within Glandore Harbour on the south coast of Ireland. The vessel had sailed from this port less than two days previously, intending to fish for five or six days, but having fished for about thirty hours an engine defect had forced the skipper to abort the intended operations and return to base. TIT BONHOMME was steaming at reduced speed of about four knots as she headed towards the harbor entrance.

There are no visual aids to navigation in the outer approaches to this harbor and a further complication was the deteriorating weather conditions. The south-easterly wind had increased to force 7, gusting to gale strength, and the reduced visibility due to rain meant a challenging approach to a lee shore. Reaching the inner sheltered waters safely would mean reliance on good local knowledge and effective use of the vessel's outfit of navigational aids, chiefly radar.

Apparently, with little or no prior warning TIT BONHOMME grounded on the rocky southern shore of Adam's Island at about 0535 local time (GMT). The vessel broke up and sank very quickly, and only one of the six persons on board survived.

1.2 *The Vessel and Crew*

The motor fishing vessel (mfv) TIT BONHOMME, C331, (Fig. 1) was an unremarkable, coastal/offshore stern trawler, rigged and equipped for bottom trawling for prawn and whitefish species. The vessel was built in 1988 in France, of steel construction, 21 meters (LOA) and powered for a service speed of about 10/11 knots. She was owned and skippered by Mr Michael Hayes, an experienced fisherman and holder of a Second Hand Special Certificate of Competency.

Figure 1. mfv TIT BONHOMME

Four other crewmen were carried, all Egyptian nationals, but two of whom seem not to have undergone the requisite basic safety training. Otherwise, this complement satisfied the appropriate safe manning regulations. In addition, one other person was embarked: Irish national, Mr Kevin Kershaw, had joined for this voyage as a student visitor.

The vessel's safety equipment (for five persons) appears to have been in order, though her radio equipment certification was found wanting. The information on TIT BONHOMME's outfit of navigational equipment is patchy but the vessel is understood to have been fitted with a comprehensive range of standard items for her class.

1.3 The Aborted Voyage

TIT BONHOMME, having departed Union Hall (in Glandore Hbr, Fig. 2) in mid afternoon on Friday, 13 January 2012, was at the fishing grounds and had commenced trawling operations (the first 'shot' of fishing gear) by 1730. They would make only five 'shots' in all, before suspending operations at about 2300 on the following day. This interruption in operations had been triggered by a defective bilge pump requiring the attention of the skipper — it appears he was the only person on board with sufficient engineering skill.

During the course of the repair, a more serious defect became obvious; lubricating oil was leaking from a pressurized pipe serving the main engine. Faced with worsening weather and with insufficient spare parts on board to properly remedy this defect the skipper decided to abort fishing operations and return to port. The main engine was restarted and run on low revs to give a boat speed of about two knots. After some 45 minutes running at these revs the skipper felt confident enough to increase speed to about four knots at around 0300 (Sunday morning 15 January). The homeward trip continued at this speed.

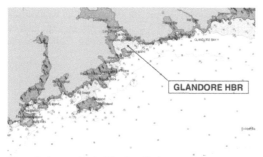

Figure 2. Approaches to Glandore Harbour

One of the Egyptian crewmen, Mr Abdelbaky Mohamed (the sole survivor), having assisted the skipper in the engine room for the previous three hours and before that, time on watch, now retired to his bunk, one of six berths in the common crew cabin. Two other crew and the supernumery visitor were also turned-in at this time, suggesting that the remaining crewman was standing watch in the wheelhouse for the remaining hours of the homeward voyage. The accident investigators could not establish who was on watch at the time of the stranding.

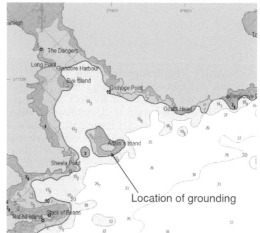

Figure 3. Casualty grounded on southern shore of Adam's I.

2 THE INVESTIGATION

2.1 Investigation and Report by MCIB

The stranding, and the circumstances surrounding the tragedy, was the subject of an investigation by the Marine Casualty Investigation Board (MCIB), who published their findings in Report No. MCIB/210 dated 27 March 2013.

It is a matter of some regret that the investigation and report are marred by significant defects and shortcomings. Much of the report addresses the search and recovery phase after the event, while insufficient attention is given to a thorough analysis of the evidence and circumstances germane to the stranding.

2.2 Reliance on 'the cause'!

Of particular concern is the conclusion that "...the single overriding causal factor is considered to be insufficient rest for the crew...". Any accident investigation report that relies on the identification of a single causal factor is, almost by definition, flawed; an experienced investigation team will invariably look for multiple factors in the quest to identify 'the causal chain' of events contributing to a serious casualty.

2.3 Fundamental Questions

The investigation report lays heavy emphasis on the issue of insufficient rest for the crew, concluding that at the time of the grounding "…all crew appear to have had at most four to five hours sleep" in the 40 hour period since departure from Union Hall. Whether this was so is open to conjecture, bearing in mind the very curtailed duration of this fishing expedition and the fact that only four tows or 'shots' and one partial tow had been completed. Crew fatigue may well have been, and most probably was, a causal factor and it is only proper that the issue should be rigorously analyzed. But whether it was the single overriding cause is certainly open to dispute.

2.4 Was Watchkeeper Asleep?

The report further concludes that because of fatigue and inadequate watchkeeping arrangements on board TIT BONHOMME "…it steamed into and stranded on Adam's Island during the hours of darkness in poor weather conditions…". Without constructing the conclusion in so many words, the report seems to be saying that those crew on watch simply fell asleep, allowing the vessel to steam ashore unchecked. Even within the scope of the report itself, there is convincing evidence to the contrary.

2.5 AIS Tracks

The report reproduces the AIS track of TIT BONHOMME, as supplied by the Irish Coast Guard. Various appendices show the vessel's earlier fishing tracks, her homeward approach and the final sequence preceding and during the grounding. The investigators' analysis of these tracks found expression in the report that "Whatever the watchkeeping arrangements put in place for steaming back to port, they resulted in the vessel steaming on an almost constant heading and speed until she was brought up short when she stranded on the southern side of Adam's Island." That the AIS track shows 'an almost constant heading' for most of the return voyage from the fishing grounds is true. But a significant, gradual and controlled change of heading during the seven minute period immediately preceding the grounding is starkly at variance with the MCIB analysis quoted above.

2.6 Navigation equipment

The report contains no information on the most likely or probable navigation equipment fitted on the vessel. Such vessels on the Irish fishing register are usually very well equipped in this regard, and there is no evidence to suggest that TIT BONHOMME was different in any way. The equipment schedule almost certainly included GPS, ENC and radar, although these are not mentioned in the report. The evidence yielded by the AIS data is most significant. While the report highlights the apparent use of the vessel's autopilot, the analysis presented, in this regard, is contentious.

2.7 Autopilot

The report notes that: "The vessel was fitted with an autopilot Navitron NT 920. The record indicates that the vessel was operating on auto-pilot up to the point of stranding. When recovered, the unit was noted with an indicated heading of 110-112°." This is followed by the further comment that: "The course setting manual adjusting control was noted rotating independently of the disc and the Mode Control noted set at *Permanent Helm*". It was also observed that "the Auto-Pilot Unit had sustained physical damage in the incident."

3 ALTERNATIVE ANALYSIS

3.1 The Final AIS Track

When TIT BONHOMME finally got under way to return home, some three hours or so, before the grounding, it is most likely that a destination waypoint was set for a convenient position in the approaches to Glandore Harbour. Common sense dictates that it would not be set on or over a hazard such as Adam's Island. But we have no way of knowing if this was the case, and the report offers no information on whether this common tactic was a routine practice on board. Thus, all the more reason to examine the AIS track very carefully.

3.2 Prudent Decisions

Everything about the skipper's decisions, (a) to abort the trip, (b) to test run his main engine for a 45minute period on low revs, (c) to return to base at slow speed, suggest responsible behavior by a cautious and prudent man.

3.3 Watchkeeping Practice on Autopilot

It seems reasonably safe to assume that once the skipper had set his course for home, he retired to his bunk at the rear of the wheelhouse. Indeed, as the report notes: "The normal watchkeeping arrangement in such circumstances where the vessel would be steaming on auto-pilot on a course previously set by the Skipper, would be for the Skipper to turn in, his cabin being in the wheelhouse, leaving instructions with the Watchkeeper as to when he was to be called. This would normally be for a specific time or a readily identifiable position to the Watchkeeper." Given the difficult circumstances and conditions that would face the skipper in navigating the entrance to Glandore Hbr, it is inconceivable that he would not have left the watchkeeper with clear instructions for his timely call, perhaps a mile or so from the entrance.

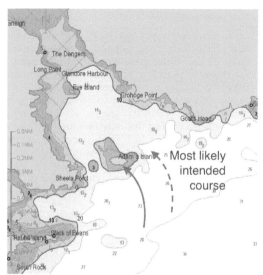

Figure 5. Comparison of actual track and likely intended path

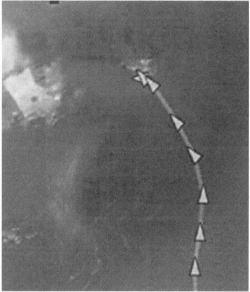

Figure 4. AIS track, from Irish Coast Guard data

3.4 Controlled Turn to Port

During the final 7-minute phase preceding the grounding, from 0528 to 0535, the AIS track (as presented at Fig. 4) paints a sequence of gradual changes in TIT BONHOMME's heading, initially from 007.5° turning to port to a heading of 326° as the vessel struck the shore. There is no significant change in the vessel's speed during this period, remaining at 4 knots, more or less. This turn, a little over 40°, is too great a deviation to be considered a random yaw from a pre-set course; it appears deliberate and controlled, and at a heading of 326° it seems to closely match a normal safe approach to pass Adam's Island.

The only reasonable conclusion to be drawn from this 7-minute sequence is that someone was exercising direct conning control of TIT BONHOMME, albeit in the mistaken belief that the vessel was making a safe approach to the harbor entrance. And it is not reasonable to believe that that person was anyone other than the skipper himself.

3.5 Use or Misuse of Radar?

If, as seems most likely, the skipper had direct control of the steering for the last 10 minutes or so, what went so tragically wrong? The report makes no mention of the possible use of other positionfixing equipment. In particular, was the use or misuse of the radar a factor in the casualty? It is simply not credible that a prudent skipper would attempt a blind harbour entrance such as this, in the prevailing conditions, without using his radar.

And if we accept that reasoning, and that radar was being used, why was Adam's Island not detected? A probable reason, given the near gale conditions, was the application of excessive sea-clutter control; an anti-clutter setting considered appropriate in the open sea would be quite incorrect if used close inshore approaching a harbor entrance with off-lying dangers. Under those circumstances, the suppression of the radar image of nearby dangers is almost inevitable, especially having switched from a longer (sea) range scale to the shorter range scales required for a port approach.

If, say, the skipper had been called when 1 mile from Adam's Island, his primary focus must have been on the steering task, and he may not have realized that the radar settings needed to be readjusted. If, in the last moments before grounding, he looked at the radar and saw no hazards immediately ahead he was probably lulled into a false sense of security and believed he was safely making the 'Most likely intended course' presented at Fig. 5.

It is not possible to say with certainty if the radar was misused in the manner just described, but the circumstantial evidence points in that direction.

Regrettably, the investigation team appear not to have considered the possibility; there is no mention in the report of any examination of recovered radar equipment. Neither does the report offer any analysis along the lines suggested in this paper.

It must also be acknowledged that crew fatigue may well have been a contributing factor in the operation of essential navigation equipment at a critical time.

3.6 Use of the Autopilot

As noted at paragraph 2.7 above, the investigators commented on the use of the vessel's Navitron NT920 autopilot. It is a very commonly fitted item on smaller vessels and within the fishing industry it is almost invariably used as the primary steering device, even when berthing and unberthing. An example of the NT920 autopilot control unit is shown at Fig. 6

Figure 6. Example of Navitron Autopilot , Model NT920

The fact that "…the vessel was operating on auto-pilot up to the point of stranding." is not surprising in the least. And the implied criticism that TIT BONHOMME simply steamed ashore unchecked on a pre-set course on autopilot is certainly wide of the mark. The report contains the curious comment that "The record indicates that the vessel was operating on auto-pilot…". What 'record'? There is no reference in the report to the identity of any such record.

In a further comment on the significance of the Navitron autopilot control unit, the report states that: "When recovered, the unit was noted with an indicated heading of 110-112°." The investigators seem to believe that this spurious or incompatible setting was the result of physical damage sustained as the wreck broke up during the hours following the grounding. This may well be so. But, there is an alternative and more rational explanation that the investigation analysis does not address. It is this: If, as suggested at section 3.4 above, the skipper was making what he mistakenly believed to be a safe entry into harbour, belatedly he must have seen and become instantly alarmed at the extensive surf and breaking water just in front of his vessel. We can readily appreciate that his immediate reaction, in those circumstances, must have been to attempt a hard-over turn by a dramatic rotation of the course control dial. The ultimate futility of any such attempt is not at issue. But the possibility, indeed probability, of such action deserved proper investigative analysis. No such analysis is offered in the report.

4 CONCLUSIONS

The MCIB report of the investigation into this casualty placed an unduly restrictive reliance on "the single overriding causal factor". It focused heavily, and almost exclusively, on 'crew fatigue' as that factor. In doing so, it gave insufficient consideration to a more probable causal chain of events. An especially serious omission has been the failure to address the possible use or misuse of radar as a causal factor, and this is regrettable.

It may never be possible to establish the final sequence of the TIT BONHOMME tragedy with any certainty, but the analysis outlined in this paper offers a more plausible error chain than that presented in the official report. It is an unfortunate consequence of a flawed investigation, where the error chain is only partially revealed, that the proper lessons from this or any serious casualty can never be fully learned. That's an added tragedy.

REFERENCES

S.I. No. 289/1988 Fishing Vessels (Certification of Deck Officers and Engineer Officers) Regulations, 1988. http://www.irishstatutebook.ie

S.I. No. 709/2003 European Communities (Workers on Board Sea-Going Fishing Vessels) (Organisation of Working Time) Regulations 2003. http://www.irishstatutebook.ie

IMO, Casualty Investigation Code, Resolution MSC.255(84) (MSC-MEPC.3/Circ.2 13 June 2008).

Marine Casualty Investigation Board (Report MCIB/210, No.5 of 2013) Report of the Investigation into the Sinking of the mfv "TIT BONHOMME", Dublin, MCIB, http://www.mcib.ie

Navigational Bridge Equipment

Radar Detection in Duct Situations in Maritime Environment.
The Vital Influence of the Radar Antenna Height

F.X. Hofele
Airbus Defence & Space, Ulm, Germany

ABSTRACT: Different meteorological conditions in maritime environment induce different optimal radar antenna heights to achieve a detection range as high as possible. Consequently the radar might be operated on an elevator moving along a mast. This offers the possibility to adjust the antenna height adaptively depending on the current atmospheric conditions. The possibility of utilizing an elevator moving along a mast and thus change the antenna position in height implies for the future for each site to require only one antenna instead of – so far – two antennas, one antenna about 40 – 60 m aloft and another so-called Duct-antenna at ground level to meet the Duct problem.

1 INTRODUCTION

Coastal radars are applied for surveillance in a specially complicated environment. The tasks of coastal radars include generally applications from identification of illegal immigration, terrorism, piracy, search and rescue up to illegal fishing and trafficking. For guaranteeing the detection of especially small objects approaching the coast it is important to know the capabilities and the performance of the deployed radar in full detail. However, the performance of radars in the maritime environment is deeply dependent on the meteorological and oceanographic conditions, the system parameters and the measurement geometry between radar and target.

In this context, it emerges the question of the influence of the antenna height on the target detection and on the detection range under various meteorological conditions. In general wording we ask: at which meteorological conditions in maritime environment which antenna height is optimal and should be chosen and how is this scheme practicable?

2 SPEXER 2000 COASTAL RADAR

Analysis with respect to these questions is done for the SPEXER 2000 Coastal Radar. It represents a member of a radar family, available for different tasks. The SPEXER 2000 Coastal Radar is a security radar with an Active Electronically Scanning Array (AESA) which provides multi-tasking and multi-mode capabilities to detect sea, land and air targets. It is a pulsed system including high Doppler resolution capabilities together with high sea clutter suppression. The main system specification relevant for the performance estimation under different meteorological conditions and sensor heights is the signal processing gain of clearly more than 20 dB. By the way, the SPEXER radars are worldwide the first radars of this type having at command AESA technology. The main task of the SPEXER radar family is the detection and tracking of threats like persons, small vehicles, small low flying airplanes like UAVs, small boats and swimmers. The SPEXER 2000 Coastal Radar can be deployed as land-based fixed-installed radar on a tower, integrated in smart vehicles, or as a transportable radar on a tripod.

Figure 1. SPEXER 2000 Coastal Radar on a tower

3 REFRACTIVITY, MODIFIED REFRACTIVITY

The terms we get to know in this chapter enable us to distinguish different propagation mechanisms in maritime environment by means of numerical values.

The Index of Refraction n describes the property of a media to change the direction of electromagnetic waves while passing through it. A measure of the diffraction is given by the quantity $n = c / v$, where c is the speed of the electromagnetic wave in free space and v the propagation velocity in the media.

For studies of propagation, the index of refraction n is not a very convenient number. Therefore, a scaled index of refraction, N, called refractivity, has been defined.

The term N is then dependent on the atmosphere's barometric pressure p in millibars, the atmosphere's absolute temperature T in Kelvin and the partial pressure of water vapor e_s in millibars according to

$$N = (n-1) \cdot 10^6 = \frac{77.6 \cdot p}{T} + \frac{e_s \cdot 3.73 \cdot 10^5}{T^2}$$

with $\quad e_s = \dfrac{rh \cdot 6.105 \cdot e^x}{100} \quad$ and

$$x = 25.22 \cdot \frac{T-273.2}{T} - 5.31 \cdot \log_e\left(\frac{T}{273.2}\right)$$

whereby rh denotes the atmosphere's relative humidity in percent.

Finally, for the physical point of view of a tactical radar operator relating to refractive gradients and propagation mechanism effects, a modified refractivity M is used.

$M = N + 0,157 \cdot h \qquad$ for altitude h in meters

h denotes the height above the Earth surface in meters [1].

4 PROPAGATION IN MARITIME ENVIRONMENT

For radars looking under low grazing angle incidence towards the sea surface the meteorological conditions can result in different propagation mechanisms.

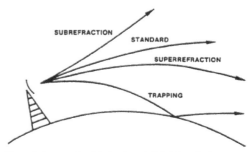

Figure 2. Propagation Mechanisms in Maritime Environment

Standard propagation mechanisms like standard refraction, free-space propagation, multipath interference (or surface reflection), diffraction and tropospheric scatter are processes that occur in the presence of a standard atmosphere. In contrast to these Standard propagation mechanisms, Subrefraction (or Anti-Duct), Superrefraction as well as Ducting (or Trapping) are Anomalous propagation mechanisms.

For subrefractive conditions the range for the detection of a target may be limited to distances less than the optical horizon. For superrefractive conditions, ducting and trapping the detection of targets may extent to distances beyond the optical horizon due to reflections of the propagating signal on the sea surface and layer boundaries in the atmosphere comparable to waveguide like propagation. For standard refractive conditions the detection range is in-between.

However, it has to be noted, that within the maritime boundary layer the propagation conditions are almost always never standard.

Now, we want to draw a bow to the previous chapter by distinguishing different propagation mechanisms using the associated refractivity gradient dN/dh and the modified refractivity gradient dM/dh. This provides the table below.

Table 1: Propagation Mechanisms with N- and M-Gradient (see [1])

Condition	N-Gradient dN/dh	M-Gradient dM/dh
Trapping/Ducting	< -157 N/km	< 0 M/km
Superrefraction	-157....-79 N/km	0....79 M/km
Normal	-79....0 N/km	79....157 M/km
Standard	-39 N/km	118 M/km
Subrefraction	> 0 N/km	> 157 M/km

5 DUCT AND ANTI-DUCT

The most interesting propagation mechanisms are undoubtedly the ones presenting the extreme gradients in Table 1. This means Ducting and Anti-Duct (Subrefraction).

A duct is a channel in which electromagnetic energy can propagate over great ranges. To propagate energy within a duct, the angle made by the electromagnetic system's energy with the duct must be small, usually less than 1°.

Non-standard atmospheres can lead to anomalous propagation. Pressure tends to be quickly restored to an equilibrium, so the most important are variations in the water vapor concentration and temperature.

Ducts tend to form when either temperature is increasing, or water vapor concentration is decreasing, unusually rapidly with height.

We differentiate between Evaporation Duct, Surface Ducts and Elevation Duct. The attributes of the different duct types see in [1], [2] and [3].

Figure 3 presents the main locations for duct occurrence. These are global seen the west coasts of the continents as well as specifically Red Sea, Arabic Gulf, Arabian Sea and in the summer months the Mediterranean Sea.

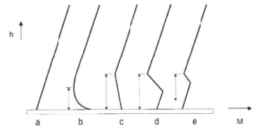

Figure 4. Idealized M profiles for various types of Duct :
a: Standard Atmosphere, no duct
b: Evaporation Duct c,d: Surface Ducts e : Elevated Duct

Figure 4 illustrates the idealized M profiles (representing altitude h versus modified refractivity M) for various types of radar Duct. The Duct height is indicated by arrows on each profile.

White areas indicate no ducting occurrence [4].

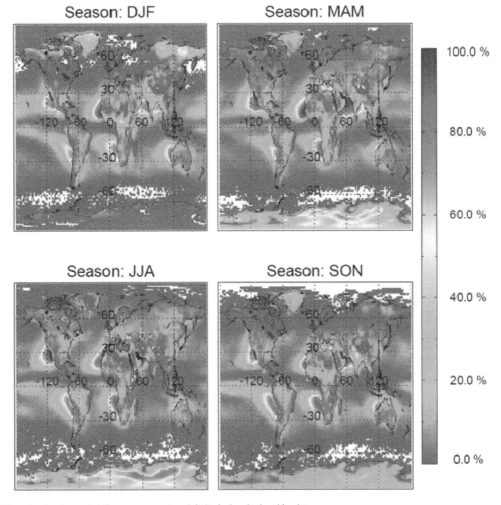

Figure 3: Ducting probability per season at each latitude, longitude grid point.

6 SOFTWARE-TOOL TERPEM

The TERPEM propagation package [5] provides a powerful tool for forecasting and analysis of refraction, ducting and terrain effects on radiowave links and radar systems. It is based on state-of-the-art hybrid models combining parabolic equation and ray-trace techniques. It is the result of on-going algorithm and software development that began in 1989 with the release of PCPEM. The computer platform of TERPEM is a PC with Windows 95, 98 or NT, 100 MHz or higher Pentium processor and at least 32 Mbytes of RAM. The frequency range is 30 MHz to 100 GHz. TERPEM specifies input scenarios including system and environmental (meteorological) parameters via a Windows interface. The propagation engine computes coverage diagrams and path loss arrays on user-specified range/height grids. Please check Signal Science Limited for getting more information.

7 IMPORTANT BASIC RESULTS

Taking first steps using the tool TERPEM, we get interesting basic results. In case of standard atmosphere, the detection range is increasing with increasing antenna height. Every duct or trapping configuration, in its level well-placed below the antenna height, will increase the radar detection range in relation to standard atmosphere. In contrast to this, subrefraction (anti-duct) will reduce the radar detection range in relation to standard atmosphere. In consequence, if the radar antenna is located above the duct or anti-duct layer, duct will benefit the detection range and anti-duct will be injurious to it.

For standard atmospheric conditions, Figure 5 and 6 show TERPEM Height-versus-Range illustrations for the coloured detection probability. The detection range refers to a False Alarm Probability $P_{fa} = 10^{-6}$, a Detection Probability $P_d = 0.9$ (yellow), an examined altitude Above Sea Level (ASL) or Above Ground Level (AGL) = 1 m and a target Range Cross Section (RCS) = κ m^2 with κ 'small'. For Figure 5 the antenna height (AH) = 65 m, for Figure 6 it is only 5 m.

As shown on the right side of each illustration, the different colours for the figures illustrate the detection probability P_d in the following way : yellow means a $P_d \geq 90\%$, red implies $80\% \leq P_d < 90\%$, blue stands for $70\% \leq P_d < 80\%$, green means $60\% \leq P_d < 70\%$, magenta implies $50\% \leq P_d < 60\%$ and white stands for $P_d < 50\%$.

Height [m]   Coloured Illustration of the Detection Probability P_d

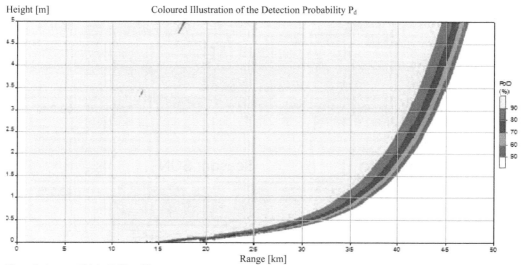

Range [km]

Figure 5. Antenna Height (AH) = 65 m

Height [m] Coloured Illustration of the Detection Probability P$_d$

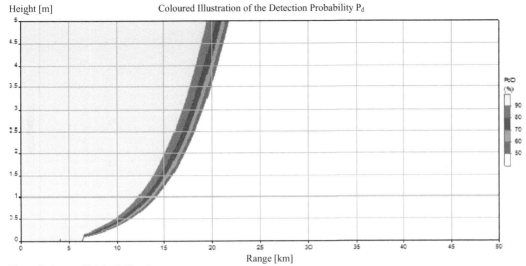

Figure 6. Antenna Height (AH) = 5 m

Height [m] Coloured Illustration of the Detection Probability P$_d$

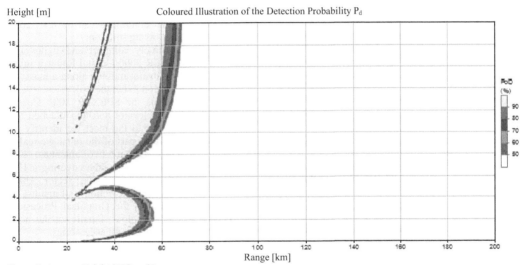

Figure 7. Antenna Height (AH) = 65 m

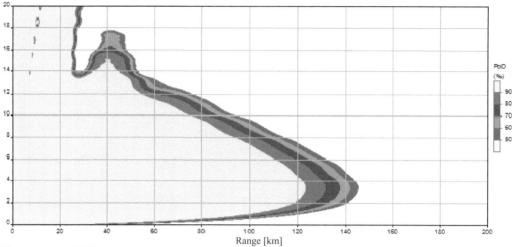

Range [km]

Figure 8. Antenna Height (AH) = 5 m

8 FURTHER IMPRESSIVE RESULTS

Usually, the detection range also increases in duct situations with increasing antenna height, if the antenna height is located higher than the duct layer. But if the antenna height is located inside the duct layer situated directly above the sea surface and if the duct layer shows an adequate strongness – for an Evaporation Duct layer this means a height of 11 m or more – then a waveguide can be formed up resulting in a clearly increased detection range in comparison to Standard Atmospheric Conditions.

For an Evaporation Duct Height (EDH) of 14 m, Figure 7 and 8 show TERPEM Height-versus-Range illustrations for the coloured detection probability, $P_d = 0.9$ (yellow), $P_{fa} = 10^{-6}$, ASL, AGL and RCS set as in the previous chapter. For Figure 7 the antenna height is 65 m, for Figure 8 it is only 5 m representing the waveguide with high range. In Figure 7, the detection range will become even smaller if there will occur a strong subrefractive layer instead of Standard Atmosphere directly above the Evaporation Duct layer. As a consequence we can state that, dependent on the given meteorological condition, the optimal antenna height for achieving a high detection range will vary.

9 CONCLUSIONS AND FUTURE SOLUTION: RADAR ANTENNA HEIGHT OPTIMISATION BY AN ELEVATOR

In the previous chapters we have seen the properties of the different anomalous propagation mechanisms, especially duct and anti-duct. We have seen that different meteorological situations result in different optimal antenna heights relating to an optimal detection range. Adapting the antenna height to the given meteorological conditions can be realized in the future by the following solution:

Due to the relatively compact and lightweight design of the SPEXER 2000 Coastal Radar it might be operated on an elevator moving along a mast and thus change the position in height.

Figure 9 represents the SPEXER 2000 Coastal Radar, affixed on an elevator and this attached to a mast of a tower. Next to the mast we identify the cable route. In the lower part of the radar we realize (in orange / yellow colour) two antennas, 120° shifted against each other and each covering a 120° sector, together covering a 240° coast sector. Above the antennas we have the protection roof against the sun in combination with the Graphical Processing Unit (GPU) in reddish colour. Finally, at the head, we see whitely the camera with accessories.

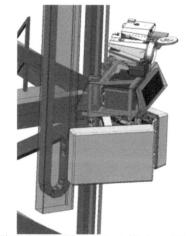

Figure 9: SPEXER 2000 Coastal Radar and Elevator

Up to now, to meet the Duct problem, an additional antenna at ground level – called Duct antenna – is used besides the traditional antenna in a height of about 40 m or 60 m. The possibility of utilizing an elevator moving along a mast and thus change the antenna position in height implies for the future for each site to require only one antenna instead of – so far – two antennas.

To control the height of the radar offers the possibility to adjust the height adaptively depending on the atmospheric conditions in the maritime boundary layer. This opens the possibility to optimize the performance of the radar depending on the actual atmospheric conditions.

Doing this, generally we fix the antenna height at an altitude of about 65 m for all meteorological conditions (combinations) without subrefraction. This way the required range - for example for a target with RCS = 1 m^2 - will be achieved. We resume our logics as follows: if there is a subrefractive layer (located below 65 m) not strong enough to reduce the detection range too much, we leave the antenna height unchanged at an altitude of 65 m. Only if the subrefractive layer is too strong and simultaneously the Evaporation Duct layer is strong enough (EDH $\geq$ 11 m), then we reduce the antenna height to 4-10 m ASL. Then the antenna will be utilized as a so-called Duct Antenna and then the suppositions are given to use the Evaporation Duct layer for the formation of a waveguide implying clearly increased detection ranges.

If we dispose of the information about the precise current Evaporation Duct Height, then we can do a vernier adjustment in relation to the Duct Antenna height. This is important, because that way it is possible to avoid detection holes (skip zones).

The knowledge whether duct or anti-duct exists as well as the altitude and the thickness of the corresponding layers, in the future this can be ascertained by radio sonde data using corresponding stations. But this would be a complex and expensive process ascertaining duct or anti-duct. An easier method would be potentially to take advantage of the clutter conditions. Clutter data could be provided by the radar on its own, without radio sonde ascent. In order to profit from clutter data in the intended manner we have to read into the future. Such an algorithm utilizing clutter information has still to be invented. In consequence, there remains work still in progress.

BIBLIOGRAPHY

[1] Merrill I. Skolnik, "Radar Handbook", chapter 26: "The Propagation Factor in the Radar Equation" from Wayne L. Patterson, Mc Graw Hill, Third Edition, New York, Chicago, San Francisco, 2008.

[2] H.T. Dougherty, E.J. Dutton, "The Role of Elevated Ducting for Radio Service and Interference Fields", U.S. Department of Commerce, Malcolm Baldrige, Secretary, NTIA-Report-81-69, March 1981.

[3] E. Brookner, E. Ferraro, G.D. Ouderkirk, "Radar Performance During Propagation Fades in the Mid-Atlantic Region", IEEE Transactions on Antennas and Propagation, Vol. 46, No. 7, July 1998, p.1056–1064.

[4] A. von Engeln, J. Teixeira, "A Ducting Climatology derived from ECMWF Global Analysis Fields", J. Geophys. Res., 109 (D18), D18 104, doi: 10.1029/2003 JD004380, 2004.

[5] Ken Craig and Mireille Levy, http://www.signalscience.com/TERPEM.htm

Signal Processing Optimization in the FMCW Navigational Radars

V. Koshevyy & O. Pashenko
Odessa National Maritime Academy, Ukraine

ABSTRACT: Frequency Modulated Continuous Wave (FMCW) or broadband radars currently realized implementation on a small unconventional vessels and at costal radar stations due to their power economy and high detection capabilities in the near-by environment. Optimization of sweep signal and correlation filter is carried out by means of applying windowing function. Compound LFM signals are suggested, which derived from compound multiphase signals, having proper ambiguity and cross-ambiguity functions.

1 INTRODUCTION

The linear frequency modulated (LFM) waveform is one of well-known and most useful radar pulse compression waveforms due to its high range resolution (depended of the waveform bandwidth) and its tolerance to Doppler shift and easy for the receiver processing.

The LFM signal with unit energy is defined as

$$x(t) = \frac{1}{\sqrt{T}} e^{i\left(2\pi f_0 t + \frac{v_0}{2} t^2\right)}, \quad -\frac{T}{2} \leq t \leq \frac{T}{2} \tag{1}$$

where f_0 – carrier frequency and sweep $v_0 = 2\pi \Delta F / T$.

Pulse compression is performed by convolving the received signal with a filter matched to the transmitted LFM. Compressed pulse yields the length, is $1/\Delta F$.

Compression ratio, defined as the ratio of the transmitted time length to the compressed pulse length, is $\Delta F T$.

Since the ambiguity function (AF) of a LFM

$$\chi(\tau, f) = \int_{-\infty}^{\infty} x^*(t) x(t-\tau) e^{i 2\pi f t} dt \tag{2}$$

has its energy is concentrated along a ridge that passes through the origin of the delay –Doppler plane with slope $v_0 = 2\pi \Delta F / T$.

According this kind of AF, uncertainty is appeared in range and velocity determination. For marine radars the compression ratios, for example, must be over 10^5, if we want to neglect of this kind of uncertainty (with consideration the maximum speeds of vessels). But such value of compression ratio is rather difficult task for practical implementation in X-band radar (main band for marine radars). The necessity of using such big compression ratio can be avoided by means compound LFM signals, which have the shape of AF differ from AF of single LFM pulse, without uncertainty in range and velocity determination. The modulation waveform of these signals may be obtained on the base of compound multiphase signals [1]. Compound multiphase signals in its tern are constructed on the base multiphase signals, which derived from LFM signals [3, 2].

Compound signals in [1] are constructed on the base of product two sequences

$$u_n^B = \exp\left\{ i \frac{\pi}{4} \alpha' [2\left(n - N_B E\left[\frac{n}{N_B}\right] + 1\right) - (N_B + \mu_0)]^2 \right\} \tag{3}$$

$$u_n^V = \exp\left\{ i \frac{\pi}{4} \beta' \left[2\left(E\left[\frac{n}{N_{B_1}}\right] - N_V E\left[\frac{E\left[\frac{n}{N_B}\right]}{N_V}\right] + 1 \right) - (N_V + 1) \right]^2 \right\} \tag{4}$$

where $n = 1 \div N$; $N = N_B \cdot N_V$; N_B – period of sequence u_n^B; N_V – period of sequence u_n^V; $E[x]$ – integer part x; $\alpha', \beta', \mu_0, N_{B_1}$ – parameters of phase modulation ($\mu_0 = [0;1]$).

AF of compound signal is the product of AF each of these signals [1]. Signals (3) and (4) have AF comb-shaped structure [2]. When they are multiplying AF multiphase compound signal has multi-peak structure. Expression for the calculation of discrete periodic cross-ambiguity function has the form

$$\chi_{SW}(k,l) = \sum_{n=0}^{N-1} w_n^* s_{(n+k)} e^{i\frac{2\pi l n}{4N}},$$ (5)

where k is the discrete values of delay and l is discrete values of frequency with the steps correspondingly T_0 and $\Delta f = \frac{1}{4NT_0}$; T_0 is elementary pulse duration; w_n is the filter coefficients for compound signal; $s_n = u_n^B u_n^V$ code sequence of compound signal (according to (3) and (4)).

The resulting AF is divided into three regions formed by multiplying:

1 The chest to the chest, i.e. the area of the central peak (CP).
2 The chest to the absence of the chest.
3 The absence to the absence of the chest.

Consider two types of multiphase compound signals each of which has:

1 $N=342$ ($N_B=18$, $N_I=19$) with the values of the co-efficients $\alpha'=1/N_B$, $\beta'=2/N_B$, $\mu_0=0$, $N_{BI}=1$.
2 $N=324$ ($N_B=18$, $N_I=18$) with the values of the co-efficients $\alpha'= -1/N_B$, $\beta'=1/N_B^2$ $\mu_0=0$, $N_{BI}= N_B$.

On Figure 1 the results of calculation AF body for aperiodic signal a) and periodic signal b), under matched filter case are represented. AF is considered in discrete time k and discrete frequency l. The upper part of the figure shows the 3D image of AF. The area around the CP will be considered. Below on the right the sections along the frequency axis AF are shown. The left part shows the section along the time axis AF. Section $l=0$ shows the correlational function of the signal

Table 1. The ratio of the CP to the maximum level of side lobes for multiphase compound signal with parameters $N=342$ ($N_B=18$, $N_I=19$), $\alpha'=1/N_B$, $\beta'=2/N_B$, $\mu_0=0$, $N_{BI}=1$ (the second region)

Type processing	$l=0$	$l=1$	$l=2$	$l=3$
Aperiodic mode	61,29	37,21	44,76	40,10
Periodic mode	perfect	54,31	39,18	56,52

Table 2. The ratio of the CP to the level of the side lobes in the cross sections $k = 2,3$ in points $l = -40$ and $l = +40$ for a multiphase compound signal with parameters $N=342$ ($N_B=18$, $N_I=19$), $\alpha'=1/N_B$, $\beta'=2/N_B$, $\mu_0=0$, $N_{BI}=1$ (the third region)

Type processing	$k=2$		$k=3$	
	$l=-40$	$l=+40$	$l=-40$	$l=+40$
Aperiodic mode	282,64	380,00	374,18	345,45
Periodic mode	$\geq 10^5$	$\geq 10^5$	$\geq 10^5$	$\geq 10^5$

The first case is the signal with $N=342$ ($N_B=18$, $N_I=19$) and the values of the coefficients $\alpha'=1/N_B$,

$\beta'=2/N_B$, $\mu_0=0$, $N_{BI}=1$ in the periodic and aperiodic modes. The results are summarized in the Tables 1 & 2.

a)

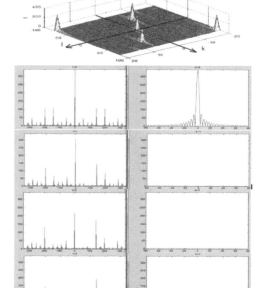

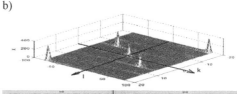

b)

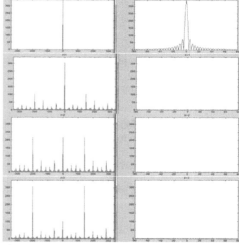

Figure 1. Body AF multiphase compound signal $N=342$ ($N_B=18$, $N_I=19$) for aperiodic (a) and periodic (b) modes of operation with coefficients $\alpha'=1/N_B$, $\beta'=2/N_B$, $\mu_0=0$, $N_{BI}=1$, (the area around the CP)

Also the signals were examined with an another set of pulses: N_B=14, N_V=23 and N_B=34, N_V=43, but with the same parameters α'=1/N_B, β'=2/N_B, μ_0=0, N_{B1}=1. The AF has multi-peak character. Free zone is formed around the CP. In all of the cases with increasing number of pulses in the signal, the area of the free zone around the CP of the AF body is grown. It must be noted, that for all the cases, considered above, the perfectness of periodic correlation functions were been reserved. So we obtained the signals, which have not only proper behavior AF on the range-Doppler plane, but the perfect periodic correlation function.

Let us consider the next type of signal with parameters N=324 (N_B=18, N_V=18) and the values of the coefficients α'= -1/N_B, β'=1/N_B^2 μ_0=0, N_{B1}= N_B. After calculation of the AF body for aperiodic signal and periodic signal under matched filter we have not enough low the side-lobes level. We may reduce it in free zone by using weighting sequence obtained from the formula (3) and formula (4). The weight function of the processing of the entire signal can be written as:

$$w_n = s_n \cdot v_n$$
$$v_m = v^B_{m-E[(m-1)/N_B]N_B} v^V_{E\left[\frac{m-1}{N_V}\right]+1}, \quad m = \overline{1,N} , \tag{6}$$

where, $v^B_{m-E[(m-1)/N_B]N_B}$ – weighting coefficients for the sequence obtained by the formula (3); $v^V_{E\left[\frac{m-1}{N_V}\right]+1}$ - weighting coefficients for the sequence obtained by the formula (4). Formula (7) was described one of the possible weighting functions for the signals (3) and (4):

$$v_n^B = sin\left[\frac{\pi(n+1)}{N_B+1}\right]$$
$$v_n^V = sin\left[\frac{\pi(n+1)}{N_V+1}\right] \tag{7}$$

We analyze the behavior of the AF of the aperiodic compound multiphase signal N=324 (N_B=18, N_V=18). The results are summarized in Figure 2 and the Tables 3-5. On the right side of figure is shown the law of changing of the phases of the compound signal.

a)

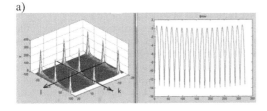

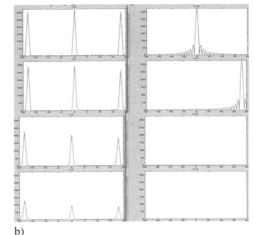

b)

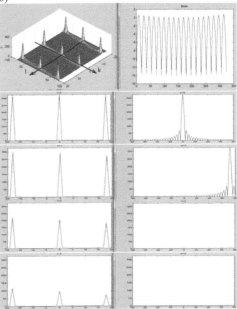

Figure 2. Body of the AF of the multiphase compound signal N=324 (N_B=18, N_V=18) for aperiodic (a) and periodic (b) modes of operation with coefficients α'= -1/N_B, β'=1/N_B^2 μ_0=0, N_{B1}= N_B (the area around the CP)

Table 3. The height of the CP for aperiodic multiphase signal with parameters α'=-1/N_B, β'=1/N_B^2 μ_0=0, N_{B1}=N_B and N=324 (N_B=18, N_V=18) (the first region)

Type processing	l=0
Matched processing	324
Weighting processing	145,64

Table 4. The ratio of the CP to the maximum level of side lobes for aperiodic multiphase signal with parameters α'=-1/N_B, β'=1/N_B^2 μ_0=0, N_{B1}=N_B and N=324 (N_B=18, N_V=18) (the second region)

Type processing	l=0	l=1	l=2	l=3
Matched processing	297,52	223,45	314,56	120,90
Weighting processing	1213,67	1693,49	291,28	280,08

Table 5. The ratio of the CP to the level of the side lobes in the cross sections $k=2,3$ in points $l=-40$ and $l=+40$ for a multiphase aperiodic signal with parameters $\alpha'=-1/N_B$, $\beta'=1/N_B^2$ $\mu_0=0$, $N_{B1}=N_B$ and $N=324$ ($N_B=18$, $N_V=18$) (the third region)

Type processing	$k=2$		$k=3$	
	$l=-40$	$l=+40$	$l=-40$	$l=+40$
Matched processing	188,37	164,47	206,37	137,87
Weighting processing	2086,53	1820,50	1471,11	1125,50

Based on these results we can see that a weighting function, which is described by (7), provides side- lobes suppression around the CP. Under weighting the multi-peaks structure cross-ambiguity function (CAF) signal has not changed.

Consider the behavior of CAF multiphase periodic signal with the following parameters: $\alpha'=-1/N_B$, $\beta'=1/N_B^2$ and $N_B=18$, $N_V=18$. Figure 3 shows the structure of CAF such signal after the weighting processing by a function sin (7).

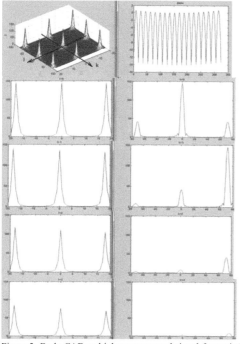

Figure 3. Body CAF multiphase compound signal for periodic modewith coefficients $\alpha'=-1/N_B$, $\beta'=1/N_B^2$ $\mu_0=0$, $N_{B1}=N_B$ and $N=324$ ($N_B=18$, $N_V=18$) after weighting, described by a function sin (7) (the area around the CP)

From Figure 3 one can see that the CAF of the multiphase signal retains the multi-peak structure under the weighting processing (7). The side-lobes level is reduced. The results of research are summarized in Table 6 & 7.

Table 6. The ratio of the CP to the maximum level of side lobes for periodic multiphase signal $\alpha'=-1/N_B$, $\beta'=1/N_B^2$ $\mu_0=0$, $N_{B1}=N_B$ and $N=324$ ($N_B=18$, $N_V=18$) (the second region)

Type processing	$l=0$	$l=1$	$l=2$	$l=3$
Matched processing	2945,45	272,27	265,57	183,05
Weighting processing	7282,00	416,11	364,10	309,87

Table 7. The ratio of the CP to the level of the side lobes in the cross sections $k=2,3$ in points $l=-40$ and $l=+40$ for a periodic signal in a multiphase parameters $\alpha'=-1/N_B$, $\beta'=1/N_B^2$ $\mu_0=0$, $N_{B1}=N_B$ and $N=324$ ($N_B=18$, $N_V=18$) (the third region)

Type processing	$k=2$		$k=3$	
	$l=-40$	$l=+40$	$l=-40$	$l=+40$
Matched processing	1022,08	952,94	1080,00	790,24
Weighting processing	7282,00	7282,00	5394,07	4854,67

Based on these results we can conclude, that by means using the weighting function (7), the side-lobes are reduced. The multi-peak structure of the CAF has not changed under weighting.

The signal-to-noise ratio losses are determined by next

$$\rho = \frac{\left|\chi_{sw}(0,0)\right|^2}{\sum_{n=0}^{N-1}\left|s_n\right|^2 \sum_{n=0}^{N-1}\left|w_n\right|^2}, \qquad (8)$$

where s_n - the complex envelopes of the signal; w_n - the complex envelopes of the filter; s/n - the ratio of signal power to noise power; $\chi_{sw}(0,0)$ - CAF value with the absence of frequency and time shift ($l=0$, $k=0$). The results of studies for periodic and aperiodic compound multiphase signal with parameters $\alpha'=-1/N_B$, $\beta'=1/N_B^2$ $\mu_0=0$, $N_{B1}=N_B$ and $N=324$ ($N_B=18$, $N_V=18$) are presented in Table 8:

Table 8. The signal-to-noise ratio losses for the signal with the phase modulation parameters $\alpha'=-1/N_B$, $\beta'=1/N_B^2$ $\mu_0=0$, $N_{B1}=N_B$ and $N=324$ ($N_B=18$, $N_V=18$)

Type processing	ρ
Matched processing	1
Weighting processing	0,725

It should be noted that the compound multiphase signal with parameters $\alpha'=-1/N_B$, $\beta'=1/N_B^2$ $\mu_0=0$, $N_{B1}=N_B$ and $N=324$ ($N_B=18$, $N_V=18$) for aperiodic and periodic modes were observed similar changes in the structure. After weighting the side lobe level is decreased. The multi-peak structure of the CAF under weighting was not changed. The signal-to-noise ratio losses is equal $\rho = 0.725$.

So, using weighing processing we get the possibility for significant suppressing the side-lobes level in the area around CP of the CAF.

We can use not only compound multiphase signals, but also equivalent compound FFM signals. As example on Figure 4 is shown the signal with saw-tooth modulation [4], which is equivalent to the compound multiphase signal from Figure 3 (derivative of the law of phase modulation gives us the law of frequency modulation). On Figure 5 is shown functional diagram of transceiver and processing unit for this kind of signal [5].

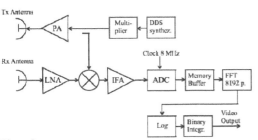

Figure 4.

Figure 5.

DDS – Direct Digital Synthesizer produces a synthesized "chirp"(compound frequency modulated) signal;

Multiplier – up-converted of modulated signal;

PA – power amplifier, which feeds X-band Compound FMCW signal to antenna;

LNA – low noise amplifier;

IFA – intermediate frequency amplifier;

ADC – analog-digital convector;

FFT – spectrum analyzer that perform Fast Fourier Transform.

2 CONCLUSIONS

Demand to the value of compression ratio of the LFM signals for Marine radar was analyzed. The way of decreasing demand to the value of compression ratio was suggested by means using compound LFM signals, which derived from compound multiphase signals. The analyze of the AF and the CAF of compound multiphase signals and filters were provided. It was shown that the size of zone with low level of side-lobs around the CP of AF and CAF is increases with increasing number of pulses in the signal. It was also shown that by means proper choosing the parameters of phase modulation in compound multiphase signals we may obtain not only good property for clutter rejection, but also perfect property of periodic correlation function. The example of transformation the modulation of multiphase compound signal to modulation of corresponding compound LFM signals was represented.

REFERENCES

[1] Koshevoy V.M., Synthesis compound multiphase signals, Izvestiya VUZ. Radioelectronika (Radioelectronics and Communication Systems), vol. 31, N8, 1988, pp. 56-58.

[2] Koshevoy V.M., Kononov A.A., Synthesis optimal one channel discrete signals and filters, Izvestiya VUZ. Radioelectronika (Radioelectronics and Communication Systems), vol. 27, N8, 1984, pp. 62-65.

[3] Lewis B.L., Kretschmer F.F., Linear frequency modulation derived polyphase pulse compression codes, IEEE Trans. on Aerospace and Electronic Systems Vol. AES-18, N5, Sept. 1982, pp. 637-641.

[4] Levanon N., Mozeson E., Radar signals, J. Wiley, NJ, 2004.

[5] Plata S., Wawruch R., CRM-203 Type Frequency Modulated Continuous Wave (FM CW) Radar, TransNav, the International Journal of Marine Navigation and Safety of Sea Transportation, Vol. 3, N3, Sept. 2009, pp. 311-314.

Autopilot Using the Nonlinear Inverse Ship Model

K. Kula
Gdynia Maritime University, Poland

ABSTRACT: In this paper the Internal Model Control (IMC) approach for marine autopilot system is presented. The inversion by feedback techniques are employed for realization of inversion such nonlinear characteristics as saturation of rudder angle and rudder rate. The extension of the model and inverse model to a nonlinear form enabled to achieve a significant improvement in the control performance.

1 INTRODUCTION

An automatic control of ships maneuvering has a long tradition. The first Proportional Integral Derivative (PID) controller was used at sea by Minorsky in 1922 to help to eliminate the steady state error as well as the disturbance rejection. That time, any helmsman could much easier and smoothly carry out a change of the ship course according to the prevailing working conditions than any existing controller. For this reason, the autopilot function was only to compensate a disturbance from the slowly-varying wave drifting force. After a manual entry of a new course the steering was typically switched on to the autopilot mode. Also today, most of the autohelm systems are mainly formulated to follow a desired course under constant speed settings. This is due to the fact that every vessel is a highly nonlinear object and its dynamic varies with the change of the longitudinal speed. The designers of course keeping autopilots as well as trajectory tracking autopilots that use the Global Positioning System (GPS) have to deal with this problem. A control system with PID regulator that does not take into account dynamic changes of the vessel model during the course maneuver requires settings that ensure large enough stability margin, which increases the settling time and as a result decreases the control performance. The control performance in the PID autopilots can be improved by use of anti-windup techniques. On the duration of the course change maneuver the integrating term can be switched off. When a new set point is reached the controller structure changes and the integral action is activated again. This way, it is possible to eliminate the steady state error resulting from the disturbances such as a wind, sea currents, and waves.

The method of switching between the two modes of work was also used in other types of autopilots such as the Sliding Mode Control (SMC). This method describes the optimal controller for a broad set of dynamic systems and is suitable for control of a non-linear plant. It does not need to be precise and it is not sensitive to parameter deviations. The SMC for ship course control is proposed by Tomera (2010), Yaozhen et al (2010).

In the ship autopilot systems there are also adopted internal model control (IMC). The difference between the IMC and single-loop systems lies in better prediction capabilities of the IMC approach.

The paper is organized as follows: the main idea and IMC control methodology are described in section 2. In the next section this methodology is applied to design to a vessel motion control of a container ship. Simulation results and comparisons of performances with PID controller are described in section 4. Section 5 is a summary of this article.

2 SHIP AUTOPILOT WITH IMC STRUCTURE

Originally, the control with the IMC was applied only for minimum phase processes. The extension of this structure for integrating systems is developed by Chia & Lefkowitz (2010). This modification could allow to use the IMC for the ship heading control. Zheng, who is focusing mainly on inland ships introduced in his work (1999) the concept of using the IMC for controlling the turning rate. In this study a design of the control system of the ship course is proposed to introduce the concept of angular speed regulation where a set point is adjusted to the deviation of the ship course as a reference signal ψ_r (Fig.1). It is a cascade control system with IMC controller in an inner loop and proportional controller in an outer loop. The structure of proposed control system is shown in Figure 1.

The main feedback prepares required value for the yaw rate, which varies depending on the difference between the desired and actual measured course.

Thus there is an additional loop for the control of the yaw rate whose function is keeping the desired yaw rate. During the course maneuver changes the longitudinal and lateral velocity heeling and drift angle, which have a very significant impact on the dynamic properties of the ship. The relationship between them are nonlinear what by employing a linear regulator leads to decreasing of the control performance.

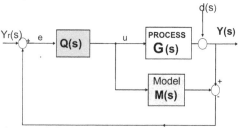

Figure 1. Block schema of proposed ship steering system

2.1 Internal Model Control

The control system structure with adopted inverse model of the plant (IMC) is shown in Figure 2. It presents a linear process $G(s)$ along with its model $M(s)$ and a regulator that is described using so-called design transfer function $Q(s)$, which should be equal to the inverse of the plant model. The internal stability is assured as long the transfer function $Q(s)$ is chosen to be any stable rational transfer function. However creating an inverse model can lead to an unstable design transfer function. To prevent this a model factorization and low-pass filter (Riviera & al 1986) are used.

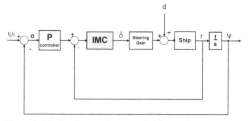

Figure 2. Block schema of Internal Model Control

$$Q(s) = \frac{u(s)}{e(s)} = F(s) M_{inv}(s) \tag{1}$$

where $M_{inv}(s)$ is inverse model of the process and $F(s)$ –the transfer function of the filter

$$F(s) = \frac{1}{(T_f s + 1)^n} \quad F(0)=1, \quad n\text{- integer number} \tag{2}$$

which has an important role to play to make $Q(s)$ biproper. It means, that n can be obtained as a difference between the order of the numerator and the order of the denominator of the inverse model

The closed-loop transfer function and the disturbance transfer function, when the closed-loop system is affected by external signals, can be derived respectively as

$$G(s) = \frac{y(s)}{y_r(s)} = \frac{G(s)Q(s)}{1 - M(s)Q(s) + G(s)Q(s)} \tag{3}$$

$$G_d(s) = \frac{y(s)}{d(s)} = \frac{1 - M(s)Q(s)}{1 + Q(s)(G(s) - M(s))} \tag{4}$$

It can be seen (3) that in the absence of uncertainties and plant modeling errors, the controlsystem is working as a open-loop system.

But let us consider that the model differs from the object and suppose that the uncertainty is of the multiplicative type. Assume that

$$G(s) = M(s)[1 + \Delta(s)] \tag{5}$$

where: $M(s)$ – is the modeled part of the process
$\Delta(s)$ is a stable multiplicative uncertainty such that $M(s) \Delta(s)$ is strictly proper

Rearranging (5) we became

$$M(s) = \frac{G(s)}{1 + \Delta(s)} \tag{6}$$

If the plant model is minimum phase the design transfer function by substituting the inverse of model (6) into (1) can be presented as

$$Q(s) = F(s) \frac{1 + \Delta(s)}{G_-(s)} \tag{7}$$

and the closed-loop transfer function will be equal to:

$$G(s) = \frac{F(s)[1 + \Delta(s)]}{1 + F(s)\Delta(s)} \tag{8}$$

Having regard to (8) we can determine the steady state output value by inaccuracy of the model

$$y_{ss} = \lim_{s \to 0} s \cdot X_0 \frac{1}{s} \frac{((T_f s + 1)^n) \cdot [1 + \Delta(s)]}{(T_f s + 1)^n + \Delta(s)} = X_0 \tag{9}$$

It follows from (9) that the model uncertainty does not change the zero steady state error, what is extremely important in the context of the use this system for the ship course control.

In a similar way as the closed-loop transfer function (8) we can derive the disturbance transfer function in the presence of uncertainties and plant modeling errors.

$$G_d(s) = \frac{1 - F(s)}{1 + F(s)\Delta(s)} \qquad (10)$$

To sum up
- the control system with internal model will be stable even if the model differs from the process, unless the $\Delta(s)$ is stable
- the closed-loop transfer function will be even more different from the filter, which acts as a reference model - the greater will be the difference between the process and the model
- the larger is time constant Tf the system is less sensitive to the impact of deviations from the correct model

In other words, when the process is perfectly modeled and $d=0$ the system is basically an open-loop. This provides the open-loop advantages. A control system design is expected to provide a fast and accurate set-point tracking, which means the output of the system should follow the input signal as close as possible. With an open loop control scheme the stability of the system is guaranteed provided that both the plant and the controller transfer functions are stable.

2.2 Nonlinear model

Zheng introduced in his work (1999) the concept of the IMC controller with the simplest linear first order model of the ship (Nomoto 1957). However, this model only applies to a single operating point and if a change occurs it greatly differs from the dynamics of the ship.

Although, the IMC system is robust to some uncertainty In practical situations where modeling errors are inevitable, the performance of the IMC system may become poor and unacceptable. A higher control performance can be achieved with the model faithfully reflecting the control process. Therefore, in this study a modified IMC strategy extended to a non-linear model has been proposed. Such properly tuned model gives static and dynamic properties of the vessel in a wide range of changes with much greater accuracy than in the case with the simple Nomoto model.

The motion of a ship can be described by using six nonlinear differential equations. Hence, ship maneuvering is treated as a horizontal plane motion and only the surge, sway and yaw modes are considered.

The following approximations (Saari & Djemai 2012) are set up:

$$\begin{bmatrix} m & 0 & 0 \\ 0 & m & mx_G \\ 0 & mx_G & I_{zz} \end{bmatrix} \cdot \begin{bmatrix} \dot{u} \\ \dot{v} \\ \dot{r} \end{bmatrix} = \begin{bmatrix} X \\ Y \\ N \end{bmatrix} + \begin{bmatrix} m(vr + x_G r^2) \\ -mur \\ -mx_G ur \end{bmatrix} \qquad (11)$$

$$\dot{\psi} = r$$

where m is the mass of the ship, I_{zz} is the inertia along the z axis, and x_G is the x co-ordinate of the center of gravity.

X, Y and N denote the hydrodynamic forces and the momentum. They result from the movement of the ship on the surface and depend on speed, weight and a profile of the hull and also on the effect of waves.

Based on (11) after linearization around a selected point of work and after elimination of the sway velocity we have a following simplified linear differential equation

$$\ddot{r}(t) + (\frac{1}{T_1} + \frac{1}{T_2})\dot{r}(t) + \frac{1}{T_1 T_2}r = \frac{k}{T_1 T_2} \cdot [T_3 \dot{\delta}(t) + \delta(t)] \qquad (12)$$

where time constants T_1, T_2, T_3 and gain k depend on derivatives of the hydrodynamic forces and momentums with respect to the sway and surge velocity and yaw rate r.

However, the change of longitudinal and lateral velocities during the maneuver leads to changes in the dynamics of the ship and thus increases the incorrectness of the linear model.

To prevent this, in the proposed control structure such a model was introduced which parameters depending on shaft velocity n and the rudder angle δ. This will ensure better quality of the model in a wider range of changes of the state variables of a system.

The equation (12) after rearranging to the new form

$$\ddot{r}(t) = \frac{k(n,\delta)}{T_1 T_2(n,\delta)} \cdot [T_3(n,\delta) \cdot \dot{\delta}(t) + \delta(t)] - $$
$$- (\frac{1}{T_1(n,\delta)} + \frac{1}{T_2(n,\delta)})\dot{r}(t) - \frac{1}{T_1 T_2(n,\delta)}r \qquad (13)$$

can be directly used to create a nonlinear model of the ship.

2.3 The inverse model structure

The determination of the inverse model encounters additional problems because the dynamic properties of the course control system are highly influenced by a steering gear. The block schema of the steering gear is shown in Figure 3

The two (or more) nonlinearities in the actuator exist: the saturation of the rudder angle and the rudder rate limitation. The maximal rudder angle, which can be generally reached in the steering gear is

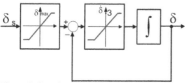

Figure 3. Steering gear scheme. δ - rudder angle

equal to 35 degrees but with higher ship velocities it will be limited to the lower values. The typical rudder rate limit in merchant vessels is within 3°/s. Because we cannot direct invert the saturation block, which in this case is a serious drawback, it is essential to apply the IMC using a different method. The outline of this method in a form dedicated to linear systems is presented below. The design transfer function which is an important part of IMC controller can be created in another way. Consider the inverse of this transfer function

$$P(s) = Q^{-1}(s) \quad (14)$$

and decompose it into a constant term p_0 and a strictly proper term $\overline{P}(s)$.

$$P(s) = p_0 + \overline{P}(s) \quad (15)$$

where: $p_0 = \dfrac{1}{\lim\limits_{s \to \infty} Q(s)}$

In the relationship between error and control signal

$$e(s) = P(s) \cdot u(s) \quad (16)$$

we substitute instead the inverse of the transfer function (14)

$$e(s) = [p_0 + \overline{P}(s)] \cdot u(s) \quad (17)$$

Then the design transfer function can be expressed as

$$Q(s) = \frac{u(s)}{e(s)} = \frac{1/p_0}{1 + \dfrac{1}{p_0}\overline{P}(s)} \quad (18)$$

and realized in form of feedback loop, what is shown in Figure 4.

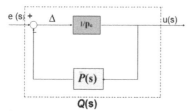

Figure 4. Design transfer function Q(s) realized in feedback loop

A strictly proper term of the inverse design transfer function which is situated in feedback we can substitute for an expression (18)

$$\overline{P}(s) = P(s) - p_0 \quad (19)$$

The inverse of design transfer function for minimum phase model can be evaluated as

$$P(s) = F^{-1}(s) \cdot M(s) \quad (20)$$

3 AUTOPILOT DESIGN

The design methodology of the proposed autopilot is shown in the example of a containership.

3.1 The object

The nonlinear 4 DOF model of the container ship is described by Nomoto (1981) and implemented in Matlab (1994) by Lauvdal. The length between perpendiculars of this ship is equal to Lpp=175 m.

Hence, this model in further consideration and simulation will be treated as a process.

The steering gear is modeled with regard to schema in Figure 3. The limitation of rudder rate is equal to 3 deg/s.

As in the other model-based control structures it is particularly important to prepare the right model of the control process.

The differential equation (12) could not been employed to create a model of the ship because the step response of the considered container ship for larger rudder angle is oscillatory. Therefore, the transfer function (21) has been selected, which step response better reflects the behavior of the ship in the responding set point.

$$M(s) = \frac{\psi(s)}{\delta(s)} = \frac{k(1 + T_3 s)}{s(T_o^2 s^2 + 2\xi T_o s + 1)} \quad (21)$$

The parameters k, T_1, T_2, T_3, as known, depend on ship velocity and it is necessary to calculate or to identify them. For this study, the parameters are identified to expend this model to a wider range of rudder angles. In Figures 5,6 are presented simulation results for $\delta_1 = 1°$ and $\delta_2 = 10°$, by $n=100$ min⁻¹.

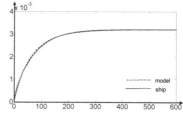

Figure 5. Step responses of ship and model, $\delta_1 = 1°$

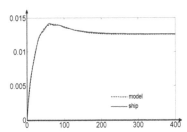

Figure 6. Step responses of ship and model, $\delta_2 = 10°$

The nonlinear functions $T_o = f_1(\delta)$, $\xi = f_2(\delta)$, $T_3 = f_3(\delta)$, $k = f_4(\delta)$ are prepared in form of polynomial of rudder angle 7th order. The polynomial coefficients are calculated using the LS Method and presented in Table 1.

$$T_o(\delta) = \sum_{i=0}^{7} a_{oi} \delta^i \qquad (22)$$

Table 1 Coefficients rounded to 11 numbers

a_{k0}	0.00341726132	a_{To0}	72.7487390248
a_{k1}	-0.00009749736	a_{To1}	-4.11356565809
a_{k2}	-0.00016716906	a_{To2}	-1.14306658459
a_{k3}	0.00004525959	a_{To3}	0.38038366768
a_{k4}	-0.00000546229	a_{To4}	-0.05350006909
a_{k5}	0.00000034658	a_{To5}	0.00394667878
a_{k6}	-0.00000001121	a_{To6}	-0.00014611649
a_{k7}	0.00000000014	a_{To7}	0.00000213119
$a_{\xi0}$	1.06302246329	a_{T30}	78.53456980163
$a_{\xi1}$	-0.00128408227	a_{T31}	-3.39944975876
$a_{\xi2}$	-0.01798662504	a_{T32}	1.37416282707
$a_{\xi3}$	0.00612658227	a_{T33}	-0.22977586218
$a_{\xi4}$	-0.00091511940	a_{T34}	0.00344625965
$a_{\xi5}$	0.00006867696	a_{T35}	0.00157226245
$a_{\xi6}$	-0.00000253244	a_{T36}	-0.00010914535
$a_{\xi7}$	0.00000003650	a_{T37}	0.00000213044

The relations between particular parameters and rudder angle are presented in graphical form in Figures 7-10.

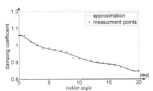

Figure 7. Function $T_o = f_1(\delta)$,

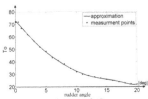

Figure 8. Function $\xi = f_2(\delta)$

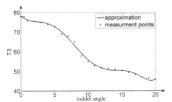

Figure 9. Function $T_3 = f_3(\delta)$

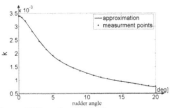

Figure 10 Function $k = f_4(\delta)$.

3.2 The inverse model design

It is apparent from the equations 18 that to determain the design transfer function it is not required to calculate the inverse model of the ship, enough for this is the knowledge of the plant model and the low-pass filter .

The filter has usually the form (2) where n is sufficiently large in order to guarantee that the IMC controller is proper. Also T_f is the only tuning parameter that has to be selected by the user to achieve the appropriate compromise between performance and robustness and to keep the action of the manipulated variable within bounds.

The expression p_0 (23)

$$p_0 = \frac{kT_3 T_f}{T_o^2} \qquad (23)$$

is not constant. It also changes depending on the speed of the vessel.

The structure of proposed IMC controller is shown in Figure 11.

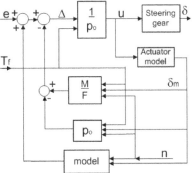

Figure 11. Schema of IMC controller.

4 SIMULATION RESULTS

The assumed time constant of the filter T_f is equal to 6 s. By computer simulation study the tests of following course change maneuvers were performed:
1 50 degrees
2 120 degrees
  The maneuvers were realized by:
– PID autopilot tuned by method for a linear model $\delta_2 = 10°$
– P and IMC controllers using a linear model $\delta_2 = 10°$
– P and IMC controller extended to nonlinear model
  Course maneuvers conducted using the PID autopilot are shown in Figure 12.

$K_p = 0.6$, $K_i = 0.012$, $T_d = 8$

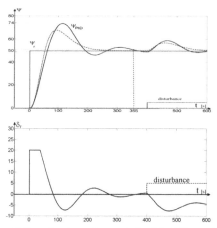

Figure 12 Control value and control signal in PID autopilot. Reference signal ψ_r = 50 deg. ----dashed line - linear model.

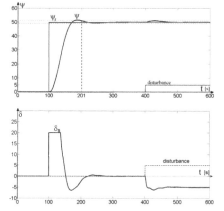

Figure 13. Control value and control signal in IMC autopilot . Reference signal ψ_r = 50 deg.

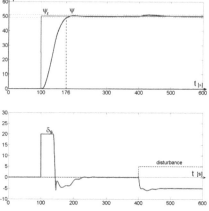

Figure 14. Control value and control signal in Nonlinear IMC (NIMC) autopilot. Reference signal ψ_r = 50 deg.

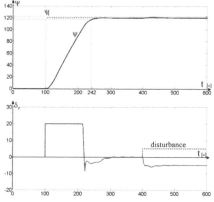

Figure 15. Control value and control signal in NIMC autopilot Reference signal ψ_r = 120 deg.

The results obtained by a computer simulation of considered container confirm the difficulties that meets autopilot with PID controller in the course-change maneuver.

5 CONLUSIONS

The PID controller is adequate for course keeping maneuvers, but is not sufficient for course changing maneuvers where nonlinear steering conditions can occur. The nonlinear steering conditions are not an obstacle to implement the IMC controller.

  The simulation results show that the use of the IMC structure in control system of the ship course allows for substantial improvement of control quality.

  By inverting the internal model of IMC by feedback technique it was possible to take into account the limitation of ruder angle and rudder rate.

As in other model-based methods the control performance depends on the accuracy of the model.

The concept of extension the linear Internal Model Control to a nonlinear form allowed to obtain a more accuracy model what also improved the control performance, which is expressed by reducing the overshoot and settling time for a wide range of work points. In this study the parametric model of the process was created by means of identification based of an open-loop step response by different rudder angle thereby the different ship velocity components. The proposed controller can be successfully used to reach a new set-point of course and also after the necessary modifications also in vessel maneuvering along a desired path.

REFERENCES

Chia T.-L. & Lefkowitz I. 2010. Internal model-based control for intergrating processes. *ISA Transactions* 49: 519-527

Minorski, N. 1922. Directional stability of automatic steered bodies. *Naval Engineering Journal* 34 (2): pp 280–309

Nomoto, K., Taguchi, T., Honda, K., Hirano, S. 1957. On the steering qualities of ships. *International Shipbuilding Progress* 4: 354–370.

Nomoto, K. 1981. On the Coupled Motion of Steering and Rolling of a High Speed Container Ship. *Naval Architect of Ocean Engineering, 20.* J.S.N.A., Japan, Vol. 150: 73-83.

Rivera, D. E., Morari, M., Sigurd, S. 1986. Internal model control.PID controller design. *Industrial Chemistry in Process Design and Development* 25(1): 252–265

Saari, H., Djemai, M. 2012. Ship motion control using multi-controller structure, *Ocean Engineering* 55: 184-190

Tomera, M. 2010. Nonlinear controller design of ship autopilot. *Appl.Math. Compt. Sci.* 20 (2) : 271-280

Tzeng, C.-Y. 1999. An Internal Model Control Approach to the Design of Yaw-Rate-Control Ship-Steering Autopilot. *IEEE Journal of Oceanic Engineering, Vol.24,No. 4, October:* 507-513.

Yaozhen, H., Hairong, X., Weigang, P.,Changshun, W. 2010. A fuzzy sliding mode controller and its application on ship course control. *International Proceedings of the 7 International Conference on Fuzzy Systems and Knowledge Discovery*: 635-63.

Reliability and Exploitation Analysis of Navigational System Consisting of ECDIS and ECDIS Back-up Systems

A. Weintrit & P. Dziula
Gdynia Maritime University, Gdynia, Poland

M. Siergiejczyk & A. Rosiński
Warsaw University of Technology, Warsaw, Poland

ABSTRACT: The article presents base issues concerning navigation in maritime transport with use of electronic chart systems. Essential aspects related to navigation by means of ECDIS system, International Maritime Organization requirements concerning safety of navigation performed with use of electronic charts, and their interpretation, has been also included. Considering "paperless" navigation by means of ECDIS and ECDIS back-up systems, reliability and exploitation analysis of such a system, plus basic outcomes, have been also shown.

1 INTRODUCTION

At the beginning of eighties of the XX century, first solutions able to indicate actual ship's geographical position onto the electronic chart, presented on computer's display (monitor), appeared in maritime transport. Development works on the solutions (named later on in general as Electronic Chart Systems), were a result of rapid technological progress in the fields of automatics, electronics and informatics, concerning in particular storing, processing, transmission and display of information. Evolution of technologies mentioned, over the previous years, caused that electronic chart was understood as not only a picture of conventional paper chart displayed on PC monitor. Electronic Chart Systems became the advanced navigational and information tools, solving important problem from the safety of shipping point of view, concerning collective presentation on one display: other sea surface objects picture from radar device, topographic and bathymetric data from electronic chart data base, and information on geographical position from external navigational positioning sensors [Weintrit, A., Dziula, P. & Morgaś, W. 2004]. The systems achieved also very wide functionality, supporting marine navigators' activities in many different fields.

Works on electronic chart systems evolving, resulted with their appearance on navigational bridges of sea going ships. It is important however to mention, that for many years from the beginning, the solutions could only be used by navigators as so

called "aid to navigation" tools. The essential navigation operations (plotting and monitoring of ship's position) could only be performed by means of traditional paper charts.

On the other side, due to obvious advantages of navigation on electronic chart, in comparison to paper one, works on developing of an electronic chart system able to substitute traditional paper charts were also started.

International Maritime Organization approved usage of electronic chart systems for the essential navigation in November 1995, by adopting the Resolution: Performance Standards for Electronic Chart Display and Information Systems (ECDIS) [IMO Resolution A.817 (19), 1995]. The resolution specified in detail the standards (requirements) vital for systems capable of substitution the traditional navigation on paper charts. It was then very important to understand properly, that IMO approval for usage of electronic chart systems for the essential navigation, was not concerning all electronic chart systems in general, but only systems (from that moment named as "ECDIS" - Electronic Chart Display and Information System) fulfilling requirements given in mentioned resolution.

The resolution defines ECDIS as a navigation information system which, with adequate back up arrangements, can be accepted as complying with the up-to-date chart required by regulation V/19 & V/27 of the 1974 SOLAS Convention, by displaying selected information from navigation sensors to assist the mariner in route planning and route monitoring, and by displaying additional navigation-related information if required.

Interesting fact is, that for a couple of years since the resolution publication, the possibility of paper chart replacement by the electronic chart, was only the theory, because there were no systems fulfilling IMO criteria (it can be said there were no ECDIS systems). The first system received an official confirmation of fulfilling all IMO performance standards for ECDIS systems (it can be then said - the first ECDIS system) in 1999 [http://www.transas.com/about/history].

Detailed requirements concerning ECDIS systems functionality, and abilities of further manufacturers to their fulfillment, caused increase of navigators' interests in these systems, resulting in growing of their usage in maritime navigation.

2 IMPORTANCE OF ECDIS SYSTEMS FOR MARITIME TRANSPORT SAFETY IMPROVEMENT

There are following key facts, proving improvement of safety of navigation performed by means of an ECDIS system [Jurdziński, M. & Weintrit A. 1992], [Kerr, A.J., Eaton, R.M. & Anderson, N.M. 1986], [Weintrit, A., Dziula, P. & Morgas, W. 2004], [Weintrit, A. 1997]:

- System, by presentation on one display: other sea surface objects from radar, topographic and bathymetric data, and information on geographical position is merging display of actual own ship movement and collision situation. This is simplifying navigation and decision making process.
- Electronic chart data base is allowing user to significantly better customize of chart content elements and their display forms. According to actual needs reflecting tasks performed, types and parameters of vessels, including their drafts, amount of data displayed is selected according to scale used (generalization), and temporary user needs (customization).
- The technology is helping in development of comparative navigation, especially in case of position determination by merging of shoreline picture obtained from radar and the same area display coming from electronic chart database (navigation system based on computer processing of marine radar images).
- The use of electronic chart database is enabling easy and quick access to additional complementary information concerning selected items of chart data (i.e. lighthouses, tides, currents, fairways, ports, pilotage etc.), that before were placed in many different nautical publications.
- ECDIS systems, by automation of significant number of functions and calculations, influence on decreasing of previous ship's navigator workload, making him able to concentrate more efforts on other tasks important for safety of navigation, i.e. visual observation.
- Implementation of electronic charts is reducing ship's costs of handling of huge, sometimes amounting with thousands of items, sets of conventional paper charts, and significant numbers of complementary publications.
- Electronic chart systems are able to support wide area of routine navigational tasks, including route planning and passage monitoring. Launch of navigational information ECDIS systems resulted with new ways and possibilities of route plan creating, and also significant changes in capabilities of planned ship's trajectory realization.
- Implementation of electronic navigational chart databases is allowing to use modern, including automatic ones, methods of navigational databases updating, also by means of radio communication means and satellite communication systems.
- Electronic chart systems are the base for further, more advanced automation of navigational process performance, plus its full documentation and recording.

Basing on above mentioned features, increasing popularity of electronic chart systems, and expanding interests in new technologies, International Maritime Organization, on the June 5, 2009, adopted amendments to the International Convention for the Safety of Life at Sea, 1974, including, among others, new regulations for the mandatory carriage requirements of ECDIS [Resolution MSC.282(86)]. Consequence of the amendments mentioned is that all ships engaged on international voyages, will be equipped in ECDIS systems after the July 1, 2018.

ECDIS system installation on board of the vessel is giving possibility to give up the traditional navigation with use of paper charts. It must be emphasized, that it is a possibility, not obligation. ECDIS systems implementation on ships as the mandatory equipment, is not meaning simultaneous obligation for paper charts usage termination. ECDIS system is able, but does not have to be used as the primary navigational tool. It can be assumed, that some part of ship owners will decide to leave paper charts on board of their vessels for some period of time, because of concerns regarding electronic chart systems reliability and safety of navigation processed by means of them. The others however will decide to cease conventional paper charts, due to at least financial matters concerned with their maintaining.

3 IMO REGULATIONS REGARDING SAFETY OF NAVIGATION PERFORMED BY MEANS OF ECDIS SYSTEM

Electronic navigation systems cannot be guaranteed to be 100% failsafe; with this in mind, there must be some form of back-up or redundancy to cover ECDIS failure. IMO performance standards require the 'overall system' to include both a primary ECDIS and an adequate, independent back-up arrangement to ensure the safe takeover of ECDIS functions without resulting in a critical situation. The independent back-up arrangement must allow the safe navigation of the ship for the remaining part of the voyage in case of ECDIS failure [Weintrit, A. 2009].

International Maritime Organization, on the 4 December, 1996, adopted the Amendments to Resolution A.817(19) – Performance Standards for Electronic Chart Display and Information Systems (ECDIS) [Resolution MSC.64(67)], including, among others, detailed technical requirements for reserve systems (named as ECDIS back-up systems), that should take over navigation in case of ECDIS system malfunction. The task of ECDIS back-up system specified in general in the Resolution mentioned, is to ensure that safe navigation is not compromised in the event of ECDIS failure. The back-up system shall allow the vessel to be navigated safely until the termination of the voyage.

The latest IMO resolution regarding performance standards for ECDIS systems: Adoption of the Revised Performance Standards for Electronic Chart Display and Information Systems (ECDIS), adopted on the December 5, 2006 [Resolution MSC.232(82)], has not significantly changed requirements for ECDIS back-up systems, specifying them as follows:
– Facilities enabling a safe take-over of the ECDIS functions should be provided in order to ensure that an ECDIS failure does not develop into a critical situation.
– A back-up arrangement should provide means of safe navigation for the remaining part of a voyage in the case of an ECDIS failure.

The resolution lists the required functions and availability requirements of back-up arrangements, including:
– Chart information using the latest official edition that are kept up-to-date for the entire voyage.
– Route planning capable of performing route planning functions, including taking over the route plan of the primary system, and adjusting a planned route.
– Route monitoring enabling take-over of the route monitoring function originally performed by the primary system, including plotting own ship's position and displaying the planned route.

– Voyage recording that keeps a record of the ship's actual track, including positions and corresponding times.

However, above rather basic statements allow considerable leeway and there are various interpretations as to what are the minimum functional requirements, or what constitute "adequate" back-up arrangements.

There are a number of possible options that could meet these requirements, including [Weintrit, A. 2009]:
– A second ECDIS connected to an independent power supply and a separate GPS position input.
– An appropriate up-to-date portfolio of official paper charts for the intended voyage.
– An ECDIS operating in the RCDS mode of operation.
– A radar-based system called "Chart-Radar" according to the IMO Performance Standards for Chart Radar.

The flag state must approve the ECDIS back-up arrangement to ensure it is in accordance with IMO performance standards; however, some flag states may delegate the ECDIS approval process to a recognised organisation.

The use of ECDIS system as a tool for primary navigation, with a simultaneous paper charts usage termination, is obviously raising questions and concerns regarding reliability of this solution. According to actual regulations [SOLAS, 1974], all ships engaged on international voyages shall carry up-to-date nautical charts and nautical publications to plan and display the ship's route for the intended voyage, and to plot and monitor positions throughout the voyage. The regulations are also saying, that an electronic chart display and information system (ECDIS), may be accepted as meeting the chart carriage requirements. The question however must be placed, if electronic devices like ECDIS system and required reserve system ECDIS back-up do guarantee 100% functioning continuity. It must be emphasized that the requirement [Resolution MSC.64(67)], is stating that ECDIS back-up system is to provide means of safe navigation for the remaining part of a voyage, meaning no disruption is allowed, causing inability of performing activities specified as plotting and monitoring of positions throughout the voyage.

IMO standards are not indicating in any way outlines for situation of ECDIS back-up system failure, that means such a situation is not predicted, consequently – meaning assumption is ECDIS back-up system is to be fully working until the end of the remaining part of an actual voyage, or eventually until ECDIS system starts working normally again. It must be then assumed, that manufacturers of ECDIS and ECDIS back-up systems are obliged to ensure solutions, that guarantee full possibility for fulfilling basic requirement of [SOLAS, 1974] –

which is ability to plot and monitor ship's positions throughout whole voyage, on its every stage.

Considering above assumption a bit risky, authors decided to conduct reliability and exploitation analysis of navigational system consisting of ECDIS and ECDIS back-up systems, allowing to analyze theoretical opportunities of appearing of failures, resulting with inability to fulfil [SOLAS, 1974] requirements regarding performing of safety navigation [Dziula, P., Jurdzinski, M., Kolowrocki, K. & Soszynska J. 2007a], [Dziula, P., Jurdzinski, M., Kolowrocki, K. & Soszynska J. 2007b].

4 RELIABILITY AND EXPLOITATION ANALYSIS OF THE SYSTEM

By conducting an analysis of the system, it can be stated, that relations taking place, concerning aspect of reliability [Epstein, B. & Weissman, I. 2008], [Verma, A.K., Ajit, S. & Karanki, D.R. 2010] and exploitation [Duer, S., Zajkowski, K. & Duer, R. & Pas, J. 2012], [Dyduch, J., Pas, J. & Rosinski, A. 2011], [Kolowrocki, K. & Soszynska-Budny, J. 2011] can be illustrated as shown in Figure 1.

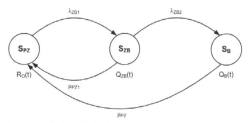

Figure 1. Relationships in the system
Denotations in figures:
$R_0(t)$ – the function of probability of system staying in state of full ability S_{PZ},
$Q_{ZB}(t)$ – the function of probability of system staying in state of the impendency over safety S_{ZB},
$Q_B(t)$ – the function of probability of system staying in state of unreliability of safety S_B,
λ_{ZB1} – transition rate from the state of full ability S_{PZ} into the state of the impendency over safety S_{ZB},
λ_{ZB2} – transition rate from the state of the impendency over safety S_{ZB} into the state of unreliability of safety S_B,
μ_{PZ1} – transition rate from the state of impendency over safety S_{ZB} into the state of full ability S_{PZ}
μ_{PZ} – transition rate from the state of unreliability of safety S_B into the state of full ability S_{PZ}.

The state of full ability S_{PZ} is one, at which both systems (ECDIS and ECDIS back-up) are functioning properly. State of the impendency over safety S_{ZB} is understood as one, where ECDIS system functioning fails. Unreliability of safety state S_B represents situation when both systems (ECDIS and ECDIS back-up) are unfit.

In case system is at full ability state S_{PZ} and failure of ECDIS system takes place, the transition

into the impendency over safety S_{ZB} state at transition rate λ_{ZB1} occurs. When being at the impendency over safety S_{ZB} state, transition into full ability state S_{PZ} is possible if appropriate actions aiming to restore full functionality of ECDIS system are undertaken successfully.

In case system is at the impendency over safety S_{ZB} state, and further ECDIS back-up system functionality fails, the transition into the unreliability of safety S_B state at transition rate λ_{ZB2} occurs.

The transition from the unreliability of safety S_B state into the state of full ability S_{PZ} is possible if appropriate actions aiming to restore full functionality of both systems are undertaken successfully.

The system illustrated in fig. 1 may be described [Rosinski, A. & Dabrowski, T. 2013, Rosinski, A. 2015b], [Siergiejczyk, M. & Rosinski, A. 2011] by the following Chapman–Kolmogorov equations:

$$R_0'(t) = -\lambda_{ZB1} \cdot R_0(t) + \mu_{PZ1} \cdot Q_{ZB}(t) + \mu_{PZ} \cdot Q_B(t)$$
$$Q_{ZB}'(t) = \lambda_{ZB1} \cdot R_0(t) - \mu_{PZ1} \cdot Q_{ZB}(t) - \lambda_{ZB2} \cdot Q_{ZB}(t) \quad (1)$$
$$Q_B'(t) = \lambda_{ZB2} \cdot Q_{ZB}(t) - \mu_{PZ} \cdot Q_B(t)$$

Given the initial conditions:

$$R_0(0) = 1$$
$$Q_{ZB}(0) = Q_B(0) = 0 \quad (2)$$

Laplace transform yields the following system of linear equations:

$$s \cdot R_0^*(s) - 1 = -\lambda_{ZB1} \cdot R_0^*(s) + \mu_{PZ1} \cdot Q_{ZB}^*(s) + \mu_{PZ} \cdot Q_B^*(s)$$
$$s \cdot Q_{ZB}^*(s) = \lambda_{ZB1} \cdot R_0^*(s) - \mu_{PZ1} \cdot Q_{ZB}^*(s) - \lambda_{ZB2} \cdot Q_{ZB}^*(s) \quad (3)$$
$$s \cdot Q_B^*(s) = \lambda_{ZB2} \cdot Q_{ZB}^*(s) - \mu_{PZ} \cdot Q_B^*(s)$$

Probabilities of system staying in a distinguished functional states in symbolic (Laplace) terms have the following form:

$$R_0^*(s) = \frac{s^2 + s \cdot \mu_{PZ} + s \cdot \mu_{PZ1} + s \cdot \lambda_{ZB2} + \mu_{PZ} \cdot \mu_{PZ1} + \mu_{PZ} \cdot \lambda_{ZB2}}{\begin{array}{l} s^2 \cdot \mu_{PZ} + s^2 \cdot \lambda_{ZB1} + s^2 \cdot \mu_{PZ1} + s^2 \cdot \lambda_{ZB2} + s^3 + \\ + s \cdot \mu_{PZ} \cdot \lambda_{ZB1} + s \cdot \mu_{PZ} \cdot \mu_{PZ1} + \\ + s \cdot \mu_{PZ} \cdot \lambda_{ZB2} + s \cdot \lambda_{ZB1} \cdot \lambda_{ZB2} \end{array}}$$

$$Q_{ZB}^*(s) = \frac{s \cdot \lambda_{ZB1} + \mu_{PZ} \cdot \lambda_{ZB1}}{\begin{array}{l} s^2 \cdot \mu_{PZ} + s^2 \cdot \lambda_{ZB1} + s^2 \cdot \mu_{PZ1} + s^2 \cdot \lambda_{ZB2} + s^3 + \\ + s \cdot \mu_{PZ} \cdot \lambda_{ZB1} + s \cdot \mu_{PZ} \cdot \mu_{PZ1} + \\ + s \cdot \mu_{PZ} \cdot \lambda_{ZB2} + s \cdot \lambda_{ZB1} \cdot \lambda_{ZB2} \end{array}}$$

$$Q_B^*(s) = \frac{\lambda_{ZB1} \cdot \lambda_{ZB2}}{\begin{array}{l} s^2 \cdot \mu_{PZ} + s^2 \cdot \lambda_{ZB1} + s^2 \cdot \mu_{PZ1} + s^2 \cdot \lambda_{ZB2} + s^3 + \\ + s \cdot \mu_{PZ} \cdot \lambda_{ZB1} + s \cdot \mu_{PZ} \cdot \mu_{PZ1} + \\ + s \cdot \mu_{PZ} \cdot \lambda_{ZB2} + s \cdot \lambda_{ZB1} \cdot \lambda_{ZB2} \end{array}} \quad (4)$$

Solution to the above set of equations in the time domain is the next step in the analysis and is not discussed here.

5 MODELLING OF RELIABILITY AND EXPLOITATION OF THE SYSTEM

Computer simulation and computer-aided analysis facilitate to relatively quickly determine the influence of change in reliability-exploitation parameters of individual components on reliability of the entire system. Of course, the reliability structure of both the entire system and its components has to be known beforehand.

Using computer aided allows to perform the calculation of the value of probability of system staying in state of full operational capability R$_O$, impendency over safety S$_{ZB}$ and unreliability of safety S$_B$. That procedure is illustrated with below example.

Example
The following quantities were defined for the system:
− test duration - 1 year (values of this parameter is given in [h]):

$$t = 8760 \ [h]$$

− reliability of basic system ECDIS:

$$R_{ZB1}(t) = 0,99995$$

− reliability of system ECDIS back-up:

$$R_{ZB2}(t) = 0,99995$$

− transition rate from the state of impendency over safety into the state of full ability:

$$\mu_{PZ1} = 0,1 \left[\frac{1}{h}\right]$$

− transition rate from the state of unreliability of safety into the state of full ability:

$$\mu_{PZ} = 0,05 \left[\frac{1}{h}\right]$$

Knowing the value of reliability R$_{ZB1}$(t), transition rate from the state of full ability into the state of the impendency over safety S$_{ZB1}$ may be estimated. Provided the up time is described by exponential distribution, the following relationship can be used:

$$R_{ZB1}(t) = e^{-\lambda_{ZB1}t} \text{ for } t \geq 0$$

thus

$$\lambda_{ZB1} = -\frac{\ln R_{ZB1}(t)}{t}$$

For $t = 8760 \ [h]$ and $R_{ZB1}(t) = 0,99995$ we obtain:

$$\lambda_{ZB1} = -\frac{\ln R_{ZB1}(t)}{t} = -\frac{\ln 0,99995}{8760} = 5,7079 \cdot 10^{-9} \left[\frac{1}{h}\right]$$

Knowing the value of reliability R$_{ZB2}$(t), transition rate from the state of full ability into the state of the impendency over safety S$_{ZB1}$ may be estimated. Provided the up time is described by exponential distribution, the following relationship can be used:

$$R_{ZB2}(t) = e^{-\lambda_{ZB2}t} \text{ for } t \geq 0$$

thus

$$\lambda_{ZB2} = -\frac{\ln R_{ZB2}(t)}{t}$$

For $t = 8760 [h]$ and $R_{ZB2}(t) = 0,99995$ we obtain:

$$\lambda_{ZB2} = -\frac{\ln R_{ZB2}(t)}{t} = -\frac{\ln 0,99995}{8760} = 5,7079 \cdot 10^{-9} \left[\frac{1}{h}\right]$$

For above initial values, by use of (4) equations and inverse Laplace transform, following results are obtained:

$$R_0 = 0,99999994$$

$$Q_{ZB} = 5,7079 \cdot 10^{-8}$$

$$Q_B = 6,516 \cdot 10^{-15}$$

Practical usage of presented above considerations allows to determine an influence of transition rate from the impendency over safety state into the state of full ability μ$_{PZ1}$, and transition rate from the unreliability of safety state into the state of full ability μ$_{PZ}$, on value of probability of the system stay at full ability state. Transition rates μ$_{PZ1}$ and μ$_{PZ}$ shall be understood as inverted times t$_{PZ1}$ and t$_{PZ}$, that determine time of restoration of both systems full functionality.

6 CONCLUSIONS

The article presents the analysis of system through the aspect of reliability and exploitation. Assuming three states (full ability S$_{PZ}$, impendency over safety S$_{ZB}$, and unreliability of safety S$_B$) and specified transitions among them, relations allowing to determine probabilities of the system stay at safety states mentioned. The relations allow to determine an influence of particular transition rates on the obtained probabilities values. Further researches are going to perform analysis concerning rationalization of intensity of restoring ability states of particular systems [Rosinski, A. 2015a].

Reliability and exploitation analysis of the maritime navigational system consisting of ECDIS and ECDIS back-up systems, introduced in the article, is the initial assumption of authors to this issue. Calculations performed are basing on reliability factors used for telecommunication equipment and systems. It must be noted, that factors for hardware used in maritime transport, produced according to sophisticated requirements, can significantly differ. However, works and calculations performed, allow to specify following general outcomes:

– There is non-zero probability of failure of both systems used for ship's navigation (ECDIS and ECDIS back-up).
– Assuming above, it can be stated, that navigation process take over by ECDIS back-up system, in case of ECDIS system failure, is not giving 100% probability of its proper functioning, for whole remaining part of a voyage, what is demanded by IMO.
– It is then recommended to perform further, more advanced researches, taking into account reliability factors for technical equipment, used in maritime transport for electronic chart systems performance.
– Detailed, more advanced reliability and exploitation analysis, basing on reliability factors describing more precisely hardware used in maritime transport, can allow to specify additional recommendations and standards regarding production, testing and surveys of the systems, lowering probability of simultaneous failure of both systems (ECDIS and ECDIS back-up) during ship's voyage.
– The probability of simultaneous failure of both systems (ECDIS and ECDIS back-up), will however never achieve zero level, despite works on its lowering. While the vessel without working properly either ECDIS or ECDIS back-up system, is not capable of continuing his voyage.
– Thus, as IMO is not giving any recommendations for proceeding in case of simultaneous failure of ECDIS and ECDIS back-up systems, despite statement, that vessel must be ensured with means allowing it to be navigated safely until the termination of the voyage, it demands deeper investigation and further works to specify eventual additional resources or procedures, that will assume ECDIS back-up system as unsatisfactory for this purpose.

REFERENCES

Duer, S., Zajkowski, K., Duer, R. & Paś, J. 2012. Designing of an effective structure of system for the maintenance of a technical object with the using information from an artificial neural network. *Neural Computing & Applications*. DOI: 10.1007/s00521-012-1016-0.

Dyduch, J., Paś, J. & Rosiński, A. 2011. *Basics of maintaining electronic transport systems*. Radom: Publishing House of Radom University of Technology.

Dziula, P., Jurdzinski, M., Kolowrocki, K. & Soszynska J. 2007. On multi-state approach to ship systems safety analysis. In *Maritime Industry, Ocean Engineering and Coastal Resources. Proceedings of the 12th International Congress of the International Maritime Association of the Mediterranean, IMAM 2007*. Paper: 1069-1073.

Dziula, P., Jurdzinski, M., Kolowrocki, K. & Soszynska J. 2007. On safety of ship systems in variable operation conditions. In *Maritime Industry, Ocean Engineering and Coastal Resources - Proceedings of the 12th International Congress of the International Maritime Association of the Mediterranean, IMAM 2007*. Paper: 1057-1067.

Epstein, B. & Weissman, I. 2008. *Mathematical models for systems reliability*. CRC Press / Taylor & Francis Group.

http://www.transas.com/about/history

International Convention for the Safety of Life at Sea, 1974, Consolidated Edition, International Maritime Organization, London, 2014.

Jurdzinski, M. & Weintrit A. 1992. *Mapa elektroniczna w nawigacji morskiej*. Gdynia: Wyższa Szkoła Morska.

Kerr, A.J., Eaton, R.M. & Anderson, N.M. 1986. The Electronic Chart – Present Status and Future Problems, *The Journal of Navigation*, Vol. 39, No. 1; and reproduced in *International Hydrographic Review*, Monaco, Vol. LXIII, No. 2, July.

Kołowrocki, K. & Soszyńska-Budny, J. 2011. *Reliability and safety of complex technical systems and processes*. London: Springer.

Resolution A.817(19), Performance Standards for Electronic Chart Display and Information Systems (ECDIS), International Maritime Organization, London, November 1995.

Resolution MSC.232(82), Adoption of the revised Performance Standards for Electronic Chart Display and Information Systems (ECDIS), International Maritime Organization, London, December 2006.

Resolution MSC.282(86), Adoptions of amendments to the International Convention for the Safety of Life at Sea, 1974, International Maritime Organization, London, June 2009.

Resolution MSC.64(67), Adoption of new and amended Performance Standards, International Maritime Organization, London, December 1996.

Rosinski, A. & Dabrowski, T. 2013. Modelling reliability of uninterruptible power supply units. *Eksploatacja i Niezawodnosc – Maintenance and Reliability*, Vol.15, No. 4: 409-413.

Rosiński, A. 2015. Rationalization of the maintenance process of transport telematics system comprising two types of periodic inspections. In Henry Selvaraj, Dawid Zydek, Grzegorz Chmaj (ed.), *Proceedings of the Twenty-Third International Conference on Systems Engineering*, given as the monographic publishing series – „Advances in intelligent systems and computing", Vol. 1089. Paper: 663-668. Springer.

Rosiński, A. 2015. Reliability-exploitation analysis of power supply in transport telematics system. In Nowakowski T., Młyńczak M., Jodejko-Pietruczuk A. & Werbińska–Wojciechowska S. (ed.), *Safety and Reliability: Methodology and Applications - Proceedings of the European Safety and Reliability Conference ESREL 2014*. Paper: 343-347. London: CRC Press/Balkema, London.

Siergiejczyk, M., Krzykowska, K. & Rosiński A. 2015. Parameters analysis of satellite support system in air navigation. In Henry Selvaraj, Dawid Zydek, Grzegorz Chmaj (ed.) *Proceedings of the Twenty-Third International Conference on Systems Engineering*, given as the monographic publishing series – „Advances in intelligent systems and computing", Vol. 1089. Paper: 673-678. Springer.

Siergiejczyk, M., Krzykowska, K. & Rosiński, A. 2014. Reliability assessment of cooperation and replacement of surveillance systems in air traffic. In W. Zamojski, J. Mazurkiewicz, J. Sugier, T. Walkowiak, J. Kacprzyk (ed.) *Proceedings of the Ninth International Conference Dependability and Complex Systems DepCoS-RELCOMEX*, given as the monographic publishing series – „Advances in intelligent systems and computing", Vol. 286. Paper: 403-411. Springer.

Siergiejczyk, M. & Rosinski, A. 2011. Reliability analysis of electronic protection systems using optical links. In Wojciech Zamojski, Janusz Kacprzyk, Jacek Mazurkiewicz, Jarosław Sugier i Tomasz Walkowiak (ed.) *Dependable Computer Systems*, given as the monographic publishing series – „Advances in intelligent and soft computing", Vol. 97. Paper : 193 – 203. Berlin Heidelberg: Springer-Verlag.

Siergiejczyk, M., Rosiński, A. & Krzykowska, K. 2013. Reliability assessment of supporting satellite system EGNOS. In W. Zamojski, J. Mazurkiewicz, J. Sugier, T.

Walkowiak, J. Kacprzyk (ed.) *New results in dependability and computer systems*, given as the monographic publishing series – „Advances in intelligent and soft computing", Vol. 224. Paper: 353-364. Springer.

Verma, A.K., Ajit, S. & Karanki, D.R. 2010. *Reliability and safety engineering*. London: Springer.

Weintrit, A. 1997. *Elektroniczna mapa nawigacyjna. Wprowadzenie do nawigacyjnych systemów informacyjnych ECDIS* (in Polish). Fundacja Rozwoju Wyzszej Szkoly Morskiej, Gdynia

Weintrit, A. 2009. The Electronic Chart Display and Information System (ECDIS). An Operational Handbook. A Balkema Book. CRC Press, Taylor & Francis Group, Boca Raton – London - New York - Leiden, 2009.

Weintrit, A. 2010. Six in One or One in Six Variants. Electronic Navigational Charts for Open Sea, Coastal, Off-Shore, Harbour, Sea-River and Inland Navigation. TransNav, the International Journal on Marine Navigation and Safety of Sea Transportation, Vol. 4, No. 2.

Weintrit, A., Dziula, P. & Morgas, W. 2004. *Obsługa i wykorzystanie systemu ECDIS. Przewodnik do ćwiczeń na symulatorze* (in Polish). Gdynia Maritime University, Gdynia.

Automatic Identification System (AIS)

An Analysis of Ship Behavior Induced by the Great East Japan Earthquake Tsunami Based on AIS

X. Liu, S. Shiotani & K. Sasa
The Graduate School of Maritime Sciences, Kobe University, Japan

ABSTRACT: The earthquake off the Pacific coast of Tohoku (the Great East Japan Earthquake) triggered extremely destructive tsunami waves. Most ports and bays in northeastern Japan were significantly affected by this earthquake. This paper studies the behaviors of vessels sailing in the northeast nearshore area after the Great East Japan Earthquake of 2011, using data obtained from the Automatic Identification System (AIS). The object of this paper is to discuss early tsunami forecasting using AIS data obtained from ships sailing in coastal areas. The conclusion obtained in the present study is that AIS data offer possibilities for the early forecasting of tsunami propagation. In this study, analyzing the cruise data and the dynamic data of a ship near wave gauges is the method proposed for recognizing potential tsunami warning signs. This requires ships to be equipped with an AIS; the author has investigated and researched the situation of a ship at the time of this tsunami.

1 INTRODUCTION

The earthquake off the Pacific coast of Tohoku (the Great East Japan Earthquake), with a 9.0-magnitude undersea megathrust, occurred at 14:46, March 11, 2011. The earthquake triggered extremely destructive tsunami waves of up to 40.5 m in Miyako, Iwate. In some cases, the tsunami waves traveled up to 10 km inland. Most ports and bays in northeastern Japan were significantly affected by this earthquake.

This paper studies the behaviors of vessels sailing in the northeast nearshore area after the Great East Japan Earthquake of 2011, using data obtained from the AIS.

The object of this paper is to discuss early tsunami forecasting using AIS data obtained from ships sailing in coastal areas. The conclusion obtained in the present study is that AIS data offer possibilities for the early forecasting of tsunami propagation. (Shiotani et al. 2013) This study is concentrated on cargo ships with lengths of between 160 m and 200 m that were sailing around the northeast nearshore area and ships of all categories in the bay that were largely affected by the tsunami. The time range has been limited to the 30 min before and after the primary crest impact of the tsunami on each concentrated nearshore area (according to the wave gauges). The targeted geolocation range of this study is the area around the Fukushima coast wave gauge and from Daini-kaiho to Keihin-Yokohama in Tokyo Bay. The tsunami data is based on the tide gauges of each nearshore area.

Programs have been developed to process information from the database and generate the AIS data as 2D and 3D KML(Keyhole Markup Language) files to show spatial variations in a virtual globe environment over time. The proposed method for recognizing potential tsunami warning signs involves analyzing the cruise data and dynamic AIS data of each ship. The author has investigated and researched the sailing situation of target ships in the time range of the tsunami's prime crest. The novelty and significance of this study lie in clarifying the behavior of ships during a tsunami wave from AIS data and analyzing the time it takes for ships to encounter a tsunami by using behavioral analysis, which allows us to ascertain the tsunami's timeline in the area of the sea where it reaches.

2 USING AIS DATA TO VERIFY THE TIME PERIOD IN WHICH TSUNAMIS AFFECT SHIPS

2.1 Studies of ship behavior patterns after a tsunami

Studies regarding ship evacuation simulations in response to pending tsunamis have previously been conducted. Murayama, Kobayashi, Kondo & Koshimura (2010) discussed the possibility of the port authority's establishing a ship evacuation policy to be used in ship evaluations. Murayama, Kobayashi, Mizunoe, Kondo, Koshimura & Osawa (2010) confirmed the possibility of using a dynamic ship traffic computer simulation to evacuate ships to sheltered areas. Finally, after the earthquake of the Pacific coast of Tohoku, analysis of ship evacuation based on the AIS data has also been made. Shigeaki Shiotani et al. (2012) investigated the evacuation actions of vessels in Tokyo Bay when major tsunami warnings were announced. The result showed behaviors unique to emergency evacuation situations and drew inferences as to the local conditions of tsunamis by analyzing the evacuation status of vessels (Iwanaga, 2012).The research of Makino presents the status of a ship at the time of a tsunami using objective navigational data (Makino, 2013).

2.2 Parameters for filtering matching ships

The Pacific coast during the Tohoku Earthquake Tsunami was observed by the ten GPS buoys of NOWPHAS, drifting on the sea surface at depths of 100–400 m, on the Pacific coast from Tohoku to the Shikoku District, Japan. The first tsunami crest of higher than 6 m reached the GPS buoys off the Tohoku Coast approximately 30 min after the earthquake. The water level rose relatively slowly in the first part of the first crest and then rose quickly in the latter part. In some locations, the GPS buoy caught the tsunami crest earlier than did the nearby coastal wave gauge and tide gauge by several minutes to dozens of minutes. The time range used as a calculating condition for filtering matching ships is 30 min before and after the time of the primary crest for each gauge. For this research, AIS data were collected from ships sailing offshore in each GPS gauge area when the tsunami occurred.

2.3 Identifying the impact of the prime crest of the tsunami on each ship

This study focused on two navy statuses of ships during the tsunami. One is ships navigating at sustained and stable speeds. The other is ships under anchor inside bay area.

2.3.1 Ships navigating at sustained and stable speeds

The time periods in which ships were most affected by the tsunami were verified by analyzing the dynamic AIS data. By comparing the wave height variance with the changes in Course Over Ground (COG) and True Heading (HDG) of each ship, we tried to find out the correlation between the tsunami wave and the motion of the ships. However, a tsunami's effect on ships can only be detected during the primary crest. Table 1 shows the current variance of the primary crest of the Fukushima wave gauge stationed in one area highly affected by the tsunami. On the top right is the current variance in the Fukushima wave gauge data on March 11, 2011. We can tell that the primary crest of the day started at 15:10 p.m., reached its maximum around 15:15 p.m., and gradually decreased until the next wave came at 15:35 p.m. Tables 2 and 3 show the COG and HDG changes of 2 ships around the wave gauge during the period of the primary crest. Both ships are ships with cargoes of 100 m to 120 m. Number 4312000632 kept a steady speed of 21 knots, and number 548632000 was moving at a speed of 8 knots before 15:40. The intervals for receiving data from the ships were 15 s and 20 s, respectively. Under usual navigating conditions, the difference between HDG and COG is steady and small (normally less than 5 degrees). When the COG changes to a different direction, the HDG changes, accordingly, under the sailors' control. However, from the sudden peak (in the red circle) in Table 3, we can tell that the HDG did not change accordingly, and the periods of sudden changes in COG are all less than 5 min before returning to the original COG degree. From the sudden and short change in the COG degree as compared to the steady HDG, the period when the ship experienced the maximum external force can be identified. Since all of the sample ships were on the left side of the tide gauge, at a parallel distance of 25 km, the times at which each ship confronted the primary crest are 3 to 5 min later than the primary crest times of the Fukushima-oki wave gauge. Figures 1 and 2 show the routes of ships 548632000 and 4312000632, respectively. The affected time period has been picked up and zoomed out, so we can clearly tell that the routes of both ships showed the same trend as the changes in COG. Usually, as a ship would navigate a linear route in a coastal area, rarely changing course, it is unlikely that each ship would intentionally change course suddenly. In this case, it is logical to suggest that the external force is the tsunami's prime crest at that local place. Unfortunately, this phenomenon was only detected during the primary crest period, with wave heights of 2.5 m. All of the following waves and the local crest wave are considered insufficiently strong to influence the ships to show such changes.

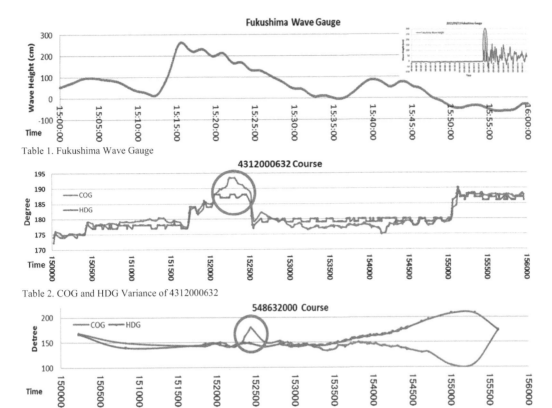

Table 1. Fukushima Wave Gauge

Table 2. COG and HDG Variance of 4312000632

Table 3. COG and HDG Variance of 548632000

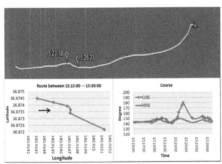

Figure 1. Track of Ship 548632000

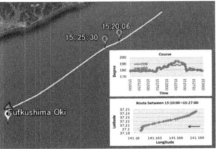

Figure 2. Track of Ship 4312000632

2.3.2 Ships' navigating situations under anchor

In contrast with ships navigating at sustained and stable speeds, ships at anchor in Tokyo Bay do not show the same obvious peak pattern in the change of COG. With sufficient wave gauge data from within Tokyo Bay, the authors first tried to analyze all ship patterns within Tokyo Bay, including the status of all naval vessels. However, since the tsunami current was not strong enough to have a major effect, there was no obvious pattern of COG change for ships sailing at sustained and stable speeds in Tokyo Bay as there was with ships in the Fukushima area.

On the other hand, anchored ships showed some variance, both on the track and angle between COG and HDG, from the effect of the tsunami current. The authors found that the changes in COG and track data were larger for vessels that were anchored during the earthquake than for those sailing at engine speed.

Although the COGs did not show any similar regular variance, as in Fukushima, the value of the course deviation of each ship's COG and HDG was calculated, and the changes in deviation showed trends similar to the current variance. Table 4 is a diagram of the prime crests of tide gauges in Keihin-Kou and Daini-kaiho. As anchored ships, the speeds of all ships in this region were under 1 knot. However, the intervals for receiving data varied. This caused similar ship tracks and chart patterns with slight differences for each ship in this area. Tables 5 and 6 show the difference in COG and HDG variance of two ships over time. We can tell that the difference of COG and HDG is smallest in the middle of the highest and first waves, and changes accordingly with the passing of the local crest and prime crest wave of both ships. Since the AIS receiver stamp of 430011005 was less than 10 s, although at the speed of 1 knot, we consider that ship 431100155 best reflected the variance of the tsunami wave. However, other ships only show the rough pattern, due to the absence of dynamic information with different time intervals from 10 s to 3 min. Even so, all ships, regardless of their size or type, show the same route variance pattern as shown in Figure 3 during the different time periods of the primary crests of the two tide gauges. Figure 3 shows the ship routes in the anchor range between Keihin-Kou and Daini-kaiho. Ships in different ranges show similar patterns in their COG and HDG differences, due to the geographic position of each ship. It is observed that the ship-course deviation time is compatible with the time of the maximum wave height of the observed tsunami. Consequently, the authors' results show that the course changes were caused by the tsunami. In three separate zones, a similar variance trend can be seen from the COG and HDG differences in each zone. Based on the time of the variance point, the motivation of the prime crest can be detected. As in ships navigating at sustained and stable speeds, this phenomenon can only be found under the influence of the prime crest.

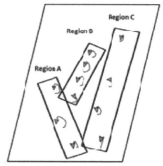

Figure 3. Anchor zone inside Tokyo Bay

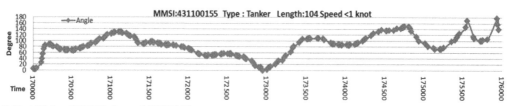

Table 4 Tokyo Bay Tide Gauge

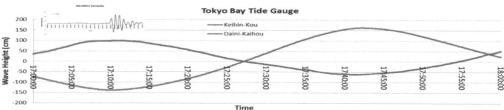

Table 5 COG and HDG Variance of 431000734

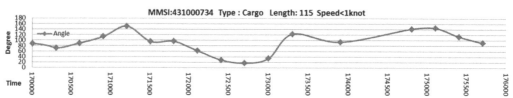

Table 6 COG and HDG Variance of 431100155

122

3 CONCLUSION

It is possible to offer early forecasting of tsunami propagation based on AIS data. By analyzing a ship's cruise data and dynamic data, the time it takes each ship to encounter the tsunami is detected. For navigated ships, a sudden and short COG degree change shows the period when the ship experienced the maximum external force. For anchored ships, both the ship track and the angle between COG and HDG show the same variance under the effect of the tsunami current. Based on the variance, the arrival of the primary crest can be detected. However, the tsunami's effect on ships can only be detected during the primary crest. To fully understand the approach of a tsunami wave, sufficient AIS data of ships in a variety of places are needed. This will require cooperation and the sharing of information by all ships in the nearsea area.

REFERENCES

Makino, H. (2013). Verification of the Time Tsunami Attacked Coasting Ships UsingAIS—Case of the 2011 Japan Tsunami. Journal of Earth Science and Engineering, 2, 126-130.

Murayama, M., Kobayashi, E., Kondo, H., & Koshimura, S. (2010, January). A Research on Ships Evacuation Simulation due to a Tsunami Attack in the Seto Inland Sea. In The Twentieth International Offshore and Polar Engineering Conference. International Society of Offshore and Polar Engineers.

Murayama, M., Kobayashi, E., Mizunoe, T., Kondo, H., Koshimura, S., & Osawa, T. (2010). Marine evacuation traffic at tsunami attack in congestion area. Journal of the Japan Society of Naval Architects and Ocean Engineers, 11, 181-188.

Iwanaga, S., & Matsuura, Y. (2012). Safety of ships' evacuation from tsunami: survey unit of the Great East Japan earthquake. Artificial Life and Robotics, 17(1), 168-171.

Shiotani, S. and Makino, H. (2012). Analysis of Ship Evacuation in Tsunami Using AIS Data. Tokyo, Japan, International Symposium on Engineering Lessons Learned from the 2011 Great East Japan Earthquake, March 1–4, 2012—Part One, pp. 475–482.

Pitana, T., & Kobayashi, E. (2010). Assessment of ship evacuations in response to pending tsunamis. Journal of marine science and technology, 15(3), 242-256.

Shiotani,S. Liu, X and Chen, C. (2013). Study on an Applicability of Early Estimation of Tsunami Propagation Using AIS Data,Japanese Association for Coastal Zone Studies、Vol.26、No.3、pp.129-139、2013

Onboard AIS Reception Performance Advances for a Small Boat

K. Tokudome, Y. Arai, S. Okuda & A. Hori
Marine Technical Collage, Ashiya, Hyogo, Japan

H. Matsumoto
National Fisheries University, Shimonoseki, Yamaguchi, Japan

ABSTRACT: Performance of AIS is not the perfect communication system according to economical transmitting or broadcasting system. Some of authors surveyed the performance of transmission (may be better to say "reception") distances which are related by antenna height, and in the case of lower antenna height it shows poor reception distance. It is often said that reception distance of a small vessel or boat for fishery should be smaller than several nautical miles because of her low antenna height. Authors, proposed to increase antenna gain with sharpening its vertical directivity and surveyed that antenna gain increased using 4 segments collinear antenna and expanded reception distance. Finally, it is concluded that reception performance affected by antenna height is clearly and application will be essential to design and develop not only AIS performance but also marine VHF digital communication for safe navigation.

1 INTRODUCTION

1.1 *Improvement of AIS Reception for a small boat*

To improve the AIS reception performance requires the improvement of signal conditions as follows:
1 Increase the transmission power,
2 Increase antenna gain,
3 Decrease interference signals and/or noise, and
4 Reject the disturbance of VHF-band radio wave.

In case of small boat, class B AIS is used, the poor reception performance is caused by not only low transmitting power, but also low antenna height.

The VHF-band propagation simulation is developed and available (SHINJI M. 1991),(HORI A. et al. 2005). Using this simulation, the simple model of influence against the antenna height is shown as equation (1).

$$L_H = -0.0019H^2 + 0.4548H - 6.2945 \qquad (1)$$

where L_H (dB) is the effectiveness of H antenna height (meters), and Standard antenna height is defined as 15 meters same as marine radar system, so $L_H = 0$ dB at $H = 15$ meters.

Effectiveness of antenna height is applied not only reception but also transmission, so reception signal level is affected the sum of both.

Considering a small boat, her antenna height is assumed as from 2 to 4 or 5 meters, and small to small boat, L_H is approximately −5 dB so total −10 dB against standard.

Transmission power is set down as class A or class B. So, this point now is not able to change, but if necessary, even small boat is able to use class B. This issue should be discussed as the problem of slot capacity, and in this paper we do not discuss.

To decrease interference signals and noises, one counter-measurement is sharpen directivity and separation from objected signals. This issue was discussed as the slot collision, and it is possible to decrease these inferences or proposed as using sector antenna (HORI A. 2005), or increase slot capacity.

Another problem concerning inference is reflection from the objected signals by obstructs in the neighborhood of reception antenna, such as mast or house, or sea surface. The effect of sea surface is larger as lower antenna height, and it is possible to sharpen the vertical directivity of reception antenna.

The effect of disturbance and inference of obstructs will be fade out by the counter-measurement of selection of installation of antenna such as not only the location but also the height of antenna.

1.2 *Antenna Gain*

In case of surface moving vehicle such as vessels, it is popular to use vertical dipole antenna for VHF communication or broadcasting. One reason is that it is horizontal omni-directivity.

Antenna gain is defined the signal level ratio against the objected direction to total power from antenna. So, horizontal and vertical omni-directivity antenna gain is 0 dBi.

Vertical dipole antenna is shown as figure 1 and its horizontal is omni-directivity, but vertical is not, so the antenna gain is 2.16 dBi.

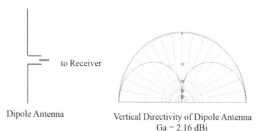

Dipole Antenna   Vertical Directivity of Dipole Antenna
Ga = 2.16 dBi
Figure 1. Dipole Antenna and its Vertical Directivity.

Vertical dipole antenna is not perfect omni-directivity, and the sensitivity toward zenith is zero.

2 COLLINEAR ARRAY ANNTENA

It is difficult to sharpen the horizontal directivity because of using surface moving vehicle and low cost, so to increase antenna gain it is possible to sharpen the vertical directivity.

To sharpen directivity, it is popular to use phase array system, and nowadays collinear array antenna for the antenna system of the mobile base station is often used, which is called collinear array antenna.

2.1 *Architecture*

The phase array system of collinear array antenna is set plural dipole antennas vertically and mechanical or electrical connected.

The mechanical connected system is shown in Figure 2, which is used four dipole antennas. Each dipole antenna are using 2/λ coaxial cable, actual length is decided by the fractional shortening of used cable and from 0.7 to 0.9.

The frequency of AIS is approximately 160 MHz, λ is 1.875 meters, so the length of antenna is 65 cm in case of FS 0.7. So, in case of using 3 elements its total length is 2.65 meters.

2.2 *Antenna Characteristics*

The most important Characteristics is the antenna directivity. This characteristic affects the antenna gain which is one of the most important parameters to decide the transmitting or receiving range. In the case of mobile communication or connecting

system, each station are movable, so ideally it is requested to omni-directivity.

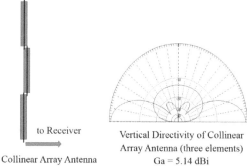

to Receiver   Vertical Directivity of Collinear
Array Antenna (three elements)
Collinear Array Antenna   Ga = 5.14 dBi
Figure 2. Collinear Array Antenna and its Vertical Directivity.

The vertical directivity is more sharpening as more number of elements and increase antenna gain. The vertical directivity shown in Figure 2 is in free space, and affected by sea surface.

Figure 3 shows the effect of antenna height for 4 elements collinear array antenna, and even if install to a small boat, antenna gain should be approximately 10 dBi. As mentioned in section 1.1, it is necessary to gain 10 dB to counter measure the effect of antenna height. In this research, we adapt the 4 elements collinear array antenna.

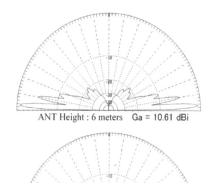

ANT Height : 6 meters   Ga = 10.61 dBi

ANT Height : 4 meters   Ga = 10.12 dBi

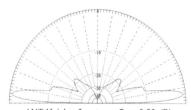

ANT Height : 2 meters   Ga = 9.06 dBi
Figure 3. Antenna Gains affected by Antenna Heights in case of Collinear Antenna with 4 elements.

3 ONBOARD OBSERVATION

Proceeding the study on the characteristics of AIS propagation, onboard observation was executed and analyzed the AIS propagation.

3.1 *Outline*

Onboard survey was executed using small motor boat belonged in National Fishery University, Table 1 shows her principals and Figure 4 shows her photograph.

Table 1. The Principal of motor boat.

LOA	6.56 meters
B	2.37 meters
D	1.22 meters
GT	3.00 GT
H.P.	110.30 kW

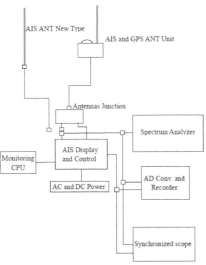

Figure 4. A small motor boat used for this research.

Figure 5 and 6 show the observation system block dialog and picture onboard. They are shown construction of observation system.

Figure 5. Observation System Dialogs.

Figure 6. Observation System Onboard.

Observe onboard, the difference of performance of each antenna, one is the conventional VHF antenna, and other is new prototype 4 elements collinear array antenna shown in Figure 7.

Conventional ANT   New Prototype ANT

Figure 7. Onboard Antenna

Conventional antenna is set up on the house of the boat and it is 2 meters high above the sea level, and new prototype one is set up on the boat deck 1 meters high starboard side.

Observation onboard was executed at northern part of Kanmon channel (Kanmon Kaikyo Traffic Route), Japan in 23rd July, 2014. Figure 8 shows the area where the observation was executed. The observation boat left the pontoon of National Fishery University at 17h00m JST and observed the number of target ships and AIS reception level at a point surrounded by a blue circle. Yellow line is trace of observation boat. Black lines mean traces of target ships by using conventional antenna from 17h00m to 17h13m40s, red ones are new prototype ones' from 17h14m to touching the pontoon. Figure 9 is a photograph of Spectrum Analyzer for observing AIS reception levels during 30 sec. (Y-axis : reception level from -120 dbm to -40 dBm, X-axis: time) for example.

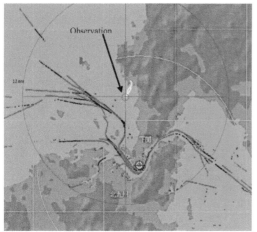

Figure 8. Observation area and traces of target ships.

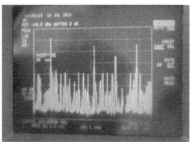

Figure 9. Monitoring Reception Level by Spectrum Analyzer.

3.2 *Result*

During the observation, Figure 10 shows traces of target ships at the observation point (black shows conventional type from 17h09m30s to 17h13m30s and red shows new prototype one from 17h14m00s to 17h19m00s).

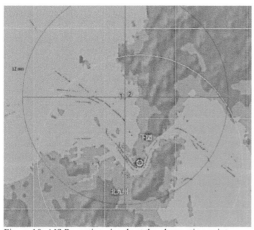

Figure 10. AIS Reception signals at the observation point.

During the observation, using the measuring system the situation of AIS receptions were observed. Figure 11 shows time series of the reception slots' numbers. First half prat shows the result using Conventional ANT and last half shows the result using new ANT (collinear ANT).

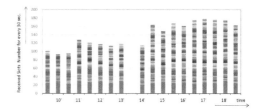

Figure 11. Number of Slots Received by AIS for every 30 seconds.

Figure 12 and 13 show the reception levels during the observation, the former is the result using conventional ANT and latter is the result using new prototype ANT.

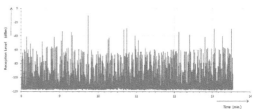

Figure 12. Reception Level of Slots using Conventional ANT.

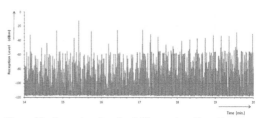

Figure 13. Reception Level of Slots using New Prototype ANT.

4 DISCUSSION

The reception performance is related to the reception level and reception rate, so two factors are discussed and evaluate the performance using new prototype antenna (collinear antenna) in this chapter.

4.1 *The Reception Level*

To discuss one of the reception performance, the reception level, we analyzed the reception level observed from each ship and the reception threshold level of AIS receiver performance.

4.1.1 *The Reception Level*

The reception level of slots are shown in Figure 12 and 13, and the difference between both of levels are not clearly because of the effects of disturbances and/or interferences of obstructs such as the construction of hull and /or mast onboard. The example of the time history of AIS slots level is shown in Figure 14 in case of 10,000DWT PCC sailing near the observation boat, and Table 2 shows the difference of reception level between both antennas.

Table 2. The Differences of Levels between Conventional and New ANT (TOKUDOME K. et al., 2014).

Range (nm)		Level (dBm)		Effect of Range		Diff.
Ave.	Sig.	Ave.	Sig.	Calc.	Comp.	(dBm)
3.15	0.04	-80.4	3.18	-1.64	-78.8	5.4
2.20	0.07	-74.6	5.43	-1.20	-72.4	

$Calc. = 10\log R - 2.1R$

In this case, difference between both antennas, so new one is approximately 5 dBm higher than conventional one.

The fluctuation of levels and some effects of drop of levels are found. The former is approximately within 5 dBm and the latter is more than 20 dBm down. It is assumed that they are affected by the effects of disturbances and/or interference of obstruct onboard such as hull, house and/or mast. According AIS information , the antenna position is very near to the main mast (approximately 4 meters ahead, 9 meters starboard side from main mast on the compass deck), and the relationship of her and observation boat position shows the effects of reflection and/or masking of main mast.

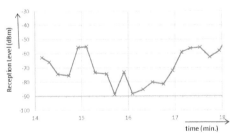

Figure 14. Reception Level of Slots using New Prototype ANT.

4.1.2 *Threshold Level*

In case of AIS transmission GMSK (Gaussian filtered Minimum Shift Keying) modulation is used, so BER (Bit Error Rate) is approximately 10^{-4} when C/N ratio (Carrier to Noise) is 10 dB (SHINGE M. 1991)(ITU-R 1988). So, in case of low noise level, it is possible to receive lower carrier and the number of received slots will increase.

In the observation the number of received slots were captured to count the AIS data recorded. So, it is possible to get the threshold level to compare the number of received slots and the number of slots of which reception level is over the simulated threshold level.

After simple simulation, the threshold levels are got -96 dBm to conventional ANT and -99 dBm to new prototype ANT. So it is possible to estimate that new ANT is approximately 3 dBm better than conventional ANT. It is possible to assume that the new ANT will be able to suppress the interference reflected from the sea surface.

In the proceeding section, the difference of level is 5 dBm and the suppression 3dBm so total 8 dBm. This means that the performance of antenna characteristics is nearly same as discussed in Chapter 1. So, it is possible to expect the development of reception performance using the new prototype antenna.

4.2 *The Reception Rate*

In the case of AIS application, the reception level is not so important, but the reception number is essential to safety navigation. Figure 15 and 16 show the number of ships of which slots were received and cllasifed by range.

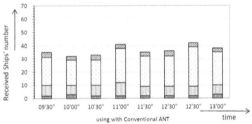

Figure 15. Number received by AIS Receiver with Conventional ANT classified by Range.

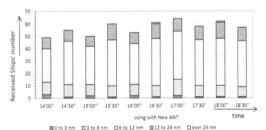

Figure 16. Number received by AIS Receiver with New Prototype ANT classified by Range.

It is clearly different that the number of ships in case of new one is greater than conventional one over 6 nm. It is roughly said that the reception performance of new one is better than conventional one.

The definition of reception rate is the ratio between the reception interval and the transmission interval (ARAI Y., et al. 2007). In case of class A, the transmission interval is decided by the ship's speed, so it is possible to calculate the reception rate using her ship's speed, turning flag and navigation mode. Even in case of class A (NAV 2005), the mooring and anchoring mode are omitted because of long transmitting interval. In case of class B, the transmitting interval is not fixed, so these cases are also omitted.

The reception rate R_{RATE} is shown in Equation (2), and the result of observation are shown in Figure 17.

$$R_{RATE} = \frac{\sum (dT_i \times n_i)}{T_o} \qquad (2)$$

where, dT_i is transmission time interval for i-th speed range, n_i is number of reception slots during the observation time T_o.

In Figure 17, ✖ and ☐ shows reception rate during conventional ANT and new prototype ANT.

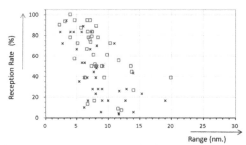

Figure 17. Reception Rate Comparing between Conventional and New Prototype ANT.

Roughly saying, new prototype ANT is higher than conventional ANT. To evaluate the improvement of new ANT, the ratio between conventional and new is calculated and classify by this ratio and show in Table 3.

Table 3. Improvement of New Prototype ANT.

Classification	Range of Rate	Number of ships
Improved	≤ 1.3	30
C:null → N:increasing	--	14
same	$0.7 \leq \leq 1.3$	8
G:some → N:decreasing	--	0
Not Improved	≤ 0.7	4
total		56

The number of improved or same is 52 (92.9%), and left 4 (7.1%) is not improved. So, almost of them are improved, one of not improved is

possibility of shat down AIS after passing Channel, two are possibility of slot collisions and one is possibility of affecting masking by land-mass, large vessels and/or large constructions.

The possibility of slot collision may increase because of improvement of reception level, but it seems not so large and the effectiveness of improvement of reception level is essential to safety navigation.

5 CONCLUSION

In this research, the possibility of application to a small boat will be clearly not only to AIS but also to VHF digital communication and data transmission, and summaries of findings as follows.
1 Onboard observation, we confirmed the improvement of reception level even in case of low antenna height applying the collinear array antenna to sharpen the vertical direction.
2 It is useful to gain the reception performance of AIS signals, and reduce the missing AIS information.

REFERENCES

HORI A., ARAI Y. and OKUDA S. 2005; "Effective Onboard Application on UAIS Information", Proc. of Asia Navigation Conference 2007, Tokyo Japan, pp. 185--190, 2005

NAV 2006; "ITU MATTERS, INCLUDING RADIO COMMUNICATION ITU-R STUDY GROUP 8 MATTERS", NAV 52/INF.2, February 2006

SHINJI M. 1991; "Radio-wave Propagation of Wireless Communication", The Institute of Electronics, Information and Communication Engineers, ISBN4-88552-102-5, pp31--pp.33, 1991

ITU 1988; Rec. ITU-R M.1371, "Technical Characteristics for a Universal Ship borne AIS TIME DIVISION MULTIPLE ACCESS in the VHF Marine Mobile Band"

HORI A., ARAI Y., and OKUDA S. 2006; "Study on Application of Real Time AIS Information", Proceedings of Asia Navigation Conference 2006, October 2006, pp.63--70

ARAI Y., HORI A., OKUDA S. and FUJIE S. 20009; "The Development of Reliability on Automatic Ship Identification System", Journal of Marine Technical College Vol.52, pp51--pp81

TOKUDOME K., NAKASHIMA K., OKUDA S., HORI A. and ARAI Y., 2014, "Study on the Development of Onboard AIS Reception Performance for a Small Boat", Proceedings of 14th Asia Conference on Maritime System and Safety Research, August 2014, pp.157—162

A Subject of Class B AIS for Small Trawler

H. Matsumoto
National Fisheries University, Yamaguchi, Japan

M. Furusho
Kobe University, Hyogo, Japan

ABSTRACT: Class B AIS is used for vessels not equipped with an Automatic Identification System (AIS). In addition to the difference in transmission systems and transmission output between Class A and Class B, the transmission interval for Class B has only two patterns. Because there is not much Distance of Closest Point of Approach (DCPA) between fishing boat and other vessels in congested area, it is important to understand the relationship between speed and displacement of fishing boats for which transmission interval is determined.

This study described herein focuses upon equipping small trawler operating in congested area with Class B, and upon applying the data obtained to DCPA between fishing boats with other vessels in the area, as well as transmission interval and displacement of Class B. The objective of this study includes identifying problems and themes pertaining to Class B used for small trawler.

1 INTRODUCTION

Problem and themes pertaining directly to the introduction of Class B AIS have been pointed out in the reports (Yamashita 2008, 2009). AIS data and radar images for fishing boats in open rough sea have already been obtained in the previous research (Matsumoto 2014), and it has been pointed out that the AIS symbol is missing from the radar images even at a short distances.

Unlike merchant ships, the maneuverability of fishing boats is limited while fishing operation, and speed is frequently increased and decreased. Fishing boats are propelled at low speed when using dragnets; the relationship with AIS, which is automatically determined by the transmission interval according to her speed, is therefore particularly important in this case.

Therefore, this study described herein involves equipping six small trawlers (less than five gross tonnages) operating in congested area with Class B AIS and studying themes and problems pertaining to Class B AIS used by fishing boats by determining the following:
1 DCPA of small trawler with merchant vessels (equipped with AIS) navigating in the area.
2 Displacement of UAIS VHF Data-link Own-vessel report (AIVDO) transmitted data by small trawlers.

2 EXPERIMENTS AND DATA ANALYSIS

The subjects of this study were six small trawler operating in the western area of the Akashi Strait. This experimental study was carried out for a period of approximately two months, spanning from September 11 to November 26, 2014. The data collected from the six small trawler was AIS data (AIVDO/AIVDM) and GPS data (GPRMC). The data for each fishing boats was stored on a cartridge SD (Secure Digital) card. Furthermore, an AIS station was set up near the operation area, and data was collected with a 24-hour observing system.

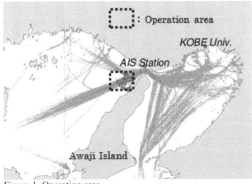

Figure 1. Operation area

As shown in Figure 1, the operation area is connected to Osaka Bay by the east opening of the Seto Inland Sea. Thus, there is a lot of maritime traffic, and it is also a popular place for catching sea bream, octopus, sand lance, etc. Recommended sea lanes have also been established, and there is concentrated maritime traffic through the area.

DCPA between small trawlers and ships navigating in the area is extracted and calculated from AIVDM for each small trawler, and for the GPRMC of small trawlers that match the reception time/date. Since there is no way for small trawlers to know the bearing of other vessels relative to them such as the satellite compass, forward direct true heading were calculated using the course over ground (COG).

Displacement of the AIVDO transmitted from small trawlers was extracted and calculated, using only the data transmitted by channel A or B.

The transmission interval of Class B AIS is given in Table 1; two patterns are divided at SOG 2 knots.

Table 1. Reporting intervals for Class B AIS

Reporting interval	Platform's condition
30 s	SOG > 2 knots
3 min	SOG ≤ 2 knots

3 RESULTS

3.1 *DCPA while trawling*

Figure 2 shows the DCPA between small trawlers while trawling and other vessels near the navigation area. As used in this study, "trawling" is defined as moving at a maximum of two knots ground speed calculated by the reported bearing of fishermen and recorded GPS data. The most frequent distance for 1 nm or less is 0.1 to 0.2 nm. There were 106 vessels that were less than 0.1 nm away.

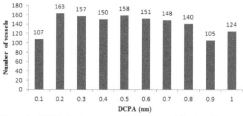

Figure 2. DCPA between small trawlers while trawling and other vessels (less than 1 nm)

Figure 3 shows the DCPA 0.1 nm or less. The most frequent DCPA was 0.05 to 0.06 nm (18 boats). There were 35 boats that were less than 0.05 nm away.

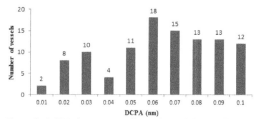

Figure 3. DCPA between small trawlers while trawling and other vessels (less than 0.1 nm)

3.2 *DCPA while not trawling*

Figures 4-5 show the DCPA when not trawling. This includes the DCPA when traveling in fishing areas. The most frequently detected DCPA was 0.1 to 0.2 nm, after which the number of vessels gradually decreases. At less than 0.1 nm (Figure 5), the most frequently detected DCPA was 0.07 to 0.08 nm (68 vessels).

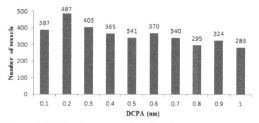

Figure 4. DCPA when not trawling (less than 1 nm)

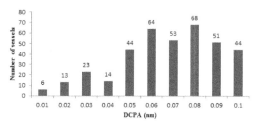

Figure 5. DCPA when not trawling (less than 0.1 nm)

The following points can be assumed in regard to the DCPA between small fishing boats and maritime traffic in congested waterways:
1. The most frequently detected DCPA was 0.2 nm both in trawling and not in trawling.
2. At less than 0.1 nm, whereas the most frequently detected distance was 0.05 to 0.06 nm when trawling (18 vessels), the most frequently detected distance when not trawling was 0.07 to 0.08 nm (68 vessels), followed by 0.05 to 0.06 nm (64 vessels).
3. The DCPA tended to be 0.1 to 0.2 for fishing boats, regardless of the operation state.

3.3 Relative bearing of DCPA

Figure 6 shows the DCPA and bearing of other vessels related to the six small trawlers. The upward direction of the vertical axis with the origin position in the center indicates the ship's bow direction and the downward direction of the vertical axis indicates the ship's stern direction. The DCPA reflecting the closest point of approach from among all AIS data received by small trawlers is plotted by distance from the fishing boat. Since the small trawlers used in the test were not equipped with a compass that could detect direct bow bearing, COG indicating the GPRMC sentence was used as the direct bow bearing. The closest approach to small trawlers during the applicable period was 0.012 nm (approx. 22 m), followed by 0.018 nm (approx. 33 m).

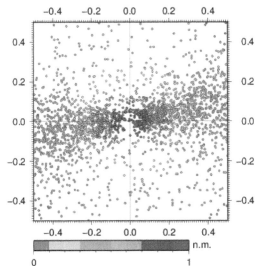

Figure 6. DCPA and bearing of other vessels relative to the six small trawlers.

Figure 7 shows the bearing distribution of Figure 6. Ships approaching fishing boats tend to be concentrated directly lateral to the fishing boats.

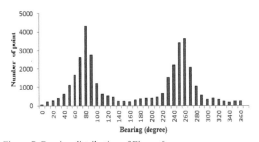

Figure 7. Bearing distribution of Figure 6.

3.4 Displacement of AIS transmission data

Figure 8 shows AIVDO displacement transmitted from small trawlers. With the position of the fishing boat as the origin point, the upward direction of the vertical axis indicates the ship's bow direction and the downward direction of the vertical axis indicates the stern direction. Table 2 gives the frequency of displacements appearing in Figure. 8. Although displacements for Table 2 converge at 0.1 nm, it suggests the possibility of displacement of at least 0.4 nm when AIS data is updated.

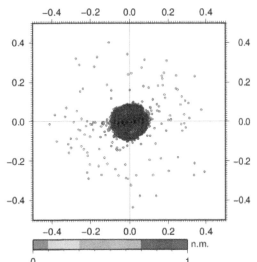

Figure 8. AIVDO displacement transmitted from small trawlers.

Table 2. Frequency of displacements appearing

Distance (nm)	Frequency	Ratio (%)
0	315	0.8
0~0.1	36,189	98.6
0.1~0.2	96	0.3
0.2~0.3	54	0.2
0.3~0.4	27	0.1
0.4~0.5	10	0.0

4 DISCUSSION

Displacement in AIVDO data was observed to exceed 0.4 nm (approx. 740 m) in the test results.

First, there are ten cases from Figure. 8 and Table 2 where displacement exceeds 0.4 nm. Movement of fishing boats will be studied by identifying applicable small trawlers and date/time, and extracting SOG. In Figures 9-10, the transmission interval of AIS is three minutes, which indicates SOG at which displacement would be 0.48 nm (approx. 887 m) during that time.

In some cases where displacement was excessive, the following three conditions were satisfied:

1 The speed of the fishing boat did not exceed 2 knots while trawling and while hoisting the net.
2 Speed increased suddenly after hoisting the net.
3 AIS data (AIVDO) was transmitted immediately prior to increasing speed.

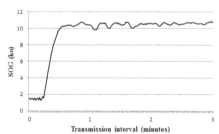

Figure 9. SOG of displacement would be 0.4 nm over

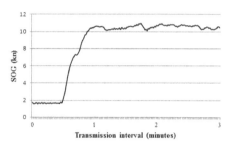

Figure 10. SOG of displacement would be 0.4 nm over

5 SUMMARY

This study focuses on the distinctive movement of fishing boats and performance of Class B AIS, and involved a study of the displacement transmitted from fishing boats by Class B AIS. A backdrop to this was the clarification of DCPA between fishing boats and maritime traffic in congested fairways. A summary of this study is shown as follows;

1 DCPA between fishing boats and other vessels is most often 0.2 nm regardless of whether or not the fishing boat is operating.
2 The closest point of approach (CPA) is concentrated directly abeam to the fishing boats.
3 Merchant vessels navigating near fishing boats tend to pass directly abeam (overtaking or head on) to fishing boats in order to maintain a safe distance from them.
4 At less than 0.1 nm, the most frequent DCPA was 0.06 nm when trawling (18 boats) and 0.08 nm when not trawling (68 boats).
5 Displacement of Class B AIS converges at 99% at less than 0.1 nm, and at 99.8% at less than 0.2 nm.
6 If speed is increased immediately after AIVDO transmission from less than 2 knots (transmission rate of 3 minutes) after hoisting the net,

displacement of the fishing boat may exceed 0.3 nm.
7 According to specifications of Class B AIS, transmission rate should be changed instantly when proceeding at least 2 knots speed.

Due to the superimposed display of AIS information on marine radar or ECDIS on merchant vessels in recent years, AIS information can be obtained in addition to the image of fishing boats. On the other hand, there are some vessels that can only reproduce AIS information without radar information using special software. In this case, movement of each vessel changes only as often as AIS information is updated.

As shown in Figures 6 and 8, determining movement of fishing boats when within the range of vision using Class B AIS only is extremely dangerous. This suggests the possibility of a fishing boat separated at a directly parallel distance at least 0.3 nm appearing directly in front of one's own vessel three minutes later. In such a condition, an AIS symbol could also exist where a fishing boat actually does not exist, thereby reducing the number of selections when maneuvering a ship, and creating a potentially dangerous situation. If a fishing boat is proceeding at less than 2 knots OG (over the ground) speed, the person steering the ship must be able to ascertain from AIS information the possibility that the fishing boat is currently engaged in fishing and that it could suddenly increase speed (excessive displacement of AIS information).

With Class B AIS, there are no standards concerning change of transmission rate; the rate differs according to the transmitter. It is necessary to understand and utilize the features of such apparatuses when equipping fishing boats that tend to increase or to decrease speed suddenly with Class B AIS.

ACKNOWLEDGEMENTS

The authors would like to thank Yasuo Arai for detailed comments, suggestions, and constant support.

REFERENCES

Yamashita, T. 2009. Extraction of Problem of VDL(VHF Data Link) When AIS Class B CSTDMA is Introduced. *Academic journal NAVIGATION (Japan Institute of Navigation)*. 168: 77-83

Yamashita, T. 2009. Potential Problems on AIS System. *Academic journal NAVIGATION (Japan Institute of Navigation)*. 170: 20-23

Hirofumi,M., Masao, F., Shimooka, N. & Ono, M. 2014. A Study of Effective Utilization AIS with fishing boats. *The journal of Japan Institute of Navigation*. 130: 69-75

Research on the Real Movement of Container Ship between Japan, China, and South Korea Using AIS Data

X. Gao
Kobe University, Graduate School of Maritime Sciences, Kobe, Japan

H. Makino
Osaka University, Naval Architecture and Ocean Engineering, Osaka, Japan

M. Furusho
Kobe University, Graduate School of Maritime Sciences, Kobe, Japan

ABSTRACT: This study analyzes the real movements of container ships. This type of study is significant for the problem of waiting ships and the creation of a smooth operating system for maritime transportation. This paper analyzes the movement of container ships sailing in the Seto Inland Sea, Japan. This ocean area represents a primary traffic route for container transportation particularly for China, Japan, and South Korea. This study utilizes an automatic identification system (AIS) data, which facilitates exact and quantitative collection of ship navigation data. The AIS data analysis reveals the total operations of container ships. This study represents core research that contributes to securing the safety and efficiency of maritime transportation.

1 INTRODUCTION

The development of international trade requires the transportation of a substantial amount of goods and materials between countries. Particularly in recent years, as the Chinese economy has rapidly grown, maritime transportation centering on Asia has compounded. The vessels responsible for the mass transportation of goods have grown in size and quantity. Consequently, ships frequently anchor offshore before entering a port. Because ships cannot sail smoothly into a port, they are typically crowded together offshore. This causes numerous maritime accidents, including collisions and offshore grounding, particularly when ships anchor because of wind and currents. These waiting ships produce negative effects far beyond their influence on maritime safety. Waiting ships have created economic and environmental problems. For example, a waiting ship incurs expenses from demurrage, fuel, and crew freight. Moreover, most ships anchor offshore without stopping their engines. Thus, hazardous substances are discharged from these ships as with sailing ships. Therefore, the problem of waiting ships requires a solution for maritime transportation.

The purpose of this study is to mitigate the need for ships to anchor offshore to construct safer and smoother operations. This paper represents core research to grasp and analyze the real movement of ships. This paper analyzes container ship movements

in the Seto Inland Sea, Japan. The shipment of containers is a principal means of international maritime transportation because container shipment is low cost and minimizes damage. Additionally, the number of container ships departing and arriving from China, Japan, and South Korea continues to increase. Seto Inland Sea is a primary traffic route for transportation. Therefore, securing the safety and efficiency of container ships in this ocean area is central to the stabilization and development of the Asian economy.

This paper is organized as follows. The following section describes the application of AIS (automatic identification system) data to an analysis of actual ship operations. Section 3 and Section 4 describe traffic flow and the operations of container ships in the Seto Inland Sea using the extracted ship data and based on examples of the real movements of container ships.

2 APPLICATION OF AIS DATA

In past years, general container ship research has focused on container volumes (Shibasaki. R, et al. 2005). Typically, containers are investigated and analyzed using port data and statistics information. Although this method provides the quantity of container ships, it does not explain the operating dynamics. The work required to gather information over a prolonged period has historically been laborious; therefore, information concerning the actual movement of container ships is limited. Moreover, studies on the contributions for maritime safety and efficiency have been conducted using only models and simulations. However, a grasp of the real movements and conditions of ships is required to create safe and smooth operations.

The method used in this study for the investigation and analysis of ship movement utilizes AIS data. It is possible to obtain the ship navigation data exactly and quantitatively. Recently, various maritime fields, such as port management (Murai. K, 2003) and traffic control using AIS data, have been brought to public attention by an increase in AIS-quipped vessels. AIS data is an optimal approach to understanding shipping trends (Takahashi. H, et al, 2007). The present study using AIS data to analyze the real movement of container services. This analysis is reliable because the majority of containers in a regular service are AIS-equipped.

The AIS technology automatically provides information that can help avoid collisions. The International Convention for the Safety of Life At Sea (SOLAS) stipulates that AIS be installed aboard all international voyaging ships of 300 gross tonnage (GT) and all non-international voyaging ships exceeding 500 GT (IMO, 2003). AIS transponders utilize high frequency (VHF) signals, over which information related to the piloted vessel and other vessels is transmitted, including static information (such as the names of the ships and the call signs), dynamic data (such as the current locations, navigation speed, and directions), and the voyage-related data (such as the draft and destinations).

3 CONTAINER SHIP TRAFFIC FLOW SURVEY USING AIS DATA

3.1 *Traffic volume of container ships in the Seto Inland Sea*

The research area for this study is the Seto Inland Sea and its oceanic waters located in the western part of Japan (32°31′N~ 34°57′12″N; 130°28′40″E~ 135°39′4″E). Figure 1 shows a map of Seto Inland Sea and the ocean area investigated. The black dots indicate the trajectory of ships according to the ships' AIS data position. Seto Inland Sea is an important route for Japan and an international route for periodical container carriers.

The traffic route of Seto Inland Sea is important as an international container liner service, especially, the East Asian economic development of recent years. Figure 1 shows that traffic exits Kanmon Strait on the westernmost side of Japan as the portals ships pass through and into the port of Japan. Seto Inland Sea also contains two Japanese international trade container ports, namely Kobe and

Osaka. Most passing ships reside in these ports; therefore, the area is frequently crowed with large ships.

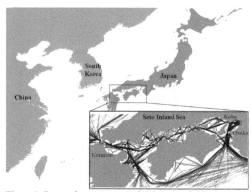

Figure 1. Research target area and trajectory of ships

The AIS data were acquired between March 1 and 7, 2012. We compiled statistics on vessel traffic and the cargo type of every ship with an MMSI (Maritime Mobile Service Identity) number based on the AIS data. A bar graph in Figure 2 shows the statistical results of the number and types of ships in the Seto Inland Sea. We verified from the MMSI number that there were 2,610 passing ships during the period March 1 to 7, 2012, and there are an average of 1,600 ships navigating this area per day. The graph shows that the number of passing ships was greatest on March 7 (Wednesday), with a slightly smaller number of ships on March 3 (Saturday). Additionally, the main type of ships in this area was cargo with a percentage contribution of approximately 63% of all passing ships.

The container ship targets in this study were normal container service. With respect to container ship size and loading capacity, we combined AIS data and the International Transportation Handbook 2013 (Ocean Commerce Limited, 2013) based on the IMO number to extract container ships. This handbook collects international transport and normal service information. From this combination, we obtained the container ship information from AIS data and information on container ship operations.

According to the statistics for the extracted container ships, we obtained 197 container ships during the research period and confirmed that 7% of all passing ships were container ships in the area. Figure 3 shows the percentage of container ships according to the operator's region. We verified that most of the operators belonged to Asian countries, particularly China and South Korea. The total percentage of the contribution to traffic from these two countries exceeds 50% of all container ships. China has the largest number of container ships in the study area.

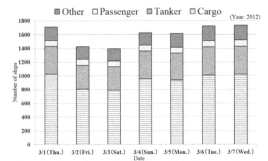

Figure 2. The number and Types of vessels in the research area

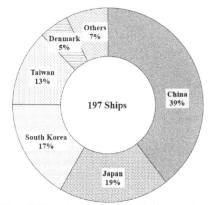

Figure 3. Percentage of container ships by operators

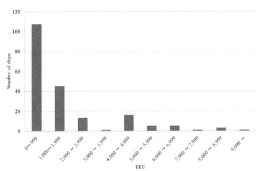

Figure 4. Number of container ships based on capacity

The loading capacity of container ships is typically described in twenty-foot equivalent units (TEUs), which is a unit of the cargo capacity of a standard container. The statistic is based on maximum TEU. Figure 4 shows the number of container ships based on TEU. We verified that the greatest number of ships was within the size of 999 TEU. Container ships in this range have an overall length of between 79 m and 150 m. A total of 107 ships were within this range. During the investigation, the largest size of container ship was 9,012 TEU, and overall length was 338 m.

3.2 Analysis of container ship operations based on tracking

We verified 197 container ships during the investigation. In this section, we analyze the real movement of these container ships based on tracking. Figure 5 shows the movement trajectories of all container ships in the research area during the investigation. The trajectories are obtained from a geographic information system (GIS) based on the ship's position included in the AIS data. From this trajectory, we observe that three general trends of container ships. Most ships reside in Osaka and Kobe, operating between Kanmon and Osaka and passing through Seto Inland Sea routes and the open sea. Some ships represent direct traffic into these ports without operating in the Kanmon Strait.

Figure 6 shows the tracking in Figure 6 (A, B, and C) based on a TEU range of 0 ~ 1,999, 2,000 ~ 4,999, and 5,000 ~ 9,999, respectively. The black dots in Figure 6 (A) indicate the tracking of container ships within the range of 0 ~ 1,999 TEU. From the trajectories, we found that these ships navigating in this area had two routes, sailing the inland sea between Kanmon Strait and Akashi Strait, and sailing the open sea between the Kanmon Strait and the Tomogashima Strait. There were 152 ships within the range. It was found that approximately 45% of all ships in this range navigated between Kanmon and Osaka passing through the inland sea, and approximately 7% of all ships navigated in open sea passing between Kanmon and Osaka. Ships have an overall length less than 200m, and the overall length of 198m was the longest ship in this range that sailed both the inland sea and open sea.

Ships with a TEU in the range of 2,000 ~ 4,999 navigated in open sea between the Kanmon and Tomogashima Strait and sailed in open sea only. The tracking of these ships are shown in Figure 6 (B). The black dots indicate the tracking of container ships within the range of 0 ~ 1,999 TEU. A total of 30 ships were in this range. We confirmed that container ships over 2,000 TEU are large container ships that operate the outward passage.

Figure 5. Trajectory of container ships during the investigation

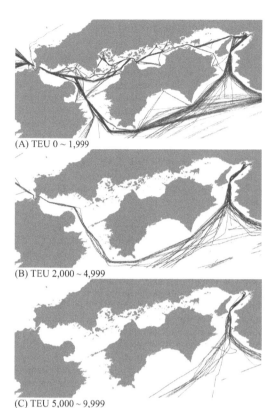

(A) TEU 0 ~ 1,999

(B) TEU 2,000 ~ 4,999

(C) TEU 5,000 ~ 9,999

Figure 6. Trajectory of container ships based on the capacity

Figure 6 (C) shows the tracking of container ships exceeding 5,000 TEU. 15 ships were confirmed from this analysis. These ships have an overall length of 300m or more, and their draft is more than 15m. Therefore, they can pier docking and undocking only to a particular pier of Kobe and Osaka. when these ships sail towards China and Korea, or dock from China and Korea, they passed in and out of the Tomogashima Strait, sailing in the open sea without sailing in the Kanmon Strait.

4 ANALYSIS OF TOTAL CONTAINER SHIP OPERATIONS

4.1 *Extraction of waiting ship for entering port and cargo handling working time using AIS data*

This section analyzes container ship navigating from their arrival in the Seto Inland Sea until port entry. This analysis obtains the total movement of container ship including ship waiting time for berth and cargo handling. Total operations were also analyzed using AIS data.

Figure 7. Flowchart of extraction of waiting ship and cargo handing time

Figure 7 shows the extraction process. The navigation status information included in the dynamic AIS data is used to indicate either a ship at anchor or sailing. However, some errors may have occurred because this information is inputted manually, and most ships drift in an area without anchoring during a temporary stay. Therefore, using navigation status only would render determining the target data difficult. In the proposed method, we extract the waiting ships and the ships during cargo handling using the position and speed data. According to the weather information during the research period, the weather did not have a significant effect on ships.

Consequently, we established whether a ship was at anchor or sailing based on its speed over ground (SOG) and sailing distance. We conducted the computations when the branch condition was satisfied. Finally, ships during cargo handling and waiting offshore were concentrated based on the ship position within and without berths. At the same time, waiting and cargo handling times were calculated and recorded.

According to the statistics for the extracted waiting ships, we found that a total of 197 container ships passed through this area. A total of nine container ships were waiting for a berth before entering the port, and the longest waiting time of a container ship was 22 hours. The time taken for ship cargo handling implies that the actual loadage of container ships can be understood based on AIS data analysis. Typically, a regular service has a regulated entering time and a dedicated berth without having to wait for a berth. The action of the container is difficult to obtain; however, AIS data made it possible to grasp the movement of ships.

Container shipping is a regular service required to arrive at a planned time. Usually, the operating plan of a ship is conducted to ensure early arrival, and a ship will anchor offshore to keep up with port entry times. Consequently, ships often wait offshore before entering the port. However, the waiting ships cause numerous offshore maritime accidents and ship congestion. Therefore, eliminating ship waiting can contribute to safe and smooth operations. We analyzed ship operating over a total navigation and including ship waiting. This type of analysis of waiting ships and cargo handling has been conducted, and data are available from each port. However, it is difficult to collect the statistics. Using AIS data, it is possible to grasp the situation of cargo handling easily and in a timely way.

We analyzed total operations using speed and sailing time. This analysis was conducted on ships navigating the three routes described in Section [3.2]. Table 1 lists the principle characteristics of the sample vessels.

Table 1. Principal aspects of the sample ships

Item	Ship A	Ship B	Ship C
Route	Seto Inland Sea	Open Sea	Open Sea
Vessel's type	Full container	Full container	Full container
Overall length (m)	148	148	338
Max. TEU capacity	1118	1118	9012

Figures 8 to 10 show the trajectories and the changes in ship speed for Ships A, B, and C. The higher illustration in each figure shows the mapping of the ship trajectory (a black dashed line) and the lower illustration shows the transition of ship speed when sailing.

Figure 8 shows the sailing analysis of ship A; this ship sailed in the Inland Sea and arrived at Kobe port. During sailing, the maximum speed was 19.4 knots, and the average speed was 14.5 knots. Navigating the inland sea requires passing through the four narrow waters of Kanmon Strait, Kurushima Strait, Bisan-seto, and Akashi Strait. The time zone in which this ship navigates each strait is indicated by a blue rectangle in the ship speed graph. The change in speed was frequently checked, and speed in particular slowed down when the ship passes through both straits. This ship rapidly decreases its speed before passing through each strait by an average of approximately five knots. We confirmed a similar tendency of all ships that sailed in the inland sea from the analytical results.

Compared with ship A, the speed distribution of ship B sailing in the open sea was significantly up or down. Figure 9 shows the tracking and speed distribution of the ship. The maximum speed of

sailing was 18 knots, and the average speed of ship B was 15 knots faster than ship A when sailing in the inland sea.

The analysis of the distribution of sailing time explains ship navigating time in the inland sea and the open sea. Ship A passed through Kanmon Strait at 8:00 on March 5 and arrived in Akashi Strait at 0:00 on March 6, taking approximately 15 hours and 229 nautical miles to navigate.

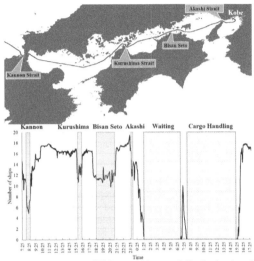

Figure 8. Trajectory of Ship A and its speed distribution based on sailing time

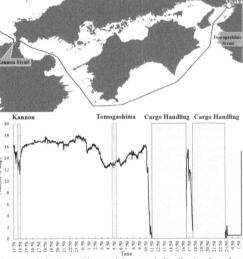

Figure 9. Ship B tracking and its speed distribution based on sailing time

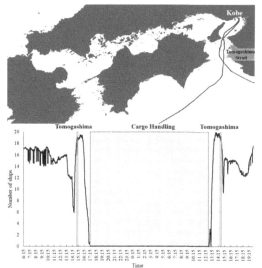

Figure 10. Tracjectory of Ship C and its speed distribution based on sailing time

Compared with ship A, ship B navigated between Kanmon Strait and Tomogashima Strait in approximately 19.5 hours and 295 nautical miles. Additionally, this analysis explains the sailing time of the ships operating in each Strait. Therefore, it is possible to estimate the time required for the ship to reach its destination, and to effectively plan the navigation.

Figure 8 follows the speed of ship A and shows a speed of 0 knots. The extracted ship waiting information confirmed that ship A waited offshore before entering the port. The waiting time was approximately 5.5 hours. Unlike ship A, ship B and ship C entered the port directly. We also verified that a speed of 0 knots was reflected in ship load handling. The time of cargo handling was approximately 7.5 hours, which is shown by the green section in the line graph in Figure 8. The 0 knot speed in Figures 9 and 10 was obtained because of ship load handling. Ship B also stopped at two ports, and ship C has the largest TEU in the investigation that took approximately 18 hours of cargo handling.

5 DISCUSSION

This study's analysis provided an overview of total ship operations for vessels navigating the Seto Inland Sea and its open sea and for container ships of the same size and length sailing in the Seto Inland Sea and the open sea, respectively, which navigate the situation differently. The results show that ships navigating the Seto Inland Sea showed frequent changes in speed. The causes of the varying speeds are fast currents and ship congestion in the straits of the Seto Inland Sea. Therefore, these factors led speed reductions and additional operating times. In such cases, the operator increases the speed to gain time entering the port. However, high speed navigation is a risk to all ships. Although ships sailing open seas are affected by Japanese currents, and the sailing distance is longer than sailing the inland sea, the operator can gain time because there is less traffic and wider navigation.

This analysis explains ship operations considering ship waiting times affected by early ship arrival and the need to wait to berth. This action causes numerous offshore maritime accidents and ship congestion. If waiting time is added to sailing time, the economic effects in fuel cost will be saved because of reductions in speed. This represents a solution to improve the safety and efficiency of maritime transportation.

6 CONCLUSIONS

This study implements an analysis on real movement using AIS data to improve efficiency of the sea transportation and safety. We successfully extracted the container ships from AIS data. Using this method, we grasped the situations of a substantial quantity of vessels. Therefore, this paper also proves that AIS data are a useful contribution to the basic data used in the field of maritime traffic.

This study analyzed container ship traffic based on loading capacity and tracking. Moreover, the actual movements of ships over the total operating were extracted including ship waiting and cargo handling. According to the analysis, the main results for the actual situation of waiting ships are the following:

– Approximately 7% of all passing ships were container ships sailing in the Seto Inland Sea and its open sea. The container ships were identified from ship operators and loading capacity. Most of the container ships in Seto Inland Sea belong to the Asia region, particularly China and South Korea.

– From the analysis based on the trajectories of ships, the ships navigated three routes in this area depending on their TEU capacity. Ships within the range of 0 ~ 1,999 TEU passing between Kanmon and Osaka navigated in the inland sea and the open sea. However, ships with a TEU in the range of 2,000 ~ 4,999 navigated in open sea between the Kanmon and Tomogashima Strait and sailed in open sea only. And ships exceeding 5,000 TEU did not navigate inland sea, just passing in and out of the Tomogashima.

– Analysis of ship waiting and cargo handling, we confirmed that nine out of all container ships waited, and we obtained the waiting and cargo handling times.

– We analyzed the total, particular according to speed and sailing time. Consequently, we identified the sailing time distribution of total operations and the passing time for each strait in the Seto Inland Sea. We found that ships navigated the inland sea with frequent increases and decreases in speed.

This study explains the real movements in total ship operations, which provides core research central to securing the safety and efficiency of maritime transportation.

REFERENCES

IMO. 2003. International Convention for the Safety of Life at Sea. London, International Maritime Organization.

Murai. K. 2003. A Few Comments on the Effects of AIS for Port Service, Navigation of Japan, No. 156, pp57-58.

Shibasaki. R, Watanabe. T, Kadono. T, Kannami. Y. 2005. Estimation Methodology and Results on International Maritime Container OD Cargo Volume Mainly Focused on the East Asian Area, Research Report of the National Institute for Land and Infrastructure Management (NILIM) Japan, No.25, pp1-47.

Takahashi. H and Goto. K. 2007. Study on Inflection to the Port and Harbour Development by AIS DATA, Technical Note of National Institute for Land and Infrastructure Management, No.420, pp1-89.

Modeling of Observed Ship Domain in Coastal Sea Area Based on AIS Data

R. Miyake & J. Fukuto
National Maritime Research Institute, Tokyo, Japan

K. Hasegawa
Osaka University, Osaka, Japan

ABSTRACT: It is well- known that an imaginary domain, which never let other ships enter, exists around a vessel. While the models of the domain in areas such as bays or harbors have been advocated and in practical use, there is little compiled data regarding the actual collision avoidance actions taken by operators under real encounter situations. The authors analyzed collision avoidance maneuvers which were extracted from AIS data recorded for a month, under one-to-one encounter situation, in order to examine collision avoidance behaviors in a coastal sea area. The results of analyses showed tendencies different from those described in the previous studies with regard to the actual offset distance between vessels. In this paper, we present some models of representative offset distances between a give-way vessel and a stand-on vessel in the coastal sea area.

1 INTRODUCTION

Determining a moment of initiating collision avoidance action and a passing distance for the action is critical in order to develop a collision avoidance algorithm. Regarding the passing distance, as is well-known, an imaginary domain which never let other ships enter, exists around a vessel. While the models of the domain in a bay or channel (Fujii 1980, 1983) and of the passing distance in a harbor (Inoue 1994) was advocated and empirical ship domain have been presented with AIS data when passing a bridge or a narrow channel (Martin 2013), there is still little compiled data regarding the actual collision avoidance actions taken by operators under real encounter situations.

The authors analyzed real behaviors under one-to-one encounter situations, which were extracted AIS data of Tokyo Bay, recorded for a month, in order to examine actual collision avoidance behaviors. In this area, a lot of encounters of various types and sizes of ships have been observed. According to the behaviors analyses based on AIS data, tendencies different from those described in the previous researches (Fujii 1980, 1983, Inoue 1994, Martin 2013, Yamasaki 2013) were seen regarding the actual offset distance between vessels in the coastal sea area.

The purpose of this paper is to specify the representative offset distances between ships in coastal sea area. Then we found that the offset distances were approximately proportional to the length of the stand-on vessels.

2 ANALYSES OF COLLISION AVOIDANCE BEHAVIORS

2.1 Area and period of AIS data

To examine an actual domain for collision avoidance around a vessel in a coastal sea area, we analyzed the real maneuvers which were extracted from the AIS data recorded for a month, under one-to-one encounter situation in southern Tokyo Bay. In this area, a lot of encounters of combination of ships of various types and sizes are constantly observed.

The AIS data recorded from 1st to 30th of June in 2013 were used for the analyses. Figure 1 shows the trajectories of AIS data only on 1st of June 2013.

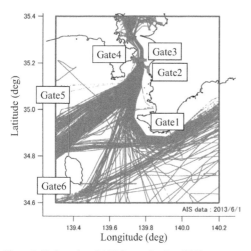

Figure 1. Trajectories of AIS data on 1st June 2013.

2.2 Procedure for extracting collision avoidance behaviors

In order to automatically extract the collision avoidance behaviors in the situation of the one-to-one encounter, six gates were set as shown in Figure 1. The vessels through these gates were extracted and their collision avoidance behaviors were analyzed. The vessels used in this analysis were satisfying all the following conditions: (1) the situation of encounter between a give-way vessel and its target ship, i.e. stand-on vessel, was head-on, overtake or crossing; (2) two vessels met on the five combinations of gates, i.e., 3-6 and 1-3, 3-6 and 1-5, 3-6 and 1-6, 3-6 and 2-4 or 3-6 and 3-5; and (3) two vessels met under the conditions that the distance of them was within 18520 m (10 miles), TCPA was within 30 minutes and DCPA was within 3704 m (2 miles). In this paper, the encounter situation was determined based on the relative position between the give-way vessel and the stand-on vessel at the moment of the aforementioned three encounter conditions.

The collision avoidance behaviors were automatically extracted by the procedure described hereafter. First, the status of each encountered vessel, such as heading, speed, DCPA and TCPA and so on, was calculated at every synchronized 10 second based on the AIS data. Next, in the case where the actions of the give-way vessel for evading the stand-on vessel, such as altering course or reducing the vessel's speed, have been observed simultaneously with the increase of DCPA or TCPA, we considered that the give-way vessel evaded the stand-on vessel. In addition, the time at which the give-way vessel started to evade the stand-on vessel, have been identified.

2.3 Extracted collision avoidance behaviors

Table 1 shows the number of extracted collision avoidance behaviors from AIS data.

Table 1. Number of extracted collision avoidance behaviors.

Gates	Avoided			Not Avoided		
	HO*	OT**	CS***	HO*	OT**	CS***
3-6 & 1-3	106	44	428	222	66	737
3-6 & 1-5	31	29	230	85	57	474
3-6 & 1-6	30	29	131	86	49	352
3-6 & 2-4	29	29	243	84	49	424
3-6 & 3-5	63	61	912	125	93	1650
Total	259	192	1944	602	314	3637

* Head-on; ** Overtake; *** Crossing

Figure 2 shows the relation between the lengths of the give-way vessel and the stand-on vessel. Figure 3 shows the types of the give-way vessel and the stand-on vessel. In these both figures, the give-way vessel passed by Gate 3&6, the stand-on vessel passed by Gate 1&5 and encountered with the give-way vessel. As shown in these figures, the extracted cases of collision avoidance behaviors include the various situations on encounters of combination of various types and sizes of ships.

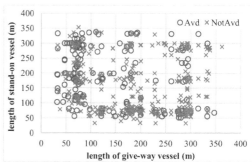

Figure 2. Length of give-way and stand-on vessels under one-to-one encounter situation.

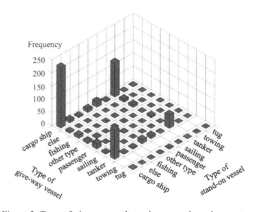

Figure 3. Type of give-way and stand-on vessels under one-to-one encounter situation.

144

3 OBSERVATION OF OFFSET DISTANCE FOR MODELING

3.1 *Examination of actual offset distance*

The individual distance on longitudinal or lateral line of a give-way vessel between the center of the give-way vessel and the stand-on vessel, is defined as an offset distance, which is minimum passing distance between two vessels in each direction. In this paper, we analyze four directions of the offset distance, i.e. "forward", "backward", "starboard" and "port" under three encounter situations, i.e. "head-on", "overtake" and "crossing".

Relative trajectory under a one-to-one encounter situation is shown in the small figure for explanation in Figure 4. A forward offset distance is defined as the length of line A on the ordinate in the figure, i.e. the distance between the origin, i.e. the center of the give-way vessel, and intersection of the ordinate and the relative trajectory. Similarly, the port offset distance is defined as the length of line B on the abscissa.

Each line in Figure 4 is a relative trajectory of a stand-on vessel to a give-way vessel. Each trajectory is plotted on a body-fixed coordinate system. The origin of coordinate is the center of give-way vessel. The relative trajectories shown in Figure 4 are those of stand-on vessels when give-way vessels evaded them under the situation of crossing on the combination of gates 3-6 and 1-5.

The respective offset distances were identified for all extracted collision avoidance behaviors, and a database of these offset distances was developed.

order to eliminate irrelevant values owing to extract the behaviors as one-to-one encounter even if a give-way vessel avoided some vessels as its target.

In our previous study (Miyake 2014), a typical sequence for collision avoidance was observed. In the sequence, the give-way vessel altered its heading toward the space behind of the stand-on vessel for evading, and after the stand-on vessel crossed in front of the give-way vessel, it returned gradually to the original course following the stern of the stand-on ship.

The forward offset distances under crossing situations are analyzed as explained below.

Each arrow in Figure 5, i.e. the result of additional analysis to the data used in the previous study (Miyake 2014), is a relative velocity of a stand-on vessel to a give-way vessel when crossing the forward longitudinal line of the give-way vessel. Each relative velocity is plotted on a body-fixed coordinate system. The origin of coordinate is the center of give-way vessel. Unfortunately, some irrelevant values are included in the figure. As shown in this figure, regarding the relative velocities at the time when stand-on vessels are crossing the front longitudinal lines of give-way vessels, the lateral components are small negative and the longitudinal components are large negative when the distance between the stand-on vessel and the give-way vessel is not less than 9260 m (5 miles). The similar tendencies of components of the relative velocity are observed in another analysis as well, at the time when the give-way vessels are starting to return to the original courses.

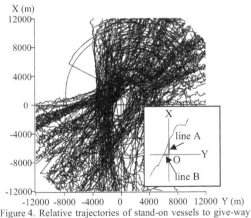

Figure 4. Relative trajectories of stand-on vessels to give-way vessels.

3.2 *Screening of offset distance on the database for modeling*

For the purpose of development of the model of the offset distances for one-to-one collision avoidance, we screened the offset distances on the database in

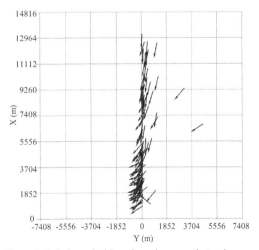

Figure 5. Relative velocities of stand-on vessels to give-way vessels when crossing the forward longitudinal line of the give-way vessels.

Therefore, from Figure 5 and the aforementioned of the observation, it can be said that the give-way vessel is not taking the evading action for the stand-on vessel which is far from the give-way vessel 9260 m or more. In other words, the stand-on vessels whose positions are 9260 m or more from the give-way vessels are deemed as out of the imaginary domain. Thus such give-way vessels can completely ensure their safety.

In this paper, the offset distances within 9260 m are extracted from the database to develop the model. The threshold of 9260 m applies only to the analysis for the forward offset distance under crossing situations.

Offset distances other than the forward offset distance under crossing situation are screened by the similar manner using different thresholds of distance, for the analysis of respective direction of offset distances under respective encounter situations.

3.3 Dependence of offset distances on collision avoidance action

We compare the offset distances on the both situations that give-way vessels evade the stand-on vessels and that the give-way vessels do not evade the stand-on vessels, in order to examine whether operators keep their offset distance regardless of evading stand-on vessels or not.

Figures 6 and 7 show the frequency and the cumulative values of the forward offset distances under crossing situations of give-way vessels, using the length of the give-way vessels as the parameter. Respective graphs in these figures show the similar tendency. Namely, the proportions of frequencies of give-way vessels of respective ranges by length are almost the same, in each offset distance range in both figures. Thus, it could be said that such proportion does not depend on whether give-way vessels take collision avoidance actions or not. This feature is observed in the cases other than forward offset distances under crossing encounter situations.

Therefore, we analyze the AIS data without distinction whether give-way vessels take collision avoidance actions or not.

3.4 Dependence of offset distances on encounter situation

We further compare the distributions of offset distances under respective encounter situations, in order to verify the dependence of offset distances under encounter situation.

Figures 8 and 9 show the frequencies and the cumulative values of the forward offset distances when give-way vessels take actions for avoiding collision with their stand-on vessels under the situations of head-on and overtake, respectively. As shown in Figures 6, 8 and 9, the distribution patterns of offset distances are quite different under

respective encounter situations. Thus, we develop the model of the forward offset distance by each encounter situation.

On the other hand, significant differences of offset distances have not been observed under all three encounter situations in the respective three directions other than the forward offset distances, i.e. backward, starboard and port. Thus, we develop the models of the offset distance of three directions regardless of the encounter situations.

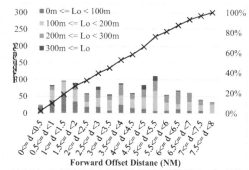

Figure 6. Frequency of forward offset distance by length of give-way vessel in crossing situation, where give-way vessels take collision avoidance actions.

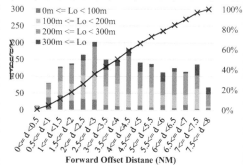

Figure 7. Frequency of forward offset distance by length of give-way vessels in crossing situation, where give-way vessels take no collision avoidance actions.

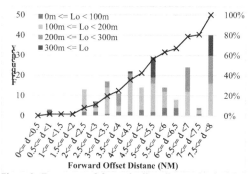

Figure 8. Frequency of forward offset distance by length of give-way vessels on head-on encounter, where give-way vessels take collision avoidance actions.

146

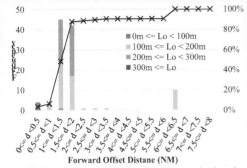

Figure 9. Frequency of forward offset distance by length of give-way vessels on overtake encounter, where give-way vessels take collision avoidance actions.

4 MODEL OF ACCEPTABL AND CRITICAL DOMAINS

4.1 Observation of tendency of offset distance

We developed models of actual representative offset distances mathematically in coastal sea area, applying to the previous research which mathematically presented models of offset distances in harbor by analysis of questionnaires (Inoue 1994).

In Sections 4.1 to 4.4, analyses of the forward offset distance under crossing situation are shown as concrete examples.

We sought to examine features of the relationship of three factors, i.e. representative offset distances, lengths of give-way vessels and stand-on vessels.

However, the direct correlation has not been examined between offset distances and give-way vessels' length. Then, the correlation between the ratio of give-way vessels' length to offset distance and give-way vessels' length was examined.

Figures 10 to 13 show the relations between the ratios and the length of give-way vessel under the respective ranges of length of stand-on vessels. A cross denotes the individual encounter. The abscissa indicates the length of give-way vessels. The ordinate indicates the ratio of length of give-way vessels to forward offset distances.

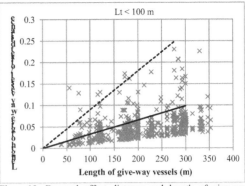

Figure 10. Forward offset distance and length of give-way vessels, where length of stand-on vessels is smaller than 100m.

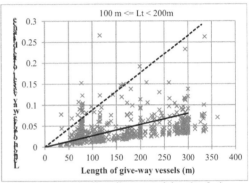

Figure 11. Forward offset distance and length of give-way vessels, where length of stand-on vessels is equal to 100m or more but smaller than 200m.

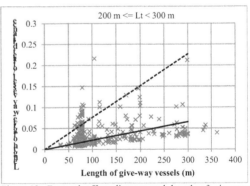

Figure 12. Forward offset distance and length of give-way vessels, where length of stand-on vessels is equal to 200m or more but smaller than 300m.

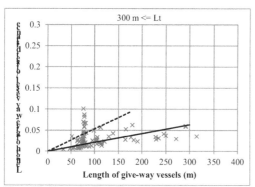

Figure 13. Forward offset distance and length of give-way vessels, where length of stand-on vessels is equal to 300m or more.

The difference of density of crosses could be observed in these figures, and the difference could be divided into three types. High, low and medium density is observed in the lower, upper and middle parts of these figures, respectively. The lines in the respective figures are drawn by one of the author in order to distinguish these parts.

4.2 *Definition of two domains*

Inoue et al. (Inoue 1994) described that two domains existed around vessels in the models at harbor. One is core domain which never let other ships enter, and the other is the area having to additional room to the core domain to maintain safer situations.

Applying the definition, two domains in coastal sea area are defined based on the analysis described in Figure 10 to 13. The left of Figure 14 shows the two domains, i.e. *acceptable domain* and *critical domain*. The *acceptable domain* is defined as the area where it is acceptable to let the vessels enter, but operators are concerning about the approaching vessels. The distances which configure the *acceptable domain* are called *acceptable offset distances*. The *critical domain* is defined as the area where it is unacceptable for operators to let the vessels enter. The distances which configure *critical domain* are called *critical offset distances*.

The *acceptable and critical offset distances* have four directions as illustrated in the right of Figure 14, and are called as mentioned in the figure with adding the words "acceptable" or "critical", e.g. "*forward critical offset distance*", respectively. In this paper, an abbreviated notation regarding the individual offset distance shown in Figure 15 is used. The superscript denotes the type of domain, i.e. "acceptable (A)" or "critical (C)". The subscript denotes directions of the offset distance, i.e. "forward (f)", "backward (b)", "starboard (s)" and "port (p)", and the encounter situations, i.e. "head-on (h)", "overtake (o)" and "crossing (c).

According to the definitions, solid and dot lines in Figures 10 to 13 correspond to the *forward acceptable offset distance* and the *forward critical offset distance*, respectively.

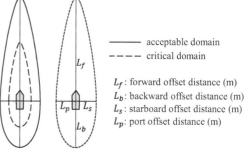

——— acceptable domain
- - - - critical domain

L_f: forward offset distance (m)
L_b: backward offset distance (m)
L_s: starboard offset distance (m)
L_p: port offset distance (m)

Figure 14. Definition of the domain around give-way vessel in coastal sea area and the respective offset distances configuring the domains.

(unit: m)

L (Domain type)

(direction)(encounter situation)

Figure 15. Definition of abbreviated notation.

4.3 *Relationship between offset distance and length of vessel*

The respective offset distances corresponding to the lines in Figure 10 to 13 are evaluated. They are expressed as the reciprocal value of the gradient of the individual lines.

Table 2 shows the offset distances of the respective range of length of stand-on vessels. It is observed that the *acceptable and critical offset distances* increase as the lengths of their target vessels increase. It could be also said that operators of the give-way vessels try to keep the offset distance irrespective of the length of give-way vessels.

Table 2. Offset distance in the respective range of length of stand-on vessels.

Ranges of length of stand-on vessels (Lt)	offset distance (m)	
	acceptable	critical
Lt < 100m	3030	1110
100m <= Lt < 200m	3700	1140
200m <= L t< 300m	4550	1320
300m <= Lt	4760	1890

4.4 *Model of offset distances*

Figure 16 shows the relations between the stand-on vessels' length and the *acceptable and the critical offset distances*. The abscissa and the ordinate indicate the length of stand-on vessels and the offset distance, respectively. Closed circles and triangles denote the *acceptable and critical offset distances* given in Table 2, respectively. In the figure, the range of the length of the stand-on vessels are representative by the median values.

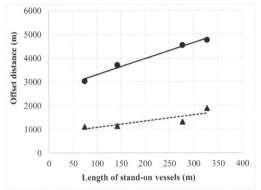

Figure 16. Forward offset distance in crossing and length of stand-on vessels.

Figure 16 shows that both circles and triangles increase linearly as the length of the stand-on vessels increases. Then it is assumed that the *acceptable and critical offset distances* are proportional to the length of stand-on vessels and the respective lines in this figure are determined by the regression analysis as shown in Equations (1) and (2). Namely, the *forward acceptable offset distance* and the *forward critical offset distance* under crossing encounter situation are:

$$L_{fc}^{A} = 6.75 \cdot Lt + 2627 \tag{1}$$

$$L_{fc}^{C} = 2.62 \cdot Lt + 826 \tag{2}$$

where Lt is the length of the stand-on vessel (unit: m).

4.5 Offset distances other than forward offset distances under crossing situation

By using the similar procedure, the *forward offset distances* for situations other than crossing can be obtained as shown in Equations (3) to (5).

Regarding the *forward offset distance* under overtake encounter situation, the two domains, i.e. *acceptable and critical domains*, cannot be distinguished, because the individual offset distances are uniformly distributed on the figures similar to Figures 10 to 13. On the other hand, the tendencies similar to those of *forward offset distance* in crossing situation are observed. Namely, the larger lengths of stand-on vessels are, the larger offset distances are. Then we draw single linier line in each figure on "forward offset distance and the length of give-way vessels" and determine the offset distance

of a range of the length of the stand-on vessels in accordance with the gradient of the linier line. Thus, we determined the *forward acceptable offset distances* based on the results of the regression analyses, for the reason that the range of the distances are similar to the *forward acceptable offset distances* in the other encounter situations. Namely, the *forward critical offset distances* under overtake situation cannot be determined.

The *forward acceptable offset distance* and *forward critical offset distance* under head-on and overtake encounter situations are:

$$L_{fh}^{A} = 17.69 \cdot Lt + 6192 \tag{3}$$

$$L_{fo}^{A} = 3.66 \cdot Lt + 2309 \tag{4}$$

$$L_{fh}^{C} = 11.40 \cdot Lt + 3347 \tag{5}$$

Furthermore, the backward, starboard and port offset distances are determined by the similar procedure for the *forward offset distances* under crossing situation. These offset distances are expressed by the individual regression equations without distinction of encounter situation as shown in Equations (6) to (11) because the significant differences were not observed in the three directions.

The *backward acceptable offset distance* and *backward critical offset distance* are:

$$L_{b}^{A} = 9.42 \cdot Lt + 1975 \tag{6}$$

$$L_{b}^{C} = 6.03 \cdot Lt + 752 \tag{7}$$

The *starboard acceptable offset distance* and *starboard critical offset distance* are:

$$L_{s}^{A} = 0.64 \cdot Lt + 875 \tag{8}$$

$$L_{c}^{A} = 0.31 \cdot Lt + 394 \tag{9}$$

The *port acceptable offset distance* and *port critical offset distance* are:

$$L_{p}^{A} = 2.05 \cdot Lt + 664 \tag{10}$$

$$L_{p}^{C} = 1.16 \cdot Lt + 308 \tag{11}$$

Based on the above mentioned analyses, we confirm the features of the domain in coastal sea as follows. There are two domains of give-way vessels in the coastal sea area. The individual offset distance depends on length of stand-on vessels, and does not depend on length of give-way vessels.

4.6 Illustration of domains

Figure 17 illustrates the models of the domains around give-way vessels under the three encounter situations. The digits denote the approximate ratios of individual offset distance in accordance with the modeled equations. Here it should be noted that the *acceptable offset distances* in the three directions, i.e. backward, starboard and port, are the same irrespective of encounter situations. Furthermore, it should be noted that the *critical offset distances* in the three directions, i.e. backward, starboard and port, are the same under the two encounter situations.

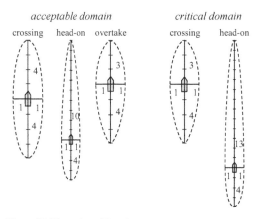

Figure 17. Illustration of domains.

5 CONCLUSIONS

We analyzed the real one-to-one collision avoidance maneuvers which were extracted from the AIS data recorded for a month, in order to examine the actual collision avoidance behaviors. Through the analyses, we find that

– The two domains around the give-way vessels can be distinguished when an operator of the vessel sails in the coastal sea area. One is the area where it is acceptable to let the vessels enter, but operators are concerning about the approaching vessels, and the other is the area where it is unacceptable for operators to let the vessels enter.

This is the same feature as the previous study on harbors by Inoue et al. (Inoue 1994). However, the individual offset distance which configures the domains is quite different from the previous study as follows:

– the individual offset distance can be determined irrespective of the length of give-way vessels;
– the forward offset distances heavily depends on the encounter situation;
– the backward, starboard and port offset distances do not depend on the encounter situations; and
– the forms of the *acceptable and critical domains* are quite different in respective encounter situations.

Meanwhile, the similar feature as the previous study regarding the offset distances is observed, i.e. the individual offset distances is approximately linear to the length of stand-on vessel.

In conclusion, we specified the representative offset distances between ships in the coastal sea area through the analyses with AIS data.

REFERENCES

Fujii, Y. 1980. A Definition of the Evasive Domain. *NAVIGATION.* Published Japan Institute of Navigation 65:17-22 (in Japanese)
Fujii, Y. 1983. Integrated study on marine traffic accidents. *IABSE Colloquium on Ship Collision with Bridges and Offshore Structures* 42:91-98, Copenhagen
Inoue, K. & Usami, S. & Shibata, T. 1994. Modelling of Mariners' senses on Minimum Passing Distance between Ships in Harbour. *The Journal of Japan Institute of Navigation* 90:297-306 (in Japanese)
Martin G. H. & et al. 2013. Empirical Ship Domain based on AIS Data. *The Royal Institute of Navigation* vol. 66 No.6: 931-940
Miyake, R. & Fukuto, J. & Hasegawa, K. 2014. (provisional title) "Analyses of the collision avoidance behaviors based on AIS" *Proc. Of Japan Institute of Navigation* vol. 2 No.2:100-103 (in Japanese)
Yamasaki, S. & Masuda, K. & Sera, W. 2013. Research on Safety Navigation of the Small Vessel within the Traffic Route. *The Journal of Japan Institute of Navigation* 128:9-14 (in Japanese)

Route Planning

Safe Ship Trajectory Planning Based on the Ant Algorithm – the Development of the Method

A. Lazarowska
Gdynia Maritime University, Gdynia, Poland

ABSTRACT: The paper presents the development of an algorithm for collision avoidance at sea inspired by a collective behaviour of ant colonies. The method developed includes also a speed alteration manoeuvre in situations, when the course change manoeuvre is not possible. An improved version of the algorithm has been introduced and results of various test cases with static and dynamic obstacles have been reported. Solutions from the perspective of every ship taking part in the navigational situations are calculated and presented. Safe trajectories of all of the ships determined by the ant algorithm do not constitute conflicting solutions, what proofs that the safe ship control system using this method can be applied on board the ships.

1 INTRODUCTION

Research on ships collision avoidance started in the 1960s and the first studies were concerned on collision avoidance manoeuvre of a two-ships encounter situation: Wylie (1960), Hollingdale (1961), Morrel (1961), Merz & Karmarkar (1976). There was a necessity to consider the COLREGs in solving a collision situation, therefore Jones (1974) developed the manoeuvre diagram to provide the possibility of including the COLREGs in determining the evasive manoeuvre of the ship. In the 1970-1980s the concept of collision risk assessment with the use of the ship domain has been introduced and developed by Fujii & Tanaka (1971), Goodwin (1975), Davis et al. (1980) and Coldwell (1983). In the 1970s the idea of determining the optimal collision avoidance manoeuvres with the use of the differential games theory has been introduced by Miloh & Sharma (1977) and Olsder & Walter (1977), and considered for the two ships encounter situation. In the 1980s Lisowski (1981) developed a differential games approach for multi-ship encounter situations. In the 2000s the ship path planning approaches have been developed. Deterministic path planning method has been introduced by Chang et al. (2003) and developed by Szłapczyński (2006). Artificial intelligence methods for ship path planning have been reported by Śmierzchalski & Michalewicz (2000), Ito & Zeng, (2001), Perera et al. (2009), Tam & Bucknall (2010) and others. However, all of these methods are not deprived of disadvantages. They are characterized by the limitations such as disregarding of ship's dynamics, static obstacles, ignoring the COLREGs.

In recent years, the growth of computing power and technology advancement led to the development of new computational intelligence methods. Among these approaches methods inspired by the behaviour of natural systems such as ant colonies, bird flocks or bacterial growth experience an increasing interest in the scientific community. The algorithms based on such biological systems are classified to the group called Swarm Intelligence (SI). The dynamic development of bio-inspired computation started in the 1990s and such approaches have been reported by Bonabeau et al. (1999), Dorigo & Stützle (2004), Blum & Merkle (2008), Solnon (2010) and Yang et al. (2013).

In this paper, the ships collision avoidance decision making process is solved with the use of an ant algorithm. Presented solution constitutes a development of an approach previously introduced by Lazarowska (2013). In the former study, solely the course alteration manoeuvre has been considered as a solution of a collision situation. In this new approach also speed reduction is implemented in the algorithm for the situations, when course change manoeuvre does not lead to a safe situation.

2 ANT ALGORITHM FOR SAFE SHIP TRAJECTORY PLANNING

The solution of the safe ship trajectory planning problem is stated as the determination of such a course or speed change, or a sequence of course or speed changes, which assures a safe passage of the ship without violation of safety areas around other ships and static obstacles. The issue formulated above is solved with the use of an ant algorithm.

Input data to the algorithm are registered with the use of an Automatic Radar Plotting Aid (ARPA) and an Automatic Identification System (AIS). These information include the course and the speed of every ship, and the distances and bearings of the target ships from the own ship. For situations in restricted waters information concerning shoals, shorelines and other static obstacles are collected from the Electronic Chart Display and Information System (ECDIS).

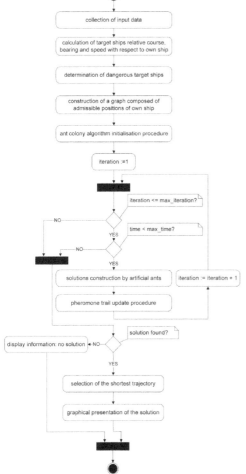

Figure 1. Flowchart of the ACO-based algorithm for collision avoidance at sea.

In the next step parameters such as the relative courses, bearings and speed of the target ships with respect to the own ship are calculated. Based upon these data every of the target ships is regarded as a dangerous or a safe object. A dangerous target ship is a ship, which intersects its course with the course of the own ship. Subsequently, a graph composed of admissible positions of the own ship is constructed. The solution is searched for by a group of agents represented by a colony of artificial ants. Every artificial ant moves on the graph, until it reaches the specified end position or the maximum number of steps.

The ant algorithm imitates the behaviour of colonies of real ants in nature. An ant colony is characterized by self-organization accomplished with the use of an indirect communication mechanism. In nature, ants deposit a chemical substance on the ground, which is called a pheromone trail. Other ants can smell this trail and that phenomenon allows them to communicate with each other indirectly, by modifying the environment.

The ant algorithm reproduces this behaviour of ant colonies in the pheromone trail update procedure. The search for a solution is terminated, when the maximum number of iterations has been reached or the maximum computational time has been achieved. Then, the shortest trajectory is selected from all of the solutions found by the artificial ants. Afterwards, the best solution is presented in a graphical form and the new course of the own ship at every segment of the determined safe, optimal trajectory is displayed. If the algorithm has not found a solution, such information is displayed after the calculations are completed. The flowchart of the ant algorithm for safe ship trajectory planning is shown in Figure 1.

The determined safe ship trajectory has to be compliant with the The International Regulations for Preventing Collisions at Sea (COLREGs). The COLREGs and their explanation can be found in the work of Cockcroft & Lameijer (2012). Rule 8 of COLREGs, defining what action should be taken by the give-way ship in order to avoid collision, allows the possibility to alter the speed in situations, when there is not enough of clear space at sea for the alteration of course.

Recent development of the ant algorithm for collision avoidance at sea includes an enhancement, which extends the problem solving capability of the previous version of the method by providing the ability to find a safe trajectory for situations, which could not be solved only by course change manoeuvres. The improved algorithm provides the possibility to determine a speed change manoeuvre for situations, when the course change manoeuvres do not solve the collision situation. The flowchart of the modified ant algorithm for safe ship trajectory planning, including the speed reduction manoeuvre, is shown in Figure 2.

The modification is the calculation of the own ship speed reduction manoeuvre, if after the maximum computational time the solution concerning the course alterations has not been found by the artificial ants. The speed is decreased by 25% and the procedure of checking, whether the new own ship trajectory with reduced speed does not collide with the trajectories of the target ships, is carried out.

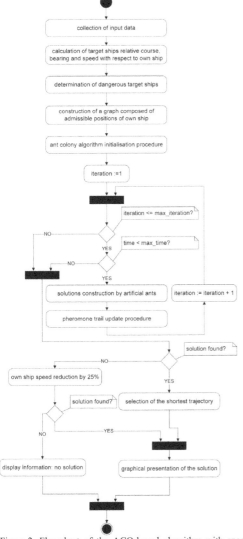

Figure 2. Flowchart of the ACO-based algorithm with speed reduction for collision avoidance at sea.

3 SIMULATION TESTS

The algorithm has been implemented in the Matlab programming language and has been tested for a number of test cases including situations both in the open sea and in restricted waters with the occurrence of static obstacles. Three representative situations have been chosen for the presentation in this paper. Situations 1 and 2 concern an encounter situation with a few target ships, while in the situation 3 also the avoidance of a collision with a landmass has to be considered.

The safe distance between the own ship and the target ships during the determination of collision avoidance manoeuvres is assured by the use of safety areas around the target ships in the form of a hexagon domain. The size of the hexagon domain used in the simulation tests is described by the parameters listed in Table 1.

Table 1. Parameters of the hexagon domain around the target ships used in simulation tests.

Parameter	Value [nm]
the length of the bow	1,0
the length of amidships	0,6
the distance towards the starboard	0,6
the length of the stern	0,25
the distance towards the port side	0,25

The static obstacles such as a landmass or a shallow are modelled in the form of convex and concave polygons described as sets of points connected by line segments. The ant algorithm parameters used for simulation tests have been tuned experimentally based upon a number of test runs and evaluation of the algorithm performance, among which the most important one was the computational time and the reproducibility of results. The parameters used in simulation tests presented in this paper are listed in Table 2.

Table 2. Parameters of the ant algorithm for collision avoidance at sea used in simulation tests.

Parameter	Value
α coefficient	1
β coefficient	2
initial pheromone trail amount	1
pheromone evaporation rate	0,1
number of ants	10
number of iterations	20

The calculations have been performed with the use of a PC with Intel Core i5 M430 2.27 GHz processor, 2GB RAM and 32-bit system Windows 7 Professional.

Situation 1 presents an encounter of 5 ships. In Table 3 the position and motion parameters of all ships including the course in degrees, the speed in knots, the bearing in degrees and the range in nautical miles are listed. Figure 3 presents the navigational situation as registered from ARPA and AIS. Figure 4 shows the point of the target ship 3 domain violation by the own ship. The collision would occur, if the ships continue to move on their trajectories without changing motion parameters. In Figures 5-7 subsequent stages of the own ship and target ships movement on their trajectories are shown for the situation, when the own ship decreases its speed by 25% from 20 kn to 15 kn, what has been determined by the ant algorithm for safe ship trajectory planning. In Figure 8 the comparison of all ships trajectories determined by the ant algorithm have been presented – t_0 to t_e indicate the positions of the ships in successive instants of time. It can be noticed that the trajectories of all of the ships do not collide with each other at any stage of the vessels movement.

Table 3. Situation 1 motion parameters of all ships.

Setting	CSE [°]	SPD [kn]	BRG [°]	RNG [nm]
OS	153	20.0	—	—
TS1	075	12.0	141	4.0
TS2	072	12.5	170	4.5
TS3	073	12.0	191	4.5
TS4	066	17.0	214	4.5

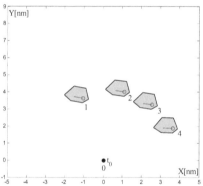

Figure 3. Graphical presentation of situation 1.

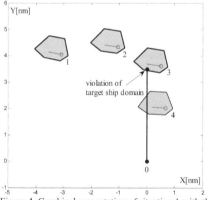

Figure 4. Graphical presentation of situation 1 with the point of collision marked.

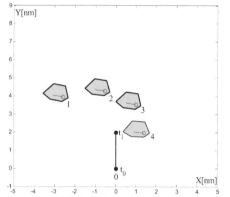

Figure 5. Graphical solution of situation 1 at the moment t_1.

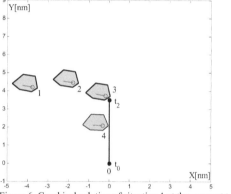

Figure 6. Graphical solution of situation 1 at the moment t_2.

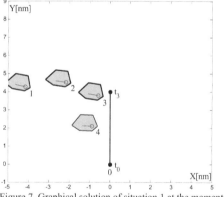

Figure 7. Graphical solution of situation 1 at the moment t_3.

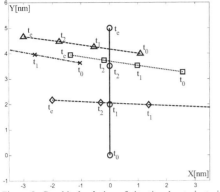

Figure 8. Graphical solution of situation 1 – trajectories of all ships.

Situation 2 concerns an encounter of 4 ships. In Table 4 the motion and approach parameters of all ships taking part in this situation are listed. In Figure 9 the navigational situation, as registered from ARPA and AIS, is shown. In Figure 10 the point of the target ship 3 domain transgression by the own ship is presented. The solution determined by the ant algorithm constitutes a speed reduction manoeuvre from 12 kn to 9 kn. Figures 11-13 show the consecutive stages of the ships movement on their trajectories. Figure 14 presents the comparison of all ships trajectories with the indication of the ships positions in subsequent instants of time. The results confirm that the algorithm determines not conflicting trajectories of all ships.

Table 4. Situation 2 motion parameters of all ships

Setting	CSE [°]	SPD [kn]	BRG [°]	RNG [nm]
OS	000	12.0	–	–
TS1	270	13.0	045	5.0
TS2	165	12.0	002	4.5
TS3	090	9.0	325	6.0

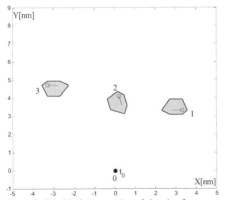

Figure 9. Graphical presentation of situation 2.

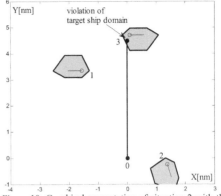

Figure 10. Graphical presentation of situation 2 with the point of collision marked.

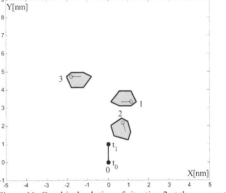

Figure 11. Graphical solution of situation 2 at the moment t_1.

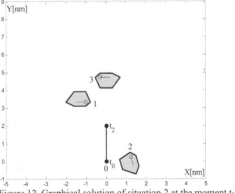

Figure 12. Graphical solution of situation 2 at the moment t_2.

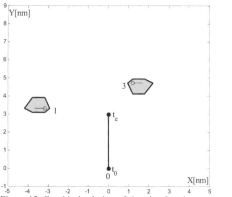

Figure 13. Graphical solution of situation 2 at the moment t_e.

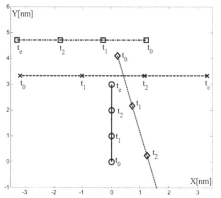

Figure 14. Graphical solution of situation 2 – trajectories of all ships.

Table 5. Situation 3 motion parameters of all ships.

Setting	CSE [°]	SPD [kn]	BRG [°]	RNG [nm]
OS	000	11.5	—	—
TS1	235	21.0	030	5.5
TS2	235	21.0	035	7.0
TS3	235	18.5	035	9.0
TS4	235	14.0	040	10.0

Situation 3 presents an encounter of 5 ships with the occurrence of two areas of land in the environment. The position and motion parameters of all ships are listed in Table 5. In Figure 15 the navigational situation, as registered from ARPA and AIS, is presented. Figure 16 shows the point of the target ship 2 domain violation by the own ship, while in figure 17 the point of the target ship 3 domain transgression is marked. In Figures 18-20 subsequent stages of the own ship and target ships movement on their trajectories are shown for the situation, when the own ship reduces its speed by 25% from 11.5 kn to 8.6 kn, what has been determined by the ant algorithm for safe ship trajectory planning. Figure 21 presents the comparison of all ships trajectories in successive

instants of time, determined by the ant algorithm. It can be observed that the trajectories of all of the ships do not constitute conflicting solutions.

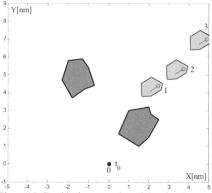

Figure 15. Graphical presentation of situation 3.

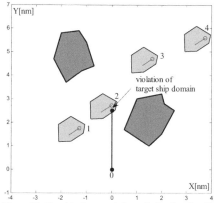

Figure 16. Graphical presentation of situation 3 with the point of collision with target ship 2 marked.

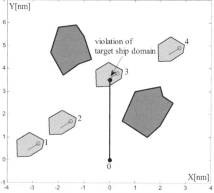

Figure 17. Graphical presentation of situation 3 with the point of collision with target ship 3 marked.

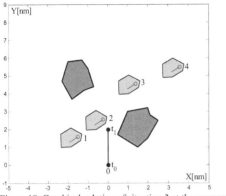

Figure 18. Graphical solution of situation 3 at the moment t_1.

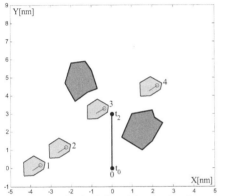

Figure 19. Graphical solution of situation 3 at the moment t_2.

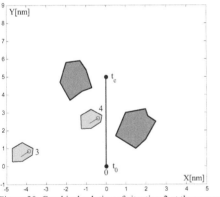

Figure 20. Graphical solution of situation 3 at the moment t_e.

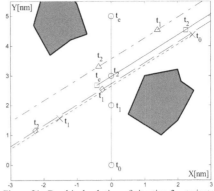

Figure 21. Graphical solution of situation 3 – trajectories of all ships.

To sum up, obtained results allow the formulation of the following conclusions:
– every solution constitutes a safe trajectory of the ship, that is a trajectory, for which at any stage of the own ship movement, its position does not exceed the target ships domains and safety areas around a landmass,
– the solutions fulfill the COLREGs,
– the results of simulation tests proof that the developed version of the algorithm provides the possibility to find a solution for situations in which the safe own ship trajectory composed of course alteration manoeuvres could not be found,
– despite the probabilistic nature of the optimisation method applied, for every repetition of calculations of the same navigational situation, an identical solution has been obtained,
– the proposed algorithm is suitable for use in the safe ship control system due to the fulfillment of all the major criteria: the safety of solution, compliance with COLREGs, short computational time and repeatability of results,
– the solutions returned by the algorithm from the perspective of all of the vessels involved in the considered navigational situation do not constitute conflicting solutions, what allows the use of such safe ship control system by all of the vessels simultaneously.

4 SUMMARY AND CONCLUSIONS

The paper presents a developed version of an algorithm for safe ship trajectory planning using stochastic global optimization method, what has allowed to include the COLREGs, a greater number of static obstacles and moving objects, dynamic properties of the vessel, the determination of a safe trajectory to the specified endpoint and consistent solutions from the perspective of all the vessels involved in the collision situation.

The use of an ant algorithm for collision avoidance at sea allows the determination of safe, optimal trajectory of a ship by the use of a collective action of agents - artificial ants, which shows that the indirect mechanism of communication between agents by modification of the environment, with the right balance between exploration and exploitation, is an effective method to solve the problem of global optimization.

The results of simulation tests show that the method of mapping the behavior of a biological system, which is an ant colony, solves complex problem of maritime transport, the protection of the marine environment and increase of ship automation by automatically determining a safe, optimal trajectory of the ship in collision situation at sea, what indicates a high potential of methods inspired by nature in practical industrial applications.

The developed method for collision avoidance at sea, taking into account a greater number of static and dynamic navigational restrictions and determining a safe, optimal trajectory of the vessel to a specified endpoint, allows its use in areas of heavy traffic, where the most conflicting situations occur.

The analysis of simulation studies has shown that despite the use of a stochastic optimisation method, the presented model of the safe ship control in collision situation at sea, using an ant algorithm, enables obtaining reproducible solutions.

The proposed algorithm is robust to changes of the target ships strategies, because it can repeatedly determine a solution in a short time, based on the current data, received from ARPA and AIS.

The ability to change the shape and size of the target ship domain allows to take into account good or restricted visibility and preferences of the decision-maker - the system operator.

The developed method for the determination of safe, optimal changes of the ship's course and speed in collision situation at sea, with the presence of other ships and taking into account the static obstacles, constitutes a process of determining a safe, optimal transition path of a moving object in a dynamic environment. The solution of this problem is widely used in robotics and military, for instance in control systems of unmanned vehicles.

REFERENCES

Blum Ch., Merkle D. 2008: Swarm Intelligence: Introduction and Applications, Springer-Verlag, Berlin Heidelberg.

Bonabeau E., Dorigo M., Theraulaz G. 1999: Swarm Intelligence: From Natural to Artificial Systems, Oxford University Press, Inc.

Chang K.-Y., Jan G. E., Parberry I. 2003: A Method for Searching Optimal Routes with Collision Avoidance on Raster Charts, The Journal of Navigation, Vol. 56, p. 371–384.

Cockcroft A. N., Lameijer J. N. F. 2012: A Guide to the Collision Avoidance Rules, 7th Edition, Butterworth-Heinemann.

Coldwell T. G. 1983: Marine Traffic Behaviour in Restricted Waters, The Journal of Navigation, Vol. 36, p. 430–444.

Davis P. V., Dove M. J., Stockel C.T. 1980: A Computer Simulation of Marine Traffic Using Domains and Arenas, The Journal of Navigation, Vol. 33, p. 215–222.

Dorigo M., Stützle T. 2004: Ant Colony Optimization. The MIT Press.

Fujii Y., Tanaka K. 1971: Traffic Capacity, The Journal of Navigation, Vol. 24, p. 543–552.

Goodwin E. M. 1975: A Statistical Study of Ship Domains, The Journal of Navigation, Vol. 28, p. 328–344.

Hollingdale S. H. 1961: The Mathematics of Collision Avoidance in Two Dimensions, The Journal of Navigation, Vol.14, p. 243–261.

Ito M., Zeng X.-M. 2001: Planning a collision avoidance model for ship using genetic algorithm, IEEE International Conference on Systems, Man, and Cybernetics.

Jones K. D. 1974: Application of a Manoeuvre Diagram to Multi-ship Encounters, The Journal of Navigation, Vol. 27, p. 19–27.

Lazarowska A. 2013: Application of Ant Colony Optimization in Ship's Navigational Decision Support System, In A. Weintrit (ed.), Marine navigation and safety of sea transportation: Navigational problems, CRC Press/Balkema.

Lisowski J. 1981: Ship's Anti-Collision Systems, Wydawnictwo Morskie, Gdańsk, (in Polish).

Merz A. W., Karmarkar J. S. 1976: Collision Avoidance Systems and Optimal Turn Manoeuvres, The Journal of Navigation, Vol. 29, p. 160–174.

Miloh T., Sharma S. D. 1977: Maritime collision avoidance as a differential game, Schiffstechnik, Vol. 24, p. 69–88.

Morrel J. S. 1961: The Physics of Collision at Sea, The Journal of Navigation, Vol. 14, p. 163–184.

Olsder G. J., Walter J. L. 1977: A differential game approach to collision avoidance of ships, Proceedings of the 8th IFIP Conference on Optimization Techniques, Würzburg.

Perera L., Carvalho J., Guedes Soares C. 2009: Decision making system for the collision avoidance of marine vessel navigation based on COLREGs rules and regulations, 13th Congress of IMAM.

Solnon Ch. 2010: Ant Colony Optimization and Constraint Programming, ISTE Ltd and John Wiley and Sons, Inc.

Szłapczyński R. 2006: A New Method of Ship Routing on Raster Grids, with Turn Penalties and Collision Avoidance, The Journal of Navigation, Vol. 59, p. 27–42.

Śmierzchalski R., Michalewicz Z. 2000: Modeling of Ship Trajectory in Collision Situations by an Evolutionary Algorithm, Transactions on Evolutionary Computation, Vol.4(3), p. 227–241.

Tam C., Bucknall R. 2010: Path-planning algorithm for ships in close-range encounters, The Journal of Marine Science and Technology, Vol. 15, p. 395–407.

Wylie F. J. 1960: The Calvert Methods of Manoeuvring to Avoid Collision at Sea and of Radar Display, The Journal of Navigation, Vol. 13, p. 455–464.

Yang X.-S., Cui Z., Xiao R., Gandomi A. H., Karamanoglu M. 2013: Swarm Intelligence and Bio-Inspired Computation: Theory and Applications, Elsevier Inc.

Ship Evolutionary Trajectory Planning Method with Application of Polynomial Interpolation

P. Kolendo & R. Śmierzchalski
Gdansk University of Technology, Gdansk, Poland

ABSTRACT: Paper presents the application of evolutionary algorithms and polynomial interpolation in ship evolutionary trajectory planning method. Evolutionary algorithms allows to find a collision free trajectory in real time, while polynomial interpolation allows to model smooth trajectory which keeps continuity of velocity and acceleration values along path. Combination of this two methods allows to find trajectory, which under some assumptions, can mimic real movement of the vessel. Paper presents the description of proposed method with simulation tests for sample collision situations at sea.

1 INTRODUCTION

The problem of evolutionary path planning is a common theme in numerous application such as: ship path planning (Śmierzchalski, 1998) (Szłapczyńska, 2009) (Fossen, 2010)(Kolendo and others, 2011), path planning for AUV's (Fogel and others, 1999), mobile robots path planning (Xiao and others, 1999)(Kurata and others, 1998) and path planning for aerial objects (Rathbun and others, 2002)(Nikolos and others, 2003)(Pongpunwattana and others, 2007)(Mittal and others, 2007). Problem is defined as a task where given a mobile object with certain dynamical and kinematical properties and an environment through which this object is travelling, one needs to plot a path between start and end points, which avoids all environments static and dynamic obstacles and meet the optimization criteria. The evolutionary path planning method is non-deterministic, based on a natural selection mechanism. Its most important advantages are build-on adaptation mechanism for a dynamic environment and reaching a multi-criteria task solution in a near-real time.

Polynomial interpolation for modeling ship trajectory was presented in (Fossen, 2010), however it concerns only the way of modeling the path from already determined waypoints. The main advantages of polynomial interpolation is ability of modeling smooth trajectory, which keeps continuous values of speed and acceleration alongside. According to (Golding 2004) it can be claimed that the trajectory modeled this way is closest to navigators

expectations. With some assumptions trajectory modeled with polynomial interpolation can mimic real movement of ship. In (Cornelliusen, 2003) the complex method combined of A* optimization algorithm (which is responsible for determine optimal trajectory) and cubic splines (for its modeling) was presented. Method allows to find smooth, collision free trajectory, however is not able to find solution in near real time, which significantly limit its application in on-line system of ship trajectory planner (according to summary of paper (Cornelliusen, 2003) author). It is due to the fact, that it has to take into consideration dynamic changes in environment.

Paper presents new method of ship path planning which combine advantages of evolutionary algorithms and polynomial interpolation for setting collision- free trajectory for marine vessel. Method allows to find a collision free, smooth trajectory in near real time with keeping continuous values of speed and acceleration alongside.

Papers is organized as follows. After the introduction the evolutionary ship path planning method is presented. Third chapter describe the application of cubic splines to path planning problem. Chapter 4 presents experimental researches of method for couple of sample environments. 5th chapter concludes paper.

2 EVOLUTIONARY PATH PLANNING

Evolutionary algorithms are one of the methods used for ship path planning. The main advantages are easy implementation in problems with large amount of constraints (independently of problem characteristics) (Nikolos and others, 2003) and ability to find final solution in near real time (Śmierzchalski, 1998). In opposition to analytic methods they allow active search of solution space (Szłapczyńska, 2009). Because of that, evolutionary algorithms are widely used in problems of ship path planning (Śmierzchalski, 1998) (Ito and others, 1999) (Śmierzchalski, 2004) (Szłapczyńska, 2009) (Szłapczyński, 2012).

According to transport plan an own ship should cover the given path in the determined time, on the other hand, it has to move safely along the planned path, while avoiding the navigational constrains and other moving objects. Path planning in a collision scenario has to stand a compromise between a deviation from a given course and ships safety. Thus the problem is defined as multi-criteria optimization task which considers safety and the economics of ships movement. Every path is evaluated based on the fitness function. In the considered case, the problem has been reduced to a single objective optimization task with weighting factors (1), (2), (3).

$$Total\_Cost(S) = Safe\_Cond(S) + Econ\_Cond(S) \quad (1)$$

$$Safe\_Cond(S) = w_c \times clear(S) \quad (2)$$

$$Econ\_Cond(S) = w_d \times dist(S) + w_s \times smooth(S) + \\ + w_t * time(S) \quad (3)$$

Figure 1 shows a single evolutionary ship path planning algorithm diagram.

Individual is one single solution (path). Population is group of paths which are in evolution process. Generation is one algorithm iteration. Fitness function determine the fitness of individual to environment. The way of encoding of chromosomes was presented in (Śmierzchalski, 1998).

In the first step a random population is being initialized. In the second, using a chosen selection scheme, a specific number of individuals is randomly selected to the temporary population. Then the genetic operators such as cross-over and mutation are working on the temporary population. In the next step a new population is established. It consists of base and temporary population best individuals. The algorithm's iterations is repeated until the termination condition is met (a certain number of iterations in this instance). Presented algorithm is a steady-state type.

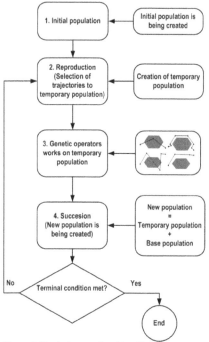

Figure 1. Evolutionary algorithm diagram

3 POLYNOMIAL INTERPOLATION

Application of polynomial interpolation for modeling ship trajectory allow to determine smooth trajectory with continuous values of speed and acceleration alongside. This kind of trajectory is desired for control system of the ship, due to the fact that it allows to control the ship along the trajectory without overshoots and with minimal difference from reference trajectory.

In considered case the most significant thing is to find smooth trajectory with continues values of speed and acceleration alongside. According to (Golding, 2004)(Fossen, 2010) 3rd degree polynomial- cubic spline is sufficient to achieve this goal. Higher degree polynomials allows assumption of more constraint conditions, however the smooth trajectory may deformed because of Runge's phenomenon. In (Corneliussen, 2003) there is comparison of cubic splines, 5th and 7th order polynomials, where this problem is pointed out.

In presented method each segment of trajectory is modeled according to equation (4). Coefficients are calculated simultaneously for all segments of trajectory.

$$x_k(s) = a_3 s^3 + a_2 s^2 + a_1 s + a_0 \\ y_k(s) = b_3 s^3 + b_2 s^2 + b_1 s + b_0 \quad (4)$$

where:

n-1 - is number of waypoints,

k - is number of segment,

Waypoints are described as follows (5):

$$s \in R = \{0,\ldots,n\}$$
$$s_0 = s(x_0, y_0) = 0$$
$$s_1 = s(x_1, y_1) = 1$$
$$s_2 = s(x_2, y_2) = 2 \qquad (5)$$
$$\ldots$$
$$s_n = s(x_n, y_n) = n$$

Sample trajectory modeled with cubic splines is presented below (Figure 2).

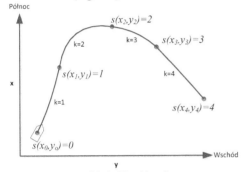

Figure 2. Trajectory modeled with cubic splines

To clarify the description, calculation of coefficients will be presented only for x- coordinate. For y-coordinate calculations will be analogical.

Due to the fact that each segment preserve continuity, the constraints for coefficient determination, should concern continuity in waypoints.

$$x_k(s) = a_3 s^3 + a_2 s^2 + a_1 s + a_0$$
$$x_k'(s) = 3a_3 s^2 + 2a_2 s + a_1 \qquad (6)$$
$$x_k''(s) = 6a_3 s + 2a_2$$

In order to determine polynomial coefficients there should be assumption of 2 constraints in start and end points, 4 constraints in rest of waypoints (Fossen, 2010). Constraints concern:

1 modeled segment k of trajectory should start in waypoint s_{n-1} and end in s_n (7),

$$x_k(s_{n-1}) = x_{n-1}$$
$$x_k(s_n) = x_n \qquad (7)$$

2 the value of derivative should be the same on both sides of waypoint (8),

$$\lim_{s \to s_n^-} x_k'(s_n) = \lim_{s \to s_n^+} x_k'(s_n)$$
$$\lim_{s \to s_n^-} x_k'(s_n) - \lim_{s \to s_n^+} x_k'(s_n) = 0 \qquad (8)$$

3 the value of second derivative should be the same on both sides of waypoint (9),

$$\lim_{s \to s_n^-} x_k''(s_n) = \lim_{s \to s_n^+} x_k''(s_n)$$
$$\lim_{s \to s_n^-} x_k''(s_n) - \lim_{s \to s_n^+} x_k''(s_n) = 0 \qquad (9)$$

To determine cubic spline there is a need to solve matrix equation (10).

$$C = AW$$
$$W = A^{-1}C \qquad (10)$$

where W is coefficient matrix for segments 1 to k:

$$W = \begin{bmatrix} a_{3,1} & \cdots & a_{3,k} \\ a_{2,1} & \cdots & a_{2,k} \\ a_{1,1} & \cdots & a_{1,k} \\ a_{0,1} & \cdots & a_{0,k} \end{bmatrix}^T$$

C is constraints matrix and A is transformation matrix. Detailed way of finding solution to this equation is presented in (Fossen, 2010) .

4 SIMULATIONS

For simulation tests 4 sample collision situations at sea, with different level of complexity, was performed. This collision situations was tested in order to find optimal configuration of parameters assignment for several initial population for each environment. The algorithm settings were found during tests and were as follows:

- algorithm with partially exchangeable population, type steady state,
- population consists of 30 individuals with 40% rate of exchange in each population,
- probability of crossover was set at 0,7,
- probability of mutation was 0,5. 5 mutation operators described in (Śmierzchalski, 1998), as standard mutation, soft mutation, adding/deletition of a gene, swap gene position, speed mutation.
- terminal condition was set for 400 generations.

In comparison to recent works the proability of mutation has substantially increased. That was the result of experiments, during which significant improvement for that setting was noticed. It helps the algorithm to leave the local extreme and better explore the solution space. It is especially important in intial phase of algorithm were there were only unfeasible trajectories. In areas, where was a domination of static obstacles, algorithm had problems with leaving forbidden areas.

Whole process of algorithm work was shown on figure 3. Figure 3 presents collision situation at sea and next set of trajectories after 50,100, 250 and 400 generations.

preserve described continiuty, movement alongside will be perfoormed theoretically without any overshoots in comparison to classic approach where trajectories are made from straight lines and ships dynamic is being approximate by circle arcs (on straight lin angular speed is equal to zero and has some defined value at circular arc). Overshoots besides energy loss caused from movement corrections are connected with unprecise steering along trajectory, what in highly congested areas may result in collision (beacsuse of significant difference beetwen reference and real trajectory).

On figure 4 second sample collision situation at similar level of complexity is depicted. Figure presents collision situation at the beginning in generation 0 and the final trajectory after 400 generations. As it can be seen on figure 4 trajectory, which was found by the algorithm allows precise control of the ship, without overshoots, from start to end point. It is very important, due to the fact that any deviation from reference trajectory may result in collision. For couple of independent runs of algorithm (for different initial populations) for both collision scenarios, final solution (trajectory) were concentrating at the similar niche. The deviation of fitness function final value was at the level of 4%.

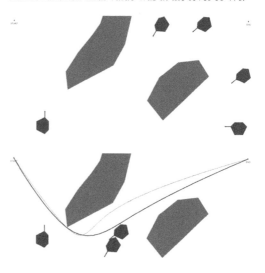

Figure 4. Ship path planning process for 2nd sample environment after: 0 400 generations

Figure 3. Ship path planning process for 1st sample environment after: 0, 100, 250, 400 generations

As it can be seen from simulations, algorithm in first stage explore the solution space and later on starts to concenrate around one dominating extreme. In classic approach without polynomial inteprolation, the situation probably would be very similar. The difference will occur during steering the vessel alongside the trajectory. Due to the fact that trajectory modeled by polynomial is smooth and

Figure 5 and 6 shows collision situation and final solution for two next environments. Analogically to previous scenarios, algorithm was able to find smooth trajectory in near real time (for couple of independent runs). The final value of fitness function differs about 6%. It can be treated as a proof, that settings of algorithms found during tests, are stable and can produce satisfying results.

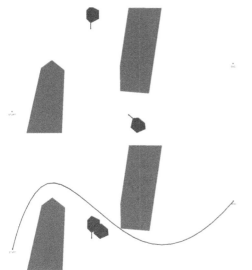

Figure 5. Ship path planning process for 3rd sample environment after: 0, 400 generations

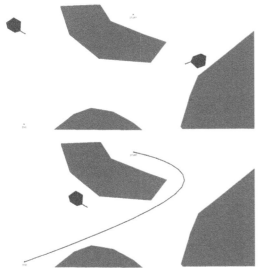

Figure 6. Ship path planning process for 4th sample environment after: 0, 400 generations

5 RESULTS AND CONCLUSIONS

Combination of evolutionary algorithms and cubic splines allows to find smooth trajectory in near real time, additionally preserving continuity of speed and acceleration of ship alongside. This trajectory allows, theoretically, precise control of the ship in congested areas and minimization of energy loss results from overshoots. It is especially important during steering the vessel in areas highly congested with static and dynamic obstacles. As it can be seen in presented situations, any deviation from reference trajectory may result in collision.

New method allows to find solution in near real time, which is significant for algorithms working in on-line mode. It makes possible to verify founded trajectory with changes in environment, during movement of a vessel along trajectory.

Moreover, trajectory set in this way, under some assumptions, can mimic real movement of the ship (takes into account dynamic properties of the ship). Due to complexity of this issue, dynamics in this method will be investigated in further works.

BIBLIOGRAPHY

1. Corneliussen J., *Implementaion of a Guidance System for Cybership II*, Trondheim, Norway : Norwegian University of Science and Technology, 2003. Master thesis.
2. Fogel D.B., Fogel L.J., *Optimal routing of multiple autonomus underwater vechicels through evolutionary programming*, Proceedings of the 1990 Symposium on Autonomus Underwater Vehicle Technology. 1999, 44-47
3. Fossen T.I., *Marine Control Systems: Guidance, Navigation and Control of Ships, Rigs and Underwater Vehicles*, Trondheim, Norway : Norwegian University of Science and Technology, Marine Cybernetics AS, 2010
4. Golding B.K., *Industrial Systems for Guidance and Control of Marine Surface Vessels*. Trondheim, Norway : Norwegian University of Science and Technology, 2004.
5. Ito M., Zhang F., Yosida N., *Collision avoidance control of ship with genetic algorithm*. Proceedings of the IEEE International Conference on Control Application, 1999
6. Kolendo P., Śmierzchalski R., Jaworski B. *Comparison of selection schemes in evolutionary method of path planning*, Lectures Notes in Artificial Intelligence: Computational Collective Intelligence: Technologies and Applications LNAI 6923, 2011
7. Kurata J., Grattan K.T.V., Uchiyama H., *Path Planning for a Mobile Robot by Integrating Mapped Information*. Madrid : 3rd IFAC Conference IAV Intelligent Autonomous Vehicle, pp. 319-323, 1998.
8. Mittal S., Deb K. ,*Three Dimensional Offline Path Planning for UAVs Using Multiobjective Evolutionary Algorithms*, IEEE Congress on Evolutionary Computation CEC. 2007
9. Nikolos, I.K., Valavanis, K.P., Tsorveloudis, N.C., Kostaras, A.C., *Evolutionary Algorithm Based Offline/Online Path Planner for UAV Navigation* IEEE Transactions on Systems, Man and Cybernetics 2003
10. Pongpunwattana A., Rysdyk R., *Evolution-based Dynamic Path Planning for Autonomous Vehicles*, Innovations in Intelligent Machines, pp. 113-145, 2007
11. Rathbun, D.; Capozzi, B.; Kragelund, S.; Pongpunwattana, A., *An Evolution Based Path Planning Algorithm for Autonomus Motion of a UAV Through Uncertain Envirnoments*, Proceedings of the AIAA Digital Avionics System Conference. 2002
12. Śmierzchalski R., *Synteza metod i algorytmów wspomagania decyzji nawigatora w sytuacji kolizyjnej na morzu*, Akademia Morska w Gdyni, 1998
13. Śmierzchalski R., *Ships domains as collision risk at sea in the evolutionary method of trajectory planning*, IEEE Transaction on Evolutionary Computation, vol.4, 2004.
14. Szłapczyńska J., *Zastosowanie algorytmów ewolucyjnych oraz metod rankingowych do planowania trasy statku z napędem hybrydowym*, Uniwersytet Zachodniopomorski w Szczecinie, 2009, Phd Thesis

15. Szłapczyński R., *Evolutionary Sets of Safe Ship Trajectories within Traffic Separation Schemes,* Journal of Navigation, vol.66, 2012.
16. Xiao, J. and Michalewicz, Z., *An Evolutionary Computation Approach to Planning and Navigation*, chapter in Soft-Computing and Mechatronics, K. Hirota and T. Fukuda (Editors), Physica-Verlag, 1999, pp.118 - 136.

Supply and Demand of Transit Cargo Along the Northern Sea Route

T. Kiiski

Turku School of Economics at the University of Turku, Finland

ABSTRACT: The allure of the Arctic region has recently been growing, in part due to its economic potential and environmental developments. In particular, the viability of shipping in Arctic waters through the Northern Sea Route (NSR) is critical in this context. This paper analyses potential transit cargo demand and throughput capacity of the NSR. The results show that currently outside the route's proximity there are a relatively limited number of potential cargoes. Moreover, this potential is vulnerable to limitations related to physical, climatic, political and infrastructural factors. The overall transit cargo potential accounts for around 50 million tonnes consisting of mostly bulk cargoes at first and later containerized and car transports, provided that conditions improve. The potential biannual throughput capacity accounts for around 660 passages. The supply of ice breakers is likely to constitute hindrance to growth due to the wearing of the ageing fleet and the slow rate of newbuildings.

1 INTRODUCTION

The Arctic region has attracted growing global interest, typically associated with economic motives related to its abundant natural resources and the provision of shortcuts to principal maritime routes. Limited accessibility is inherent due to its harsh climate and the perennial presence of ice. Recent developments have subjected the Arctic to profound changes driven by factors outside the region (Arbo et al. 2013). One of the main catalysts has been climate change that appears to decrease the ice extent of the Arctic Sea. This process is expected to lead to an almost ice-free Arctic Sea in the autumn by mid-century (IPCC 2014).

Shipping is the primary method of transport in the area. Furthermore, thawing ice improves the viability of maritime shortcuts across Arctic Sea, most notably the Northern Sea Route (NSR) (Stephenson et al. 2013). The NSR is a central part of the Northeast Passage, a maritime connection between the Pacific and Atlantic Oceans through Russian Arctic areas. In principle, the shorter distances of the NSR may lead to savings in transit time and related shipping costs. Other benefits relate to the avoidance of the politically volatile Middle East region and piracy-troubled African coasts. Over the past five years, the NSR has drawn growing attention outside Russia, resulting in a number of

trial-based transit sailings. So far, concerns incorporated with predictability and short navigational season have ruled out large-scale shipping.

Strategic and political aspects are an inherent part of the Arctic since the Cold War. Together with the receding ice, competition among Arctic nations has been provoked to secure ownership of the yet undivided Arctic areas, with prospects of natural resources and control over strategic maritime passages. According to estimates of the United States Geological Survey (2008), 22 % of the world's undiscovered hydrocarbons are to be found in the Arctic. Russia has been especially active, due to its presence in the area dating back to the Soviet era. China's economic and political interest in the Arctic has also grown (Alexeeva & Lasserre 2012). Intensified political tensions between both the EU and the USA towards Russia due to the Crimean crisis and the events in the Ukraine have repercussions to the Arctic, e.g. by dampening the investment impetus.

A widespread use of the NSR in the global shipping could, in theory, become a game changer for global trade patterns, analogous to the upcoming expansion of the Panama Canal. Put simply, the NSR could generate both exogenous and endogenous growth. Exogenous growth would occur if cargoes on Europe-Northeast Asia and Europe-

Northwest America trade were diverted to the NSR instead of using the conventional canal routes. Endogenous growth occurs when new cargo flows would emerge within the proximity of the NSR, i.e. destination-based shipping.

This paper focuses on transit shipping, where transcontinental shipping traverses the entire length of the NSR, but does not call at any port along the route. Destination-based shipping in the NSR is excluded from the analysis.

To our knowledge, research regarding the capacity of the NSR has been scarce with the only exception of Ragner (2000), whereas its cargo potential has attracted a reasonable number of studies (e.g. Arpiainen 1994, Ramsland 1999, Ragner 2000, Ship and Ocean Foundation 2001, Tavasszy et al. 2011). The range and scope of these estimates varies considerably.

Further research is needed to establish a holistic picture of the shipping potential of the NSR. In order to investigate the topic, this paper will quantitatively assess two objectives. First, the magnitude of the demand of the NSR transit shipping services will be examined. Second, the annual throughput capacity of the NSR under contemporary conditions will be computed. The research methods include capacity modelling and cargo analysis. Capacity analysis employs a model that takes into account the NSR's operational constraints such as the supply of ice-classed fleet and ice breakers, length of the navigational season and related conditions. Demand potential is evaluated by using trade analysis.

This paper provides insights for potential short to medium-term impacts of the NSR into the global shipping market. The overall viability of the NSR is a sum of costs and operational requirements related to reliability and cargo-specific items. This paper concentrates on the operational capacity and cargo potential issues. For more detailed analyses on the feasibility of the NSR (see, e.g. Liu & Kronbak 2010, Kiiski 2014b, Lasserre 2014). The main data sources are statistics from canal and port authorities, Eurostat, NSR administration and Clarkson's World Fleet Register.

The paper consists of six sections. Section two elaborates on the aspects related to shipping in the NSR. Section three discusses the demand potential of transit shipping. Section four analyses the supply issues related to NSR shipping. In section five the NSR throughput capacity is computed through capacity modelling. Finally, section six discuss the results and conclusions.

2 SHIPPING IN THE NORTHERN SEA ROUTE

The NSR is a central part of the Northeast Passage, a link between the Atlantic and Pacific Oceans. It is a complex myriad of routes between Novaya Zemlya and Bering Strait through five epicontinental seas: the Barents, Kara, Laptev, East-Siberian and Chuckhi Seas. The actual navigation through the NSR is subject to ice conditions, which in turn vary on an annual, seasonal and regional basis (east-west and north-south axis). Accordingly, the route's length varies between 2200 and 2900 nautical miles.

By and large, the most sheltered ice conditions prevail typically in western parts of the NSR and on the southernmost routes near the continent, which are also burdened by shallow waters. In particular, it has narrow straits (Dmitry Laptev and Sannikov) with depths of around 6.7 and 13 meters respectively, which impose draft restrictions. These and other physical constraints limit the applicable vessel size to around 50,000 deadweight tonnes (Ragner 2000), equivalent to 4500 TEU in containerships (Liu & Kronbak 2010). In contrast, northernmost routes, which have opened just recently, have greater water depths but more challenging ice conditions. Thus, offering a relaxation in the former vessel size restrictions. Brief ice-free periods occur at the end of the annual melting season in the autumn.

Given the location of the NSR in ice-covered waters, navigation imposes requirements for the ships as well as the crews. At the operational level, the ice causes among other things reduced sailing speeds, poorer fuel economy, detours, as well as damage to ships. In addition, navigation is usually conducted by ice breaker escorted convoys. An ice convoy is formed when the ice breaker leads and breaks the path for the escorted vessels. The composition of the convoy depends on ice conditions. In principle, the more severe the ice conditions the fewer number of escorted ships there are at a time and in extreme cases the ice breaker must tow the escorted ship. An alternative means to organize the ice breaker escort is to setup a zone system, whereby each ice breaker is nominated to cover a certain part of the route (Ragner 2000).

As a result of its location primarily in Russian territorial waters, the needs of the Russian and former Soviet economy has been an essential premise for the development of the NSR and its cargo flows. Since the 1950s, the NSR was strongly engaged as part of the Soviet maritime network. It conveyed cargoes related to the supply of Arctic colonies as well as the transportation of extracted minerals and hydrocarbons for further processing back to the South. The peak year of the NSR was in 1987, when 381 ships transported 6.6 million tonnes (Mt) of cargo in their 1306 voyages (AMSA 2009).

In 1991, the NSR was officially opened for international shipping. In the Soviet era, very few commercial transits took place until 1989 when positive currency rate levels stimulated operations (Ragner 2000). Numbers remained modest and reached a peak level of 0.2 Mt in 1993. Eventually, interest faded in together with worsening freight market conditions. It took nearly twenty years for international interest to capitalize. Only in 2009, motivated by favorable ice conditions, two trial-based transit sailings of German project cargo carriers commenced. In 2012 and 2013, activity stabilized to a level of around 40 annual sailings carrying approximately 1 Mt of cargo. In 2014, the modest momentum gained appeared to be in decline due to geopolitical tensions since only 31 sailings commenced, carrying just approximately 0.3 Mt of cargo (NSRIO 2015).

The bulk of transits convey natural resources to markets outside the Arctic. During the past two years, relatively few of them are made by ships flagged or based in European countries. Instead, they have consisted primarily of Russian transits. Consequently, destination-based shipping is expected to constitute the primary activity in Arctic waters in the short to medium-term (Lasserre 2014), in turn utilizing bulk shipping instead of containers (Schoyen & Bråthen 2011). However, its viability is sensitive to commodity price fluctuations (Comtois & Lacoste 2012) given the higher extraction costs of Arctic resources.

The potential of the NSR as a transcontinental shipping route relies primarily in shortcuts provided by its favorable geographical position. Under the present conditions, distance advantage is not equivalent to equal savings, at least on regular basis, since route deviations and speed reductions are common. Table 1 depicts the sensitivity of the NSR distance advantage in varying ice conditions conceptualized by the extent of the ice cover (ranging from 0, 50, 75 and 100 % of the route's length), with the consequence of reduced speeds (from 14 knots to 6 knots). Assuming that the NSR is entirely ice-free, its viability distance area relative to the Suez and Panama canals is confined ranging between Northwestern Europe and Northern Pacific in the grey areas in Figure 1.

3 TRANSIT SHIPPING POTENTIAL OF THE NSR

The global trade market does not exist in a vacuum but is instead subject to emerging shocks and trends, which in turn shape the patterns of trade. Maritime transport services are propelled by the stimulus of global economic activity and the consequent need to convey goods by sea. In principle, the demand for these services can be considered inelastic for a number of reasons: lack of close substitutes; short-term inelastic nature of the demand for bulk commodities; and relatively small proportion of the freight on the total costs (Button 2010).

The integrated global shipping network cost-efficiently carries goods from all over the world. In 2013, 9.6 billion tonnes of cargo was carried by sea making it the most common mode of transport in volume terms typically accounting for about 85 % of the totals (UNCTAD 2014). The shipping market is highly dynamic and cost competitiveness is a key factor to lure additional cargoes, provided that safety and predictability requirements are fulfilled. By and large, operating cost levels in the NSR are relatively higher compared to open-water routes. In particular, it is burdened by fees paid for ice breaking escorts, incremental cost of specialized fleet, as well as insurance (Lasserre 2014). However, high bunker levels are found to offset the relative cost difference (Liu and Kronbak 2010).

Figure 1 Northern Sea Route and principal maritime routes (potential trade areas for the NSR in Europe, Northeast-Asia and Northwest America highlighted in grey)

Table 1. Distances via the Northern Sea Route, Panama Canal and Suez Canal from Hamburg to various destinations

| Port | From Hamburg | | | | | |
	Yokohama	Shanghai	Hong Kong	Dutch Harbor	Vancouver	San Francisco
	Distance (NM)					
Suez/Panama	11,796	11,041	10,330	10,345	9080	8320
NSR	7087	7825	8505	5205	6643	7016
Advantage (%)	*−40*	*−29*	*−18*	*−50*	*−27*	*−16*
	Transit time (d)					
Suez/Panama	28	27	25	25	22	20
NSR ice-free	17	19	21	13	16	17
NSR iced 50 %	24	26	27	20	23	24
NSR 75 %	27	29	31	23	26	27
NSR 100 %	31	33	34	26	30	31

The demand potential of the NSR shipping is two-fold. On the one hand, the shorter distances of the NSR could attract seasonal time-sensitive cargoes, but on the other hand, the seasonality incorporated with the high variation of sailing conditions in low temperatures outweigh the benefits. A key factor in determining the trade potential of the NSR is the length of the annual navigational season. In the last couple of years, it has remained level at around 140 days (NSRIO 2015). Arpiainen (1994) estimated that maximum interval of two months between navigational seasons would be tolerable in order to sustain viability and accommodate the required investments. Most of the potential trades require year-round service. For example, bulk commodities may be stockpiled, but reengineering the whole supply chain in order to accommodate the requirements for seasonal traffic questions its feasibility.

Russia has the largest export potential of the NSR transit cargoes while China has the largest import potential. Albeit that the distance advantage of the NSR extends up the Northwest Pacific, this trade seems invalid not only for political reasons. From the geographical point of view another emerging Arctic maritime route, the Northwest-Passage (NWP), appears to be more compelling (Somanathan et al. 2007). In addition, the demand potential of the NSR regarding Europe-Northwest Pacific trade through the Panama Canal is marginal. In 2014, the total seaborne trade through the Panama Canal between Europe and the West Coast Pacific (USA and Canada) and respectively Europe to Asia was around 10 Mt, accounting for about 5 % of the total volumes (Panama Canal Authority 2015). Out of these, the extent of cargoes viable for NSR trade by origin-destination regions accounted only for 6.5 Mt. Furthermore, the upcoming expansion of the Panama Canal, which was initially intended to be ready by 2014, but has since then encountered a number of delays proposed to last until 2016 (Panama Canal Authority 2014), is likely to intensify the competition. The next sub-section will analyze the potential of the NSR transit shipping by shipping sectors.

3.1 Containers

The prevalent China-centric market setting supports utilization of the NSR. The global containerized trade volumes accounted for 160 million TEUs in 2013 of which Fareast Europe trade accounted for 13.1 % (UNCTAD 2014). The traffic characteristics confine the potential demand of liner shipping operated sectors, such as containers. The industry relies on among other things, fixed schedules and itineraries, leaving no room for unexpected delays. Currently, these liner shipping principles cannot be plausibly applied to the NSR (Schoyen & Bråthen

2011). This explains in part the industry's current lack of interest (Lasserre & Pelletier 2011), and why to date no trial-based sailings have taken place. Another problem for containerized cargoes, e.g. electronics, is that these types of cargoes are temperature sensitive. The internal and external icing of containers may not only damage goods but also cause problems to the stability of the ship. Application of thermal solutions (e.g. reefers) in turn incurs extra costs.

The potential transfer of a substantial part of the transcontinental container flows has been a controversial issue (Granberg 1998). Accordingly, estimates vary considerably. Arpiainen (1994) estimated that 2 Mt (0.2 million TEUs) of containerized cargoes could be shifted on the NSR. A more recent estimate by Tavasszy et al. (2011) estimates that demand on container transport through the NSR may reach up to 3.3 million TEUs. The most optimistic estimate projects 5-15 % of China's international trade, mostly container traffic, would use the NSR by 2020 (Doyle 2013). In the extreme case of an ice-free Arctic Sea up to 110 Mt of containerized cargo could be transferred to the NSR (Laulajainen 2009).

Just the sheer container volumes between Northern Europe and Northeast Asian ports indicate that there would be an influx of containers to be transferred to the NSR. In 2013, the combined Asia trade volumes of Hamburg, Antwerp and Rotterdam ports accounted for around 10 million TEUs. However, due to the inconsistent data on trade parameters and cargo contents, it is fair to assume that the actual number is considerably lower. Taking into account the biannual nature of the traffic as well as the magnitude of projections in the literature, a fair estimate would be that about 3 million TEUs may be potential for the NSR if operational requirements are met. If only the non-temperature sensitive cargoes are counted, the number is closer to 1 million TEUs.

3.2 Pure Car Carriers

Pure Car Carriers (PCC) are dedicated ships, which transport new automobiles between production locations and overseas markets. The world's seaborne trade in cars amounted to around 22 million units in 2013 (UNCTAD 2014). The industry relies on speed and schedule reliability similar to container shipping due to the high value of the cargo. It is thought to be as one of the most promising trades for the NSR (Furuichi & Otsuka 2014). The potential of which is estimated to be around 1-2 Mt (Arpiainen 1994). Based on the Eurostat data (Table 2) the 2010-2013 combined average car trade between Northeast Asian countries (China, South Korea and Japan) and Northwestern European countries (Germany, Denmark, Great

Britain, Belgium, Estonia, Finland, Latvia, Holland, Poland, Sweden and Lithuania) accounted annually for around 2 Mt with relatively balanced trade.

Table 2 2010-2013 average trade between Northwestern Europe and various countries by SITC (Standard Industrial Trade Classification) groups (thousand tonnes) (Eurostat 2014)

SITC group	Trade	Country			
		China	Japan	Canada	South Korea
oil products	Import	80	153	631	1489
(334)	Export	774	120	2690	80
coal	Import	161	190	2247	0
(32)	Export	58	58	44	44
iron and	Import	2006	229	101	556
steel (67)	Export	931	110	351	314
automobiles	Import	862	922	18	337
(78)	Export	1305	314	215	221

In light of this, it can be concluded that a fair estimate of car trade potential of the NSR is approximately 1 Mt.

3.3 Bulk transports

Bulk transported cargoes (wet and dry) consist mainly of raw materials, with low weight-to-value ratio, such as crude oil and iron ore. Large-scale utilization of the NSR to convey such cargoes can be considered to be relatively limited, since these cargoes are typically shipped directly from extraction site to processing plants usually without any trans-shipments. Neither Europe nor Northeast Asia constitutes the origin for any of the major global dry bulk trade flows. As for refined products, some potential seems to exist, e.g. oil products trade between Western Europe and Northeast Asian countries could provide potential up to 2.7 Mt (Eurostat 2014).

In the literature, the high variation indicates dissonance with regard to the bulk transport potential. This may be a result of the mixed definitions used in relation to the divergent traffic types (transit and destinational). For example, Arpiainen (1994) projects 4-5 Mt of dry bulk trade, whereas Ship and Ocean Foundation (2001) projects a total 16 Mt of trade including petroleum products, minerals, metal products and fertilizers. Potential traffic is also most likely to be unbalanced. Ragner (2000) has compiled various estimates, ranging from a modest estimate that eastbound transit traffic would be 2-3 Mt up to an optimistic 6-8 Mt, while the westbound potential is less than 1 Mt and 2-4 Mt, respectively. If the Arctic ice melts entirely then even higher number have been proposed. Laulajainen (2009) estimates that year-round ice-free Arctic Sea routes (both NSR and NWP) could have around 30 Mt of dry bulk cargoes.

It would be fair to argue, that in the near to mid-term, most of the growth generated by the NSR in the bulk sector will be of endogenous origin given Russia's large potential due to its abundant resource base. Hence, shipping activity in this sector will most likely comprise primarily of destination-based traffic.

3.4 LNG

LNG carriers are specialized liquid bulk vessels that provide the necessary refrigeration and hold pressure for the transportation of liquefied gases. Prior to the shale boom, the outlook for the Arctic LNG trade looked very promising, which resulted a number of investments made in the Russian Arctic. For example, Yamal LNG project, scheduled for completion by 2016, with an annual capacity of 15 Mt of LNG (ship-technology.com 2014). However, in the late 2000s the North-American shale gas boom altered the picture, among other things by transforming the US as an exporter of gas.

Therefore, the main destination for the Russian LNG exports is likely to be in Asian markets. In 2013, Russia already exported 14.8 billion cubic meters of LNG (BP 2014b). In 2014, the combined LNG demand of Japan, South Korea and China comprised 62 % of the totals (Goncalves 2014). In contrast, due to political disputes Europe is most likely trying to lessen its reliance on the supply of Russian gas, which currently accounts for approximately 30 % of its supply (Goncalves 2014). In addition, it is expected that energy demand in the EU is going to decline by 6 % by 2035 (BP 2014a).

LNG trade could be suitable for seasonal utilization of the NSR and provide robust volumes. AMSA (2009) projected the volume of hydrocarbon (oil and gas) related traffic on the Western part of the NSR to reach up to 40 Mt by 2020. However, as the LNG production sites are predominantly located along the NSR, this traffic may not be categorized as transit traffic but instead destination-based, similar to bulk transports. Therefore, its impact on the transit traffic volumes is likely to be modest.

4 SUPPLY OF ICE-CLASSED FLEET AND ICE BREAKERS

A fundamental feature in shipping markets is the interplay between demand and supply that creates a cyclical pattern when supply reacts with a delay to the impulses of the demand. The supply component of the NSR shipping consists of two parts: ice-classed fleet and ice breakers. A third, pertinent component, would be the supply of ice-experienced crews, but due to lack of data, it is excluded from the analysis.

Table 3 World ice-classed fleet and order book by ice-class and width of vessel as of January 2015 (CRSL 2015)

| | Breadth (m) | Ice-class | Vessel type | | | | | | |
			Tanker	Bulk carrier	Container	LNG	PCC	Other	Total
Ice-classed fleet	> 30	1AS	23	0	0	0	0	6	29
		1A & 1B	221	29	15	9	2	18	208
	< 30	1AS	14	8	13	0	0	135	170
		1A & 1B	651	56	338	2	18	1944	2128
Total by vessel type			909	93	366	11	20	2103	3502
Order Book	> 30	1AS	6	0	0	20	0	4	20
		1A & 1B	0	22	6	2	2	6	13
	< 30	1AS	0	0	0	0	0	6	6
		1A & 1B	6	2	5	0	0	48	45
Total by vessel type			12	24	11	12	2	64	125

4.1 Ice-classed fleet

At given conditions, high ice-class and winterization of vessels are basic requirements for ships to operate in the NSR. Table 3 illustrates the composition of the world's ice-classed fleet in higher ice-classes (1B and above) and order book as of January 2015 categorized based on the vessel's width and type. Vessels are categorized upon their capability to operate in the Arctic waters and the consequent number of ice breakers needed.

It is assumed that vessels over 30 meters require two ice breakers since their width exceeds the width of the largest ice breaker (30 meters). Using two ice breakers to escort a ship in turn may lead to capacity problems in peak seasons. There are a total of 3502 of vessels that have 1B or higher ice-class notation. A notable feature is the dominance of group other (60 %) (including e.g. supply ships and ferries) and tankers (26 %).

Given that the trend of increasing vessel size is prevalent in the market, it is apparent that the vessel sizes in the NSR are also likely to increase thus creating an incremental challenge to ice breaking capacity. The share of vessels requiring assistance by two ice breakers comprises 15 % of the totals (if group other is excluded).

The existing fleet is primarily suited to accommodate the needs of destination-based traffic, which may be inferred by the large share of the ships related to bulk shipping (tankers and bulk carriers). This may not directly concern Arctic waters, as there are other areas in the world where ice-classed fleet may be deployed, e.g. the Baltic Sea.

The number of highly ice-classed newbuildings in the order book looks relatively modest (around 4 %) compared to the numbers of total fleet. The investments in hydrocarbon extraction projects in the Russian Arctic have already generated large orders for LNG and tankers, capable of independent operations during the NSR most of the year. These observations agree with Lasserre & Pelletier (2011) and Kiiski (2014a) who concluded that there is no evident boom of building highly ice-classed vessels due to emerging Arctic routes.

4.2 Ice breakers

As long as ice is present, ice breakers can be considered as a key resource for successful and safe shipping along the NSR. They are designed to plough through the ice and break the path for the escorted vessels. Traditionally, ice breakers have sustained the growth of NSR cargo volumes, for example, in the early 1990s, the NSR ice breaker fleet comprised of 7 nuclear and 9 diesel-powered ice breakers (Granberg 1998).

In January 2015, the world's active fleet of ice breakers (including ice breakers and offshore supply vessels equipped with ice breaker capability) consisted of 86 ships, with an average age of nearly 35 years (CRSL 2015). Of these 44 % were Russian flagged. Russia has a fleet of the world's powerful ice breakers consisting of six nuclear-powered ice breakers (4 with 54 megawatt (MW) capacity and 2 with 35 MW) while the remainder are diesel-powered. Nuclear-powered ice breakers are normally used in escort operations, whereas diesel-powered are deployed mainly in ports and other offshore supporting duties. The supply of ice breaking services in the years to come has raised concerns (e.g. NSRIO 2013) as the average age of the fleet is approximately 30 years old and three of the nuclear powered ice breakers are coming to end of their mileage within couple of years.

There are a total of 14 ice breakers under construction, of which all except one are Russian-based orders (CRSL 2015). A peculiar feature of the newbuildings is the considerable increase in vessel size. For example, the width of vessels currently under construction has grown by approximately 25 %, with the average width reaching over 27 meters. Three of the newbuildings are nuclear powered 60 MW capacity ice breakers with a width of 34 meters and plans have been made for the construction of 110 MW ice breaker (OKBM 2014). The schedule and funding of these projects is very much uncertain and subject to change (Staalesen 2014). It seems evident that due to wearing of the existing fleet and the modest level of newbuildings, Russia is not able to maintain its current ice breaking capacity in the future. The reliance solely on diesel-electric ice breakers offers no remedy to the problem, as their operational performance including their range is limited, for example due to bunkering requirements.

5 THROUGHPUT CAPACITY

The annual throughput capacity (TC) of the NSR is a sum of following parameters: distance of the route (NSR$_{dist}$); number of ice-breakers (IB); the number of vessels that one ice breaker can escort at one convoy (EV); monthly sailing speed (v_t) determined by particular month (t). The final buffer parameter (B) represents the slack for the time needed in formation of ice convoys and waiting for the ice breaker. The following Equation 1 approximates the annual throughput capacity of the NSR.

$$TC = \sum_{t=1}^{6} \left(\left[\frac{NSR_{dist}}{v_t \cdot 24} \right] \cdot IB \cdot EV_t \cdot B \right),$$

$$NSR_{dist} = \{2200, 2900\} \quad (1)$$

$$v_t = \{6.4, 8.3, 11.0, 10.6, 9.6, 9.7\}$$

The following assumptions are inherent concerning the throughput model. We assume that there occurs a one day delay per voyage while waiting for an ice breaker to arrive and the formation of a convoy of ships. For the sake of simplicity, every ship is assumed to operate under icebreaker escort and sails along the NSR for its entire distance. The NSR's distance is assumed as either 2,200 or 2,900 NM subject to applied scenario. The navigational parameters (including speed and navigational season's length) are based on the combined monthly averages of transit statistics of years 2012, 2013 and 2014 (NSRIO, 2015). Final variable is the number of escorted ships in a convoy per ice breaker. It is assumed that it varies in accordance with the ice conditions: 5 ships in interval of easiest conditions (August to October) and 2 for the rest of the season.

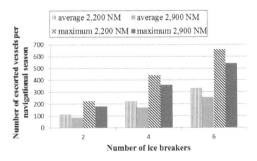

Figure 2. Biannual throughput capacity of the NSR by various route lengths (2200 NM and 2900 NM) (average = average speed and navigational season length of 2012, 2013 and 2014 transit speeds; maximum = fixed sailing speed of 14 knots speed with 180 days navigational season)

Figure 2 illustrates the throughput capacity of the NSR in average conditions (based on monthly average speeds of 2012, 2013 and 2014 transits) and optimum conditions (fixed at 14 knots average speed over a six month period) by varying the number of ice breakers (2, 4 and 6).

The number of ice breakers has a direct influence on the throughput capacity. The maximum throughput capacity of the NSR under optimum conditions (route length 2200 NM) is approximately 660 sailings per biannual navigational season. Whereas at the bottom end, the respective number, with two ice breakers, accounts for only 220 sailings. As for the details based on actual conditions (average of 2012, 2013 and 2014), the throughput capacity is 330 sailings with 6 ice breakers and respectively 110 with 2 ice breakers. Therefore, the tonne-wise capacity accounts for between 13 Mt and 26 Mt, depending on the scenario (assuming 40,000 tonnes cargo size).

6 DISCUSSION AND CONCLUSIONS

Large-scale utilization of the NSR is closely intertwined with the socio-economic development of the Arctic. The NSR constitutes a primary conveyor of cargo flows to and from the region due to undeveloped transport networks. Traditionally, the vessel sizes in the NSR have been relatively small compared to their open-water counterparts. For example, in 1987, the hitherto peak year of the NSR, the average cargo size was just over 5000 tonnes (AMSA 2009). This undermines the NSR's overall competitiveness compared to conventional canal routes, which depend on economies of scale. However, continued newbuildings of LNG and tanker sectors appear to challenge this trend.

In the early 2010s, the NSR's economical potential was appealing. Towards the end of 2014, the radical changes in the world market price of oil and gas, together with Western sanctions on Russia due to developments in Ukraine have, however, radically altered the picture. In the short-term, the outlook for large-scale utilization of the NSR as a bulk cargo transport route is relatively unclear. The level of uncertainty regarding the Arctic hydrocarbon extraction as a whole is very high. Among other things, the offshore extraction technologies required by Russia have been embargoed due to joint EU and US sanctions. In addition, commodity price levels of extracted commodities influences largely on the demand for Arctic hydrocarbons. For example, during the second half of 2014, the price of oil has fallen by nearly 40 % to around USD 60 a barrel. For oil exporting countries, namely Russia, this has had devastating effects as the ruble has considerably lost its value against the dollar, causing deficits to the federal budget (Financial Times 2014).

It seems that, the demand prospects of the NSR transit shipping in the short-term look relatively modest given the current conditions. In the mid-term, if conditions improve, a potential exists, particularly in liner-traded cargoes up to 3 million TEUs and 0.5 Mt of car transports. The overall transit potential is around 50 Mt accounting for less than 1 % of the world totals and around 6 percent of the Suez Canal totals. Most of the activity in the NSR is likely to have endogenous origin related to Arctic resources and therefore not transit traffic. Under current conditions basic characteristics of viable transit cargo is port to port based time and temperature tolerant cargo, such as bulk cargoes.

This paper has raised a number of issues mostly related to the throughput capacity of the NSR infrastructure. Russia's ambitious plans regarding the Arctic hydrocarbon production projects may also be hindered severely by the lack of ice breakers in the foreseeable future. Furthermore, the future level of ice breaking service provision to international transit shipping is a debatable issue particularly in cases when destination-based Russian transports utilize the capacity. In the 1990s, at the close of the route's peak years, it was estimated that the contemporary NSR transit shipping capacity would accommodate only 2–3 Mt of cargo (Granberg 1998). New ice breakers are being constructed, but given their modest numbers together with the slow-paced building process and uncertainties of funding, it seems obvious that the capacity offered cannot sustain growth. Building a highly ice-classed fleet with the capability of independent operations may relieve the problem to some extent. Future research should explore how the capacity of ice breaking services could be improved through different organization methods, for example by applying the zonal system.

REFERENCES

Alexeeva, O. & Lasserre, F. 2012. China and the Arctic. *Arctic Yearbook 2012*.

AMSA. 2009. *Arctic Marine Shipping Assessment 2009 Report*. Arctic Council.

Arbo, P., Iversen, A., Knol, M., Ringholm, T. & Sander, G. 2013. Arctic futures: conceptualizations and images of a changing Arctic. *Polar Geography* 36 (3): 163-182.

Arpiainen, P. 1994. *The Northern Sea Route: A Traffic potential study*. Publications of the Center for Maritime Studies at University of Turku.

BP, 2014a. Energy Outlook 2035:EU, BP PLC. January 2014.

BP, 2014b. Statistical Review of World Energy. BP PLC.

Button, K. 2010. Transport Economics 3 Edition. Edward Elgar Publishing Limited, Cheltenham, UK.

Comtois, C. & Lacoste, R. 2012. Dry bulk shipping logistics. In Song, D-W and Panayides, P. (eds), *Maritime logistics: a complete guide to effective shipping and port management*: 163-176. London, UK: Kogan Page Publishers.

CRSL, 2015. Clarkson's World Fleet Registry. Clarkson Research Services Limited.

Doyle, A. 2013. China plans first commercial trip through Arctic shortcut in 2013, Reuters. http://www.reuters.com/article/2013/03/12/shipping-china-idUSL6N0C4F9720130312.

Financial Times, 2014. Winners and losers of oil price plunge. http://www.ft.com/cms/s/2/3f5e4914-8490-11e4-ba4f-00144feabdc0.html#axzz3M5eYObCC.

Furuichi, M. & Otsuka, N. 2014. Economic feasibility of finished vehicle and container transport by NSR/SCRcombined shipping between East Asia and Northwest Europe. Proceedings of the IAME 2014 Conference. 15-18 July 2014, Norfolk, USA.

Goncalves, C. 2014. Breaking rules and changing the game: Will shale gas rock the world? *Energy Law Journal*, 35 (2): 226-262.

Granberg, A. G. 1998. The northern sea route: trends and prospects of commercial use. *Ocean & Coastal Management* 41:175-207.

IPCC, 2014. Summary for Policymakers. In: *Climate Change 2014: Synthesis Report*. Intergovernmental Panel on Climate Change.

Kiiski, T. 2014a. The Dynamics of World Ice-Classed Bulk and Containership Fleet in view of Arctic Shipping. Proceedings of Nordic Logistics Research Network (NOFOMA) Conference, 12–13 June 2014, Copenhagen, Denmark.

Kiiski, T. 2014b. The economic viability of Northern Sea Route as a seasonal supplement for container shipping between Europe and Asia. Proceedings of 6th International Conference on Maritime Transport, 25–27 June 2014, Barcelona, Spain.

Lasserre, F. & Pelletier, S. 2011. Polar super seaways? Maritime transport in the Arctic: an analysis of shipowners' intentions, *Journal of Transport Geography* 19: 1465-1473.

Lasserre, F. 2014. Case studies of shipping along the Arctic routes. Analysis and prof-itability perspectives for container sector. *Transportation Research Part A* 66: 144-161.

Laulajainen, R. 2009. The Arctic Sea Route. *International Journal of Shipping and Transport Logistics*, 1 (1): 55-73.

Liu, M. & Kronbak, J. 2010. The potential economic viability of using the Northern Sea Route (NSR) as an alternative route between Asia and Europe. *Journal of Transport Geography* 18: 434-444.

NSRIO, 2013. Russia expects "Icebreaker pause". Northern Sea Route Information Office. http://www.arctic-lio.com/node/189.

NSRIO, 2015. Transit statistics. Northern Sea Route Information Office. http://www.arctic-lio.com/nsr_transits

OKBM, 2014. Breaking the Ice of the Arctic Region. http://www.okbm.nnov.ru/english/publications/671-2014-06-20-17-29-08.

Panama Canal Authority, 2014. Panama Canal Updates Maersk Line on Expansion Program. 20.8.2014. Press release. http://www.pancanal.com/eng/pr/press-releases/2014/08/20/pr519.html.

Panama Canal Authority, 2015. Transit Statistics. http://www.pancanal.com/eng/op/transit-stats/index.html

Ragner, C. 2000. Northern Sea Route Cargo Flows and Infrastructure - Present State and Future Potential. FNI Report, Lysaker.

Ramsland, T. R. 1999. Cargo Analysis, North West EuropeThe Far East / Canada US West Coast-NW Europe. The Potential for Transit traffic on the Northern Sea Route. INSROP Working Paper No. 145-1999, III.10.2 & III 10.3.

Ship and Ocean Foundation 2001.The Northern Sea Route. The shortest sea route linking East Asia and Europe. Tokyo, Ja-pan: Ship & Ocean Foundation.

Ship-technology.com. 2014. Port of Sabetta, Yamal Peninsula, Russia. http://www.ship-technology.com/projects/-port-sabetta-yamal-peninsula-russia.

Somanathan, S., Flynn, P. C. & Szymanski, J. K., 2007. Feasibility of a Sea Route through the Canadian Arctic, *Maritime Economics & Logistics*, 9: 324–334.

Staalesen, A. 2014. Yard delays delivery of icebreaker. Barents Observer. http://barentsobserver.com/en/arctic/2014/11/yard-delays-delivery-icebreaker-21-11

Stephenson, S., Smith, L., Brigham, L. & Agnew, J. 2013. Projected 21st-century changes to Arctic Marine access. *Climate Change* 118 (3-4): 885-899.

Tavasszy, L., Minderhoud, M., Perrin, J-F. & Notteboom, T. 2011. A strategic network choice model for global container flows: specification, estimation and application. *Journal of Transport Geography* 19: 1163-1172.

UNCTAD, 2014. *Review of Maritime Transport 2013*. Publications of UN, New York

175

Northern Labyrinths as Navigation Network Elements

A.N. Paranina & R. Paranin
Herzen State Pedagogical University, St-Petersburg, Russia

ABSTRACT: The authors of the article consider the stone labyrinths as solar calendars. In the center of these structures there are usually already installed gnomons – vertical objects that give shade. Midday shadow points to north, and the change of its length during a year is correlated with the diameter of the arcs of the labyrinth. Points of sunrise / sunset at the equinoxes and solstices are very often fixed in the pattern of the labyrinth, as well as the beginning of the annual cycle. In general, patterns of labyrinths are of the same type, the differences reflect the regional characteristics of illumination, the differences in latitude and topography (shape of the horizon). The uniformity of the technology and the location on the waterways give an opportunity to consider stone labyrinths as ancient elements of local and regional navigation networks.

1 INTRODUCTION

Northern labyrinths can be found in England, Iceland, Norway, Denmark, Sweden, Finland, Estonia and Russia. They are located on isles, peninsulas, near harbors and in river mouths. Their picture is complicated but organized. In terms of structure, there are unispiral, bispiral, concentric and radial types. In terms of outer shape: circles, ovals, rarely squares (Yeliseyev 1883, Vinogradov 1927, Gurina 1948, Kuratov 2008, Kern 2000).

Hypotheses about the designation of stone labyrinths can be divided into two groups: calendar and non-calendar. It should be noted that despite all the diversity of facts of non-calendar use, most part of them is often associated with time and stage of life.

Hypothesis of calendar designation of labyrinths are mainly based on the assumption of a direct projection of the trajectory of space objects on the Earth's surface (Herman Wirth, Daniel Svyatskiy, Sergei Yershov) or consider pattern of labyrinth as a record of the results of direct sight of the annual variation of the Sun (Yuri Chekmenev). However, direct sight cannot explain the technology for using the labyrinth: 1) it is impossible to explain the quantitative ratios of the trajectory of a celestial object and its reflection in the stone pattern; 2) it is even more difficult to imagine the use of the pattern - with a diameter of 20-30 meters, it is impossible "to read" from the human height; 3) the problem of

monitoring the trajectory of the sun is that bright light dazzles eyes, just after sunrise its movement takes off from landmarks.

The proposed concept of the labyrinth-gnomon – a tool of back sight of the sun - from the shadow set in the center of the object, opens the possibility of its use as a sundial compass and calendar (Paranin & Paranina 2009, Paranina & Paranin 2009a, b. 2014, Paranina 2009, 2010, 2011a, b. 2012a, b, c, 2013, 2014). The shadow of the object is easy to observe, record, measure, and its movement reflects and a form - encodes all the movements of the sun and is consistent with the position of the elements of the structure of the labyrinth; landscape orientation on the horizon (mountains, valleys) are not necessary, on the contrary, water environment is the optimal without creating distortions of azimuths of sunrise/sunset (their normal location is on the island or cape).

Author's concept of a labyrinth -gnomon technology solves the problem of the calendar use, and is consistent with all elements of a wide range of symbolic interpretation of the signs of the labyrinth and Labrys.

2 OBJECTS AND METHODS

The objects of study were the monuments of ancient material culture of European Russia (siedis, menhirs, stone labyrinths, petroglyphs). From 2009 to 2013 the objects located on coast of the White Sea are investigated: in the archipelago of Kuzova, in the archipelago Solovki, in the gulf Kandalaksha, in the mouth of the river Vyg and etc. (Figs 1-4).

The applied field research methods (survey, description, observation, work with maps) and Earth remote sensing, as well as methods of mathematical, conceptual modeling and mapping. Theoretical analysis is based on the theory of reflection and systemic and chorological approach, methodological statements of historical geography by V.I. Paranin (Paranin 1990, 1998, Paranina 2012c).

Figure 1. Area of research - coast of the White Sea.

Figure 2. Labyrinth No. 1 in the archipelago Solovki.

Figure 3. Labyrinth in the archipelago of Kuzova.

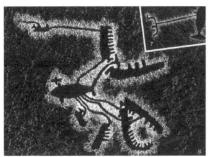

Figure 4. Petroglyphs (Lobanova, 2005).

3 RESULTS AND DISCUSSION

For interpretation of a northern labyrinth the gnomon - the elementary astronomical tool was used (Figs 5, 6). The shadow of a gnomon codes a trajectory of movement of the Sun on a firmament. In 2009 the authors proved that drawing of a labyrinth fixes astronomically significant points: 1) the provision of a midday shadow in days of winter and summer solstice corresponds to extreme arches of spirals, 2) the ends of spirals correspond to azimuths of risings/calling, 3) the entrance to a labyrinth notes the beginning of an annual cycle (in an equinox or a solstice).

The sketch of a shadow of a gnomon in days gives the schedule similar to a pitchfork, horns, wings, a fish tail. The shadow schedule in a year fills the space whose shape form represents labris - a bilateral two-horned axe of god of light (Figs 6, 7).

Figure 5. Labyrinth No. 1, the plan (Skvorzov 1990).

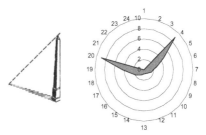

Figure 6. Gnomon and geometry of its shadows per day.

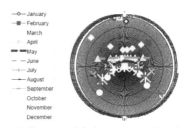

Figure 7. Geometry of shadows per year (Paranina 2010).

The basic units of the information model of the world (IMW) reveal different aspects and levels of modeling of space-time (Fig. 8): the first basic level - the navigation, creates a spatial and conceptual framework of IMW; second modeling level - reflects semiotic, linguistic, cartographic, toponymic, mythological units that encode, duplicate and replicate vital navigational information; and the crown of the model - a tradition that serves as the selection and storage of proven information to maintain the continuity of Life, including the Renaissance. Basic processes and phenomena form the reference benchmarks of fundamental concepts, the meaning of which is priceless, and therefore sacred, and their shape is less exposed to other transformation. Structure flow, which maintained sustainability of this model throughout the history of our civilization, was a continuous practical use of sunlight to ensure the order (Paranina 2010, 2011, 2012a, b, 2013, 2014).

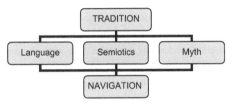

Figure 8. The structure of the information model of the world – navigation concept (Paranina 2014).

4 NORTHERN LABYRINTHS AS NAVIGATION NETWORK

Stone labyrinths are located, as a rule, on a plot of sea coast estuaries (at the source of fresh water) - it's convenient for rest and orientation, waiting for the desired date of astronomical calendar, in which marks of important phenological events of the area (cycles of fishing animals, climate and hydrological mode, lighting) can be made.

Key elements of the picture calendar are diameters of arcs and azimuths of entrance and end spirals - reflect the effect of two factors: the latitude and discrepancies of physical horizon (surface relief) with the astronomical horizon.

Polar regions differ from moderate latitudes in terms of azimuths of sunrise/sunset in the solstice that vary considerably in adjacent parallel (Tab. 1). If latitudes 40-50° rise at the summer solstice and shift by only 6.92°, and at latitudes of 50-60° only twice – 13.42 °, then advancing further at only 5° (60-65°) to the north - rise shifts at 17.37°, and latitudinal range of 1° 33" (65° - 66° 33", i.e. B. Zayatsky Island to the Arctic Circle) - to 20.03°. It is obvious that planetary space conditions of astronomical observations in the polar latitudes become the main reason for specific features of drawings.

The distorting influence of the physical horizon line on measurement of astronomical azimuths can be levelled by locating the instrument on the beach, whose calm surface coincides with the astronomical horizon, this explains the location of the labyrinth near water. This fact partly explains the abundance of labyrinths in a small area of the Big Zayatsky Island (more than 30 items on 1.25 km²): firstly, the labyrinths are located on parts of the shore, open to different sectors of the horizon, which provides accurate measurements for different astronomical dates and various astronomical objects (objects in the light of the moon cast a shadow as well); secondly, the construction of new labyrinths is associated with the retreat of the shoreline; thirdly, arranged compactly enough, they form a local network.

An equally important reason for the construction of new labyrinths is variability of subpolar latitudes of astronomical targets not only in space but also in time - here the change in slope of the Earth's axis is most visible; being observed according to displacement of the position of the Arctic Circle at other latitudes, these changes are not as dramatic (Tab. 1). The table shows that 5,000 years ago, the azimuth of the summer solstice (SS) was significantly less than modern, therefore, the line of the Arctic Circle was located closer.

Table 1. Dynamics of astronomically significant directions in space and time

N (°)	2010		3000 BC	
	WS*, 22.12	SS**, 22.06	WS, 07.01	SS, 02.08
65	160	20,03	165,31	15,46
60	142,86	37,40	144,82	35,48
50	128,41	51,82	129,55	50,71
40	121,29	58,74	122,25	57,81
30	117,39	62,74	118,20	61,97
20	115,05	64,97	115,83	64,29
10	113,85	66,19	114,50	65,56
0	113,44	66,56	114,09	65,91

*WS - winter solstice;
**SS - summer solstice.

Most labyrinths are located in the most dynamic area approximately from latitude 57° to 66° 33", which primarily determines the differences in their pattern.

At the latitude of the Arctic Circle azimuths of solstices coincide with the meridian, and the boundaries of the astronomical seasons are in the shape of direct cross. In some cases the center corresponding to the polar day, is marked by a closed circle or spiral, as in a labyrinth in Iceland.

North of the Arctic Circle, only equinoxes can be reliably determined by azimuth of sunrise/sunset. To divide the year into periods between the polar night and polar day, you can use the azimuths of sunrise/sunset, which, depending on latitude, more or less rapidly move in the range of 0° +/- 180°. When the sun does not set over the horizon, length of midday shade - diameters of arcs - become the only way to divide time into days.

5 CONCLUSIONS

The simplest astronomical tools – Gnomon, allows to read pattern of northern labyrinth as a solar calendar. Phenomena of polar days and white nights contribute to development of solar navigation in the north.

Labyrinths and other megalithic sites in Northern Europe should be considered as elements of local and regional navigation networks. This approach will provide: 1. an integral understanding of the technology of solar navigation in its space and time dynamics; 2. a more precise dating of labyrinths, using astronomical criteria; 3. additional information about the rhythm of natural processes.

Studies of northern labyrinths-gnomons fill with concrete knowledge a well-known truth that astronomy, geography and navigation are the oldest scientific knowledge. The concept of navigation modeling of the world allows to create new vision of culture of prehistoric man, and, therefore, to understand the fundamentals of the achievements of modern civilization.

REFERENCES

Gurina, N.N. 1948. *Stone labyrinths of Belomorya.* Soviet archeology. CX: 125-142.

Kern, G. 2000. *Through the Labyrinth.* Munich-London-New York: Prestel Verlag

Kuratov, A.A. 2008. *Stone labyrinths in sacral space of Northern Europe.* Arkhangelsk: Pomor University of Russia.

Lobanova, N.V. 2005. *Secrets of petroglyphs of Karelia.* Petrozavodsk: Karelia.

Paranin, V.I. 1990. *Historical Geography of the Chronicle of Russia.* Petrozavodsk: Karelia.

Paranin, V.I. 1998. *The history of the barbarians.* Ed.. St. Petersburg: RGO.

Paranina, G.N. 2009. Heritage of the Stone Age – a basis of geo-cultural space. *Balanced development of the North-West Russia: current problems and perspectives; Materials of socio-scientific conf. Pskov, November 18-19, 2010.* Pskov: State Pedagogical University.

Paranin, R.V. & Paranina, G.N. 2009. The Labyrinth: orientation in geographic space and the evolution of the mark. *Space geocultural European North: the genesis, structure and semantics.* Arkhangelsk: State Pomor University of Russia.

Paranina, G.N. & Paranin, R.V. 2009a. The ancient system of orientation of the European North. *Sustainable development and geo-ecological problems of the Baltic region; Intern. scientific and practical conf. dedicated to the 1150th anniversary of V. Novgorod 23-25 October 2009.* Novgorod: Novgorod State University,

Paranina, G.N. & Paranin, R.V. 2009b. Northern labyrinths as astronomical instruments in relation to patterns of mythology and symbols of culture. *Society. Environment. Development.* 4 (13): 120-134.

Paranina, G.N. 2010. *Light in the labyrinth: time, space, information.* St. Petersburg: Asterion.

Paranina, G.N. 2011a. Northern Labirinths – gnomon and models of geographical space. *Elsevier, Procedia Social and Behavioral Sciences.* 19: 593-601.

Paranina, G.N. 2011b. Northern labyrinth – gnomon: compass, clock, calendar. *The Quarternary in all of its variety. Basic issues, results and major trends of future research; Proc. of the VII All-Russian Quaternary conf. Apatity September 12-17 2011.* Apatity - St. Petersburg: Geological Institute KSC RAS.

Paranina, G.N. 2012a. Navigation in space-time-a basis of information models of the world. *Montreal Internat. Engineering Forum February 9-10 2012.* Montreal.

Paranina, G.N. 2012b. Stone labyrinths in the ancient navigation. *Divnogorskiy Collection: Works Museum "Divnogorie"* Vol. 3: 209-221

Paranina, G.N. 2012c. System principles of historical and geographical studies in the works of V.I. Paranin. *Problems of regional development. Finno-Ugric studies in geographic space:* 31-35. Saransk: Publishing house of the Mordovian university.

Paranina, A.N. 2013. Northern Labyrinth – A Key to Time, Space, Information. *Scientific Research Publishing (Eastern Connecticut State Univ., USA) Natural Resources (NR).* 4: 349-356.

Paranina, A.N. 2014. Navigation in Space-Time as the Basis for Information Modeling. *Scientific Research Publishing (Eastern Connecticut State Univ., USA), Vol. 2, N3, July 2014, Archaeological Discovery (AD),* 2: 83-89. doi: 10.4236/ad.2014.23010.

Paranina, A.N. & Paranin, R.V. 2014. Interaction of the nature and ancient persons on the coast of the White Sea. *Wetlands Biodiversity (JWB).* 4: 131-140.

Skvorzov, A.P. 1990. How many monuments on the Solovki? *Problems of studying of the historical and cultural environment of the Arctic:* 282-300. Moscow: Nauka.

Vinogradov, N.N. 1927. *New labyrinths of the Solovetsky islands. Island B. Zayatsky's labyrinth. Solovki, Solovki, Materials, issue XII.* Island Solovki: Society of Study of local lore.

Yeliseev, A.I. 1883. A bout so-called the vavilonakh in the north of Russia. *News of Imperial Russian geographical society.* 19: 12-16.

Ship's Route Planning in Ice Infested Areas of Northern Svalbard Following Ice Charts Made by Remote Sensing Methods

T. Pastusiak

Gdynia Maritime University, Gdynia, Poland

ABSTRACT: The data visualized on various sea ice maps are not the same. The discrepancies of the information provided by each source will result from the error of the measurement method, which can be as high as 15% of the concentration of ice floes. It should also be borne in mind that the more generalized information about the state of the ice cover, the lower probability of detection of ice floe patches of a high concentration and spatial extent.

The results of analysis allow dividing sources of information according to their applications in voyage planning according to vessel's ice class. Some sources enable navigators to plan the route of the ship avoiding areas with high concentration of ice floes. Other sources seem to meet the needs of preliminary voyage planning and scheduling the next expedition of vessels, even with low ice classes. Some sources may be misleading during preparation of voyage plan. Yet other sources allow determining the limits of areas available for navigation of the vessels of average ice class. Each vessel that is planning voyage in ice should take into consideration inaccurate estimation of concentration of ice floes by means of satellite remote sensing methods.

1 INTRODUCTION

Planning a voyage of a ship in ice-covered areas in the Arctic differs from the preparation of a standard one. The main threats to shipping are complex, variable in time and space ice conditions. They may, in extreme cases, completely disable the voyage in the whole route or part of it and potentially may pose a threat to its safety all the time.

A significant number of new sources of information characterizing the conditions for the Arctic ice appeared in recent years. They are obtained by means of satellite remote sensing methods. Some of them are compiled automatically, other are prepared with the support of highly qualified specialists. Properties and formats of these sources are very diverse. These sources are planned first and then analysed in detail in terms of their compliance with terrestrial observations. For this purpose, the results of visual observations of ice cover and recorded table components of voyage plan, collected during the voyage of the ship "Horyzont II" from Longyearbyen (78° 13' N, 015° 38' E) to Kinnvika (Murchisonfjorden, 80° 02' N, 18° 30' E) on the North of Svalbard on 10th of August 2009, and from Kinnvika to Longyearbyen

on 15-16th of August 2009 were used. Routeing was determined on the basis of previously executed voyage. The map contents of the current ice conditions were not taken into account. The next step was to compare the consistency of data between the selected sources.

The results of the analysis should answer a few questions:

1. What is the precision of the information provided by each sea ice data source?
2. What is the usefulness of particular sources for planning of route and schedule consecutively appointed after each trip?
3. Is the usefulness of particular sea ice data sources the same for each ice class of vessels?
4. Could vessel determine more favourable route if the content of sea ice data sources was taken into account?
5. Could vessel avoid moving through the field of higher concentration of ice floes at the initial stage of planning a route?

2 PRELIMINARY ASSUMPTIONS AND RESEARCH METHOD

Concentration and age (thickness) of ice were adopted as basic parameters of ice to choose optimal route of the vessel (Arikajnen & Tsubakov 1987, Arikajnen 1990). Canadian administrative method of assessing the feasibility of the ship passage in ice is also based on the concentration and age of ice (Timco at all 2005). Russian (AARI) publishes ice concentration maps in the summer time. Maps of ice thickness (age) are published during the winter time. Concentration of ice was adopted as a criterion of evaluation because comparative analysis relates to the summer season. Vessels are built according to the planned conditions of navigation in ice. AARI criterion was adopted to assess the ability of a vessel with specified reinforcements of the hull structure for navigation in ice (Table 1).

Table 1. Capability of 'ice-free' navigation in Arctic waters for various ice class vessels (following Arikajnen, 1987)

Ice Class	Criteria of "ice-free" navigation capability
Icebreakers or ULA with icebreaker's assistance	Ice pack concentration CT > 70-80%
ULA (without Norilsk class ships)	Ice pack concentration CT ≤ 70-80%
UL, L1	Ice pack concentration CT ≤ 40-60%
L2, L3, L4	Ice pack concentration CT ≤ 10-30%

At first, compliance of the information presented on maps derived from remote sensing methods with visual and radar observations made on the vessel was verified. Then, the compliance of ice floe concentration edges from various sources received by remote sensing methods was verified. For this purpose available online files were used and related to the concentration of ice floe in JPG and GeoTIFF raster formats and in GRIB and NetCDF grid formats as well as in Shapefile / SIGRID-3 vector format. The raster format maps were calibrated. Routes of the vessel and ice floe field boundaries obtained from visual and radar observations have been saved in Shapefile format. This allowed the data contained in all files to be visualized on a single screen in a georeferenced system.

Displaying all examined maps from remote sensing and observed ice floe field boundaries found by the vessel on a single screen was designed to facilitate the comparison of results from different sources of information and to increase the accuracy of the position and parameters of the ice cover. On this basis, it was intended to determine which sources of information more accurately visualize the actual limits of ice floe concentration fields as to the value of concentration, position and spatial distribution. In this way it was intended to determine

which files (maps) are best suited on board of conventional vessel, non-conventional vessel, a yacht or a boat or ashore in the owner's office or planner's office in voyage planning in ice-covered areas.

The work is a scientific analysis. That is why sources of information were not divided into the official (derived from authorized nautical providers) and unofficial (for scientific climate research). It was expected that the qualitative diversity of information sources will result from the correctness and accuracy of georeferenced files, resolution and precision of concentration scale used. This diversity should also comprise the width of filtered band around land (omitted when making the map), the value added to automatic results by the scientists (who develop maps). Finally, this diversity should verify information from remote sensing data by means of visual and radar observations made by the author of the work carried out during the voyage of the vessel in the north of Svalbard.

Conditions observed on board vessel were recorded by taking video of the radar screen and by taking pictures of visual view of sea surface ahead of the vessel. The most difficult ice conditions were identified by means of recorded waypoints position, speed of the vessel and concentration of ice floe. Positions and parameters of the detected edges of ice floe patches of higher concentration (pictures 1a, 1b) were also reconstructed. The speed of the vessel in the ice was the smaller, the greater the concentration of ice floe was. Full speed of 11 knots reflected lack of ice in the vicinity (0.5 - 1 Mm). Reduced speed of 6 knots indicated navigation in the area of ice floe concentration below 35%. Reduced speed below 3.5 knots was connected with navigation in close proximity of ice. The speed was below 2.5 knots when passing through the field of ice floe of concentration above 50%.

3 CHARACTERISTICS OF THE DATA SOURCES USED IN THE STUDY

3.1 *The NIS maps on ice concentration in raster JPG format with geographic coordinates*

The producer of these maps is the Norwegian Ice Service (met.no). The maps are developed on the basis of SAR images of European and Canadian RADARSAT and ENVISAT satellites at a resolution of 75-150 meters. Spatial resolution of these charts is 1.000 m (NIS, http://polarview.met.no/documenta-tion.html). In this way they allow determination of ice conditions in the fjords and straits which have a width of a few kilometres only. The files adopt a standard name c_map3.jpg. Examined files were downloaded from the website http://polarview.met.no/. The scale of ice concentration is discrete and includes the following information: Open Water (CT = 0-10%), Very Open Drift ice (CT = 10-40%), Open Drift ice (CT = 40-70%), close Drift ice (CT = 70-90%), Very Close Drift ice (CT = 90-100%), Fast ice.

3.2 *The NIC vector maps on the ice concentration, ice age and floe size in PDF or Shapefile format*

These maps are actually discontinued. The producer of these maps was the US National / Naval Ice Center. Examined files were downloaded from the website http://www.natice.noaa.gov/products/weekly_products.html and tab "Barents Sea NW". These maps are available in vector PDF format (files named barnwYYMMDDcolor.PDF with plotted geographic coordinates) or in vector Shapefile format (file named barnwYYMMDD.zip which contains SHP, DBF, SHX, and PRJ files). The files contained particular data on weekly or bi-weekly basis. They represent the ice conditions for the week in which they were published. Data for the analyses went back 96 hours from when they were completed. They were dated with the week they were published. Additional information included in the scale determined: Fast ice, Ice shelf, Undefined. They were based on an analysis and integration of all available data on ice conditions, including weather and oceanographic information, visual observations from shore, ship and aircraft, airborne radar, satellite imagery (RADARSAT, ENVISAT, MODIS, and GMM) and climatological information. These products were mainly used for climate analysis, climate change studies and as input to the Global Digital Sea Ice Data Bank (GDSIDB) but they can also provide ice information to marine community to enhance the safety and the efficiency of marine operations in ice covered waters.

3.3 *The AARI maps of concentrations, age and of ice forms in vector Shapefile format*

The producer of these maps is the Ice Center AARI. They are compiled on the basis of satellite information (in the visible, infrared and radar bands) and reports from the Arctic and coastal stations same like ships. The data are collected during the period of 2-5 days and after averaging are issued every Thursday which is the reference date. An example name of file is aari_bar_YYYYMMDD_pl_a.ZIP (which contains SHP, DBF, SHX, PRJ files) elaborated for the Barents Sea. Maps of concentration, age and forms of the Arctic Ocean in vector SIGRID-3 format with a sample name aari_arc_YYYYMMDD_pl_a.ZIP (which contains SHP, DBF, SHX, PRJ files) are taken into account as equivalent maps. Examined files downloaded from the website http://www.aari.ru/projects/-ecimo/index.php?im=100. Scale of the map is 1:

5,000,000 (adopted on the basis of its raster equivalent). It should be noted that the raster equivalent map for aari_arc_YYYY-MMDD_pl_a.ZIP file is drawn up on a scale of 1: 10,000,000.

3.4 *The IUP maps of the concentration of ice in raster GeoTIFF format*

The producer of these maps is the Institute of Environmental Physics, University of Bremen. The maps are developed on the basis of AMSR-E images (currently AMSR2) using the ASI algorithm. The resolution of these maps is equal to 3.125 meters. An example name of file is asi-n3125-YYYYMMDD.tif. The examined files were downloaded from the website http://iup.physik.uni-brmen.de:8084/amsredata/asi_daygrid_swath/l1a/n3 125/. The scale of visualized ice concentration is available on website http://iup.physik.uni-brmen.de:8084/amsredata/asi_daygrid_swath/l1a/n3 125/2009/aug/Svalbard/asi-n3125-0090810_nic.png and on website http://iup.physik.uni-brmen.de:8084/amsredata/asi_daygrid_swath/l1a/n3125/README.TXT. Colours of the scale reflect increments of concentration for every 10% for a value between 0 and 80% and for every 5% of values between 80 and 100%. Global Mapper software displays digital value increments of concentration of 0.5%. However, the unit of the displayed parameter is not displayed.

3.5 *The NIC "daily products" maps on the MIZ ice concentration scale in vector Shapefile format*

The producer of these maps is the National Ice Center (US). They are compiled from a variety of sources with a resolution better than 50 meters per pixel. Sources of information include (but are not limited to) ENVISAT, DMSP OLS, AVHRR i RADARSAT (http://www.natice.noaa.gov/products/-daily_products.html). The NIC analysts carry out the necessary interpretation of images that improves the value of these sources for the correct identification of the extent of the ice edges. An example name of files is nic_mizYYYYDDD-nc_pl_a.zip (which include SHP, DBF, SHX, PRJ files) drawn up for the entire Arctic Ocean. Examined files were downloaded from the website http://www.natice.noaa.gov/products/daily_-products.html and tab "MIZ Shape". The scale of ice concentration is available on website http://www.-natice.noaa.gov/produts/products_on_demand.html. Field CT81 means concentration above 80%, CT18 (Mariginal Ice Zone) means concentration between 18% and 80%, "Open water" means concentration of ice floe from zero to 17%.

MIZ maps issued by NIC (products on demand) on the concentration of ice for the Arctic Ocean in vector Shapefile format show identical edge lines. An example name of files is arctic_daily_MMDDYYYY. Examined files were downloaded from the website http://www.natice.-noaa.gov/products/products_on_demand.html. Information concerning these files is consistent with the data described in the related files nic_mizYYYYDDD-nc_pl_a.zip. The research made use of nic_miz-YYYYDDDnc_pl_a.zip files.

3.6 *The NCEP maps on ice concentration in gridded GRIB format in a resolution of 5 minutes of arc*

The producers of these maps are the NWS, NOAA, NCEP, NOMADS. Spatial resolution of these charts is 5' of latitude x 5' of longitude (geographical greed). Examined files were downloaded from the website ftp://polar.ncep.noaa.gov/cdas/archive/. An example name of files is ice5min.YYYYMM.grb. This file contains particular data for each day of the month. The scale of ice floe concentration is continuous in the range between 0 and 100% with increments of one percent. Additional scale includes the following information: Land, Weather, Bad data, Coast, No data (Grumbine R, 1996, ftp://polar.ncep.noaa.gov/pub-/pub/mmab/papers/ssmi120.ps.gz).

3.7 *The IFREMER maps of ice concentration in gridded NetCDF format without geographical coordinates plotted*

The producer of these maps is the CERSAT that is part of IFREMER. They are compiled on the basis of satellite SSMI and QUIKSSCAT images. Spatial resolution of these charts is 12.500 meters (Erzaty and others, 2007). An example name of file is YYYYMMDD.nc. Examined files were downloaded from the website ftp://ftp.ifremer.fr/ifremer/cersat-/products/gridded/psi-concentration/data/arctic/daily/ net-cdf/2009/. In order to eliminate the pixels associated with the land, grid masks have been enlarged up to two pixels (25 miles) away from the land. In connection with the use of weather filter, an area considered to be free of ice is determined by the 15% ice concentration limit. The scale of ice floe concentration is continuous in the range between 0 and 100%.

3.8 *The NIC maps on the sea ice extent and sea ice edge boundary in vector Shapefile format*

The producer of these MASIE maps is the NIC. These charts use a wide variety of data sources such as MODIS, AVHRR-VIS, GOES, SEVIRI, MTSAT, AMSR-E, SSM/I, AMSU, SAR imagery from RADARSAT-2, ERS-2, ALOS, PALSAR, ASAR.

The ice charts and ice edge products from ice charting agencies in the US, Canada, Norway, Denmark, Russia, Germany, Sweden and Japan also serve as data sources in the absence of direct satellite data. These charts are constructed by analysts trained in remote sensing imagery interpretation and sea ice climatology. Spatial resolution of these charts is 4 km. An example name of file is masie_ice_r00-_v01_YYYYDDDD_4km.ZIP (containing SHP, SHX, DBF and PRJ files). Examined files were downloaded from the website http://nsidc.org/data/-docs/noaa/g02186_masie/index.html. The analysts integrate all data sources for the best estimate of spatial coverage of ice cover. A cell is considered ice covered if more than 40 percent of the 4 km cell is covered with ice. This is regardless of the ice thickness or ice type. It is worth mentioning that the daily ice edge product is used to warn navigators and others in the Arctic where ice exists or is likely to form at any concentration. The primary users of the ice charts and ice edge products are marine transportation interests. The input product for MASIE is IMS product that is designed primarily for modellers. It is produced relatively consistently in comparison with chart and edge products. It also benefits from the same careful manual analysis that is used for chart and edge products.

4 VISUAL OBSERVATIONS

In August 2009, the ship "Horizon II" performed a return voyage from Longyearbyen to Kinnvika within the ongoing project IPY-58 KINNVIKA and special project 111/IPY/2007/01. The hull of the vessel meets the criteria for ice class L1. Due to the lower main engine power the ice class of the ship is reduced to L2. During this voyage visual and radar observations of hydro-meteorological and ice conditions were made, as well as the records of ship motion parameters from safety of navigation point of view. These conditions documented through photographs and videos of the vessel's surroundings and the radar screen. The identified parameters are presented separately in the tables for the voyage from Longyearbyen to Kinnvika (Table 2) and for the voyage from Kinnvika to Longyearbyen (Table 3). High values of concentration are related to the places where the vessel passed through or passed in vicinity of ice edge. Three locations of higher concentration of ice floe were found on the way to Kinnvika. Only one place was found during way back. The edges of a higher concentration of ice were sequentially assigned name, date and time: Edge-1 - 2009.08.10 12:00 UTC, Edge-2 - 2009.08.10 13:25 UTC, Edge-3 - 2009.08.10 15:45 UTC and Edge-4 2009.08.14 14:30 UTC. For each of these places sketch of the ice floe edge was made.

They were then stored in a Shapefile file. The spatial distribution of the observed ice edge and implemented vessel routes are shown in Figure 1.

Table 2. Voyage plan table from Longyearbyen to Kinnvika

Way-Point	Longitude E [°]	Latitude N [°]	Speed [knots]	Concentration [%]	Floe size [meters]
22	80.1405	14.1171	11.1	0	0
23	80.1475	14.3073	11.1	70 (Edge-1)	110
24	80.1285	14.4828	11.1	40 (Edge-1)	70
25	80.1285	14.5555	11.1	30 (Edge-1)	70
26	80.1295	14.8633	11.0	20 (Edge-1)	110
27	80.1208	15.0516	11.1	1	70
28	80.1298	16.5333	11.2	0	0
29	80.0391	17.4273	8.4	1	70
30	80.0310	17.5203	5.1	70 (vessel out of Edge-2)	110
31	80.0308	17.5750	9.0	70 (vessel out of Edge-2)	110
32	80.0096	17.6745	10.9	5 (vessel out of Edge-2)	20
33	80.8333	17.7133	11.1	40 (vessel out of Edge-2)	90
34	80.0053	17.7190	8.7	1	40
35	80.0053	17.7461	10.4	1	40
36	80.0048	17.8013	5.1	5	40
37	80.0103	17.8558	1.8	60 (Edge-3)	70
38	80.0130	17.8851	3.2	60 (Edge-3)	70
39	80.0096	17.8920	1.3	40 (Edge-3)	70
40	80.0103	18.0640	3.1	40 (Edge-3)	70
41	80.0156	18.0800	1.4	40 (Edge-3)	70
42	80.0135	18.1215	0.5	40 (Edge-3)	70
43	80.0148	18.1323	1.3	50 (Edge-3)	54
44	80.0168	18.1531	3.4	40 (Edge-3)	50
45	80.0233	18.1648	1.5	30 (Edge-3)	50

Table 3. Voyage plan table from Kinnvika to Longyearbyen

Way-Point	Longitude E [°]	Latitude N [°]	Speed [knots]	Concentration [%]	Floe size [meters]
1	18.2007	80.0335	6.0	2	50
2	18.1229	80.0130	11.3	4	33
3	18.0283	80.0093	11.3	4	33
4	17.8818	80.0117	11.3	4	55
5	17.7750	80.0075	11.3	4	40
6	16.4837	80.1311	11.3	1	33
7	15.4168	80.1260	11.3	0	0
8	15.1554	80.0843	11.3	25 (Edge-4)	33
9	15.0587	80.0910	10.4	60 (Edge-4)	90
10	15.0139	80.0941	4.0	65 (Edge-4)	83
11	14.9870	80.0965	11.0	2	50
12	14.9215	80.1036	11.0	2	50
13	14.8945	80.1069	6.0	2	70
14	14.8624	80.1093	1.2	6	70
15	14.8329	80.1115	8.2	2	70
16	14.7701	80.1148	11.3	1	50
17	14.7392	80.1165	11.3	1	50
18	14.6918	80.1180	11.3	1	70
19	14.5586	80.1205	11.3	1	50
20	14.3460	80.1222	11.3	1	50
21	14.0571	80.1206	11.3	1	50
22	13.9682	80.1159	11.3	1	50

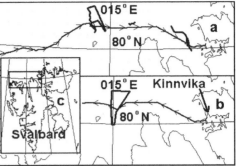

Figure 1. The vessel's routes north of Svalbard and observed ice floe edges: a – route towards Kinnvika, b – route towards Longyearbyen, c – area covered by the research. Symbols: —▶— route towards Kinnvika, ◀— route towards Longyearbyen, ——— edge of ice pack.

5 DISCUSSION OF RESULTS

5.1 Verification of information from remote sensing using visual observations

5.1.1 NIS maps on the concentration of ice in raster JPG format with geographic coordinates plotted

Map named c_map3.jpg was assessed. The exact time of the NIS map was not specified. Therefore the date indicated on the map equivalent sarmap2.jpg - 2009.08.10 08:47 UTC was adopted. The ice edge (CT = 40-70%) shown on the NIS map is observed from the vessel offset of ice edge (Edge-1) of 4.0 Nm in the direction 000°. The map did not demonstrate the existence of an ice floe wedges with a width of less than 1.0 Nm. The ice edge (CT = 40-70%) shown on the NIS map is observed from the vessel offset of ice edge (Edge-2) of 4.4 Nm in the direction 046°. The ice edge (CT = 40-70%) shown on the NIS map is observed from the vessel offset of ice edge (Edge-3) of 2.6 Nm in the direction 113°. The map did not demonstrate the existence of an ice floe wedges with a width of less than 0.3 Nm. It seems that NIS map is offset, in relation to all three observed from the vessel, of ice floes edges of higher concentrations in different directions. Average offset distance was 3.7Nm. The time difference between successive ice edges abeam of the vessel was 1.4 hours and 2.3 hours. They showed an increasing delay in relation to the publication of the NIS map by 3.2 hours, 4.6 hours and 6.9 hours.

Map named c_map3 dated 2009.08.14 14 06:30 UTC used for comparison with the edge of the ice floes observed from the vessel (Edge-4) was dated 2009.08.14 22:30 UTC. The ice edge (CT = 40-70%) and field of ice floe of concentration CT = 10-40% that is shown on the NIS map were offsets from the vessel observed ice edge of the concentration of 25-60% (Edge-4) of 2.4 Nm in the direction 042°. It seemed that the scale of the ice concentration of the NIS map depicted accurately the spatial distribution of ice observed from the ship. The undetected field of ice floe of higher concentration with a width of 1.5 Nm are presented in Figure 2. It was included inside the field with a concentration of CT = 10-40%. It was assumed that NIS map in an accurate way depicted the state of the ice cover at the time of reference for the source data.

5.1.2 *Vector NIC map on the ice concentration, ice age and floe size in PDF format*

These maps were published at intervals of 14 days. The closest map corresponding to the observation date was ice edge map dated 2009.08.17. For this reason, only Edge-4 referenced to the date 2009.08.14 22:30 UTC was analysed. The time difference was 49.5 hours (equal to 2.06 days). Map of 2009.08.03 (mean time of source data dating 2009.08.04 0:00 UTC) showed a general ice drift in the direction of 253 ° with an average distance of 20 Nm in seven days.

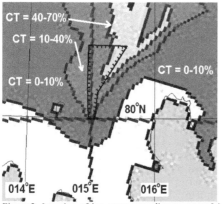

Figure 2. Location of ice cover according to c-map3.jpg map dated 2009.08.14 06:30 UTC and observations from ship

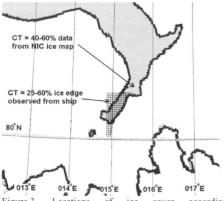

Figure 3. Locations of ice cover according to barnw090817color.PDF map and observations from ship after approximate adjustments of time difference and drift of ice cover.

Map of 2009.08.17 (average time of data source dating 0:00 UTC) showed a general ice drift in the direction of 245 ° at an average speed of 35 Nautical miles in seven days. This means drift ice equal to 10 Nautical miles of the time difference between visual observations and the dating of the ice edge map source data. After moving, the ice edge position on the map of a concentration of 40-60% for estimated resultant drift ice noted the consistency of data in comparison with the observed Edge-4 (Figure 3).

It was assumed that the NIC map accurately reflects the state of the ice cover for reference time of the data source. It was assumed that remote sensing method did not detect the ice pack field with a width of less than 1.5 Nm. When only the information about the ice drift from the map dated 2009.08.17 was considered, the position of the edge of the ice floe would be shifted 16 Nm in the direction of 130 ° with respect to the observed ice floe edge from the vessel.

5.1.3 *AARI maps of concentration, age and ice forms in vector Shapefile format*

The aari_bar_20090811_pl_a (SHP, SHX, DBF, PRJ) file was used for comparison with the observed from the vessel ice floe field edges dated respectively Edge-1 (2009.08.10 12:00 UTC), Edge-2 (2009.08.10 13:25 UTC) and Edge-3 (2009.08.10 15 45 UTC). It was assumed that the map represented ice conditions dated 2009.08.10 12:00 UTC because it reflected the average ice cover for the last three days. Another assumption was made that the time difference equal to 0-3.75 hours is negligible. Edge-1 field had no equivalent on the AARI map. The closest edge of the ice floe field of concentration CT = 13% on AARI map, that was coincident with shape of the Edge-1, was located at a distance of 6 Nm in the direction of 350°. The closest edge of the ice floe field of concentration CT = 78% on AARI map resembling the shape of the Edge-2 was found at a distance of 4 Nm in the direction of 014°. The Edge-3 field generally corresponded to the position of ice floe field of concentration CT = 13% situated at a distance of 1-5 Nm in the direction of 090°. It was assumed that there was moderate harmony of data.

The aari_bar_20090818_pl_a (SHP, SHX, DBF, PRJ) file was used for comparison with the observed from the vessel edge of ice floe field Edge-4. It was assumed that the map represented ice conditions dated 2009.08.17 12:00 UTC because it reflected the average ice cover for the last three days. The time difference was 2.56 days. The closest edge of the ice floe field (CT = 91%) was offset, in relation to the edge of the ice edge of concentration 25-60% observed from the vessel (Edge-4) at a distance 27 Nm in the direction of 312 °. It was assumed, therefore, that there was no relationship between these data. Mean offset position with respect to the edge of the ice floe that was observed from the vessel was 10 Nautical miles.

5.1.4 The IUP maps of ice floe concentration in raster GeoTIFF format

The asi-n3125-20090810.TIF file was used for comparison with the observed from the vessel ice floe field edges dated respectively Edge-1 (2009.08.10 12:00 UTC), Edge-2 (2009.08.10 13:25 UTC) and Edge-3 (2009.08.10 15 45 UTC. It was noted that the Geo-TIFF map was offset from the edges of the ice fields and land masks 6 Nm in the direction of 120°. It was found out that, after taking this offset into account, the nearest edge of the field of concentration CT = 10% on GeoTIFF map was coincident with the shape of the Edge-1 located at a distance of 17 Nm in the direction of 350° and was coincident with the shape of the Edge-2 at a distance of 23 Nm in the direction of 346°. At the same time the edge of ice floe field of concentration CT = 20-40% consistent with the shape of the Edge-3 was lying at a distance of 4 Nm in the direction of 065°. Thus configuration of the ice fields edge on the IUP map was distinctive and consistent with the edges of the ice floe seen from the vessel but at relatively large distance.

The asi-n3125-20090815.TIF file was used for comparison with the observed from the vessel ice floe field Edge-4 dated 2009.08.15 10:43 UTC. It was noted that this GeoTIFF map was offset from the edges of the ice field and land mask 6 Nm in the direction of 120°. It was found out that, after taking this offset into account, the nearest edge of the field of concentration CT = 20% on GeoTIFF map was coincident with the shape of the Edge-4 located at a distance of 34 Nm in the direction of 013°. Configuration of the ice floe edge on IUP map is not very clear.

5.1.5 The NIC maps on the ice concentration in a MIZ scale in vector Shapefile format

The nic_miz2009222nc_pl_a.zip (SHP, SHX, DBF, PRJ) file dated 2009.08.10 was used for comparison with the observed from the vessel ice floe field edges dated respectively Edge-1 (2009.08.10 12:00 UTC), Edge-2 (2009.08.10 13:25 UTC) and Edge-3 (2009.08.10 15 45 UTC. The ice floe field Edge-1 was located like border CT = 18-81% presented by MIZ file. This limit was also the beginning of the route in the area covered by ice. The MIZ wedge of a high concentration of ice floe shown on NIC map was offset 4 Nm in the direction of 028° with respect to Edge-2. Field of Edge-3 was found to be entirely located in the area of CT = 18-81%concentration as shown by the MIZ map. It was assumed that data were in general harmony.

The nic_miz2009227nc_pl_a (SHP, SHX, DBF, PRJ) file dated 2009.08.15 was used for comparison with the observed from the vessel ice floe field

Edge-4 dated 2009.08.15 10:43 UTC. The nearest edge of ice (CT = 18-81%) on NIC map passed exactly through the area of the observed Edge-4field. Instead of the expected wedge with a higher concentration of ice extended to the South this map showed the general ice limit on NW-SE direction. This was consistent with the observed field of low ice floe concentration on the route from Kinnvika to the area of Edge-4 (see Table 3). The closest edge of ice floe field of CT > 81% concentration was offset from the observed edge of ice of 25-60% concentration (Edge-4) 20 Nm in the direction of 000°. It might be assumed that there was a general correlation between data. However, it was low in detail due to the different edges concentration presented on map nic_miz2009227nc_pl_a and the edge observed from the vessel.

5.1.6 The NCEP maps of ice concentration in gridded GRIB format in a resolution of 5 minutes of arc

The ice5min.200908.grb file for the date 2009.08.10 was used for comparison with the observed from the vessel ice floe field edges dated respectively Edge-1 (2009.08.10 12:00 UTC), Edge-2 (2009.08.10 13:25 UTC) and Edge-3 (2009.08.10 15 45 UTC. Field of ice floe Edge-1 was situated in the area of 40-50% concentration; Edge-2 in the area of 0-5% concentration, Edge-3 was located in the area of 0-40% shown on GRIB map. Data consistency is correct for Edge-1 and Edge-3. However, the lack of continuity presented by the GRIB map raised concern that the results of the comparison may be largely random. I was assumed that there was only very general harmony of data.

The ice5min.200908.grb file for the date 2009.08.15 was used for comparison with the observed from the vessel ice floe field Edge-4. The closest corresponding to the shape of the edge of ice floe concentration field CT = 25-35% was offset from the vessel observed 25-60% ice edge (Edge-4) 8 Nm in the direction of 013°. It might be assumed that there was a general consistency between data. However, it was low in detail due to the different concentration edges being on map ice5min.2009-08.grb and the edge observed from the vessel.

5.1.7 The IFREMER ice concentration maps in gridded NetCDF format without geographical coordinates

The 20090810.nc file was used for comparison with the observed from the vessel ice floe field edges dated respectively Edge-1 (2009.08.10 12:00 UTC), Edge-2 (2009.08.10 13:25 UTC) and Edge-3 (2009.08.10 15 45 UTC. Field of ice floe Edge-1 was in the area of 0-11% concentration, Edge-2 in the area of 11-22% concentration, Edge-3 was in the area of 11-44% visualised by NetCDF map. Thus it has been assumed that this grid map does not recognize the local significant changes in ice floe concentration.

The 20090815.nc file was used for comparison with the observed from the vessel ice floe field Edge-4 dated 2009.08.14 22:30. The closest NetCDF map area of the CT = 22-44% concentration corresponding to the shape of the ice floe edge observed from the vessel 25-60% (Edge-4) was offset 32 Nm in the direction of 354°. It might be assumed that there was a general consistency between data. However, it was low in detail due to different concentration edges being on map 20090815.nc and the edge observed from the vessel.

5.1.8 *The NIC maps of sea ice extent and sea ice limit in vector Shapefile format*
The masie_ice_r00_v01_2009222_4km.ZIP (SHP, SHX, DBF, PRJ) file dated 2009.08.10 was used for comparison with the observed from the vessel ice floe field edges dated respectively Edge-1 (2009.08.10 12:00 UTC), Edge-2 (2009.08.10 13:25 UTC) and Edge-3 (2009.08.10 15 45 UTC. The shape of MASIE line was not correlated with the observed from the vessel ice floe edges. The closest corresponding to the MASIE line of ice floe CT> 40% concentration is located with respect to the observed from the vessel 25-60% ice edge (Edge-1) at a distance of 25 Nm in the direction of 003°. It might be assumed that there is no correlation between the data.

The masie_ice_r00_v01_2009227_4km (SHP, SHX, DBF, PRJ) file dated 2009.08.15 was used for comparison with the observed from the vessel ice floe field Edge-4. The closest MASIE line of CT > 40% corresponding to the observed ice edge (Edge-4) of 25-60% concentration was offset 28 Nm in the direction of 354°. It might be assumed that there was a general consistency between data. However, it was low in detail due to the different concentrations of edge provided by the MASIE map and ice floe edge observed from the vessel.

5.2 *Verification of ice data presented by remote sensing sourcesNIS maps named c_map3*
The first the content of the NIC map named barnwYYMMDDcolor and NIS map named c_map3 was compared. The analysis included offset of ice edge position on the NIC map barnwYYMMDDcolor described in section 5.1.1. A 30-50% concentration boundary on the map barnwYYMMDDcolor was consistent with the edge of 40-70% concentration on c_map3 map. The concentration limits of 70-90% and 90-100% on the map c_map3 coincided with 50-70% edge on barnwYYMMDDcolor map and 90% edge on barnwYYMMDDcolor map where the edge of 50-70% was not shown on the map.

The edge of the concentration of 81% on the NIC map named nic_mizYYYYDDDnc_pl_a reflected approximately 40% concentration edge on the NIS map named c_map3. The limit concentration of 18% on the map nic_mizYYYYDDDnc_pl_a reflected approximately 0-10% concentration edge on the map c_map3.

The edge of the ice concentration of 90 (91) % on the AARI map named aari_bar_20090818_pl_a clearly reflected the edge of the concentration of 90% on NIS map named c_map3. Fields with a concentration of 13-78% on the aari_bar_200908-18_pl_a map reflected 10-40% and 40-90% concentration field on the map c_map3. Field of concentration of 40-70% on the map c_map3 was lying inside the field of concentration of 46-78% on the map aari_bar_20090818_pl_a. Fields with a concentration of 78-90% on the map aari_bar_20090818_pl_a reflected field of concentration of 40-70% on the map c_map3. The edges of the ice concentration of 0% were consistent on both maps. The edge of the ice concentration of 40% on the NIC map named masie_ice_r00_v01_2009227_4km mostly reflected the edge of the concentration of 40% on NIS map named c_map3.

5.2.1 *NIC maps named barnwYYMMDDcolor*
The analysis included offset of ice edge position on the NIC map named barnwYYMMDDcolor described in section 5.1..2. The edge of the ice concentration of 81% on the NIC map named nic_mizYYYYDDDnc_pl_a was consistent with the edge of 50-70% and 90% on the map barnwYYMMDDcolor where the edge of the concentration of 50-70% was not shown on the map barnwYYMMDDcolor. The edge of the concentration of 18% on the map nic_mizYYYYDDDnc_pl_a coincided with the edge of the concentration of 50-70% on the map barnwYYMMDDcolor. In this case, the edge of 50-70% was the limit of data related to sea ice.

Fields with a concentration of 46% and 78% from the lower limit of the field edge and concentration of 90% as the upper limit on aari_bar_20090818_pl_a maps were consistent with the edge of concentration of 90% on the map barnwYYMMDDcolor. The edge of the ice coverage (concentration of 0%) on the map barnwYYMMDDcolor north of the Nordaustlandet Island was the same as on the map aari_bar_YYYYMMDD_pl_a but the edge of 0% on the North and Northwest of Svalbard on the map barnwYYMMDDcolor was located more to the South and contained more details than aari_bar_YYYYMMDD_pl_a map.

The edge of ice concentration of 40% on the NIC map named masie_ice_r00_v01-_2009227_4km roughly coincided with the edge of 30-50% on the NIC map named barnwYYMMDDcolor. Just like in

188

the case of AARImaps, the edge of the ice concentration of 30-50% on the map barnwYYMMDDcolor was located more to the South than on the map masie_ice_r00_v01_2009227_4km. It also contained more details than MASIE map.

5.2.2 *IFREMER maps named YYYYMMDD*

The fields with the specified concentration on the IFREMER map represented fairly the ice concentration field on the NIS map named c_map3. The limitation was the relative large size of the grid (cell) on IFREMER map. Similar difficulties for comparison were due to lack of information in the fields appearing on the IFREMER map.

The edge of the ice concentration of 81% on the NIC map named nic_mizYYYYDDDnc_pl_a approximately reflected the distribution of concentration fields of 44-78% on the IFREMER maps named YYYYMMDD.nc. However, the edge of the ice on the NIC maps ran parallel through the fields with smaller values of concentration of IFREMER maps.

The edge of the concentration of 78 (90)% on the AARI maps named aari_bar_YYYYMMDD_pl_a reflected 78% concentration field on the IFREMER map named YYYYMMDD.nc. The edge of the concentration of 13% on the AARI maps reflected fields of the concentration of 33% on the IFREMER map.

The edge of the ice concentration of 40% on the NIC map named masie_ice_r00_v01_2009227_4km runs through or near the fields of concentration of 44-56% on the IFREMER maps. The spatial distribution of these edges was very consistent.

5.2.3 *NCEP maps named ice5min.YYYYMM*

The edge of the concentration of 81% on the NIC maps named nic_mizYYYYDDDnc_pl_a generally ran along the edge of the concentration of 40% on the NCEP map named ice5min.YYYYMM. There were discrepancies. The edge of the ice on the nic_mizYYYYDDDnc_pl_a map ran through lower concentration values on the NCEP map. The edge of the concentration of 18% on the nic_mizYYYYDDDnc_pl_a maps generally ran along the edge of the concentration of 0-20% on the ice5min.YYYYMM map but there was derogation. The fields on both maps were consistent.

The edge of the concentration of 90% on the AARI map named aari_bar_YYYYMMDD_pl_a mostly ran through the fields of concentration of 70-80% on the map ice5min.YYYYMM but there was derogation. The edge of the ice concentration below 13% on the AARI map edge reflected approximately edge of 20% concentration on the NCEP map. The edges of the concentration of 46% and 78% on the AARI map reflected data on the NCEP map only in a very general way.

The edge of 40% concentration on the NIC map named masie_ice_r00_v01_2009227_4km mostly ran through the fields of concentration of 40% on the ice5min.YYYYMM map. There was derogation. In such cases, the NCEP map showed mostly lower concentration values than MASIE map.

Fields with 85% concentration on the NCEP map named ice5min.YYYYMM quite accurately reflected edge of 90% concentration on NIS map named c_map3. The edge of the concentration 70% on the NCEP map was less coincident with the field of 50-70% concentration on NIS map. The limit of ice 0% concentration on the NCEP map corresponded roughly the edge of concentration of 0-10% on NIS map.

5.2.4 *IUP maps named asi-n3125-YYYYMMDD*

The edge of 40% concentration on NIS map named c_map3 reflected roughly the fields of concentration 20% or 70-80% on the IUP map. It was noted that the edge of 0% concentration on the IUP map corresponded to the edge of the 40% concentration on the NIS map. The higher the concentration, the smaller is discrepancy of position for each ice floe concentration.

The edge of 81% concentration on the NIC map named nic_mizYYYYDDDnc_pl_a accurately reflected edge of 0% concentration on the IUP map named asi-n3125-YYYYMMDD.

The edge of 90 (91) % concentration on the AARI map named aari_bar_YYYYMMDD_pl_a accurately reflected fields of 80-90% concentration on the IUP map. The edge of 78% concentration on the AARI map accurately reflected 10-20% concentration field on the IUP map.

The edge of 13-46% concentration on the AARI map took place in areas of 0% concentration on the IUP map. The edge of 40% concentration on the NIC map named masie_ice_r00_v01_2009227_4km mostly reflected edge of concentration of 0-10% on the IUP map.

6 EVALUATION OF USEFULNESS OF REMOTE SENSING DATA FOR ROUTEING PURPOSES

NIS maps named c_map.jpg accurately reflected the edges of the ice floes field observed from the ship at the moment of reference. Mean offset position was 3.7 Nm. The ice floe patches of a width of less than 1.0 Nm were not detected by remote sensing. Also NIC map named barnwYYMMDDcolor accurately reflected condition of the ice cover observed from the ship at the moment of reference. It was assumed that wedge of ice field of a width of less than 1.5 Nm was not detected by remote sensing. Based on observations made from the vessel, the two maps seemed to be appropriate for voyage planning and routeing of the vessel in ice for every ice class vessel (see section 2). It is also possible to optimize the routeing using these maps in accordance with the criteria specified by Kjerstad (2011), Arikajnen and Tsubakov (1987) and CCG (1992). It is routeing of the vessel along the lightest ice conditions.

NIC maps named nic_mizYYYYMMDDnc_pl_a show general compliance with the shapes of the edge of the ice fields observed on the vessel. Mean offset position for the characteristic shapes of the edge of the ice fields was 6.5 Nm. AARI maps named aari_bar_YYYYMMDD_pl_a seemed to be highly generalized and did not reflected the state of the ice cover observed from the vessel. Mean offset position for the characteristic shapes of the edge of the ice fields was 5.8 Nm. It was assumed that ice field of a width less than 5 Nm remained undetected by remote sensing. Both of these maps showed a consistent location of MIZ lower limit of 0-18% and MIZ upper limit of 70-90% visualised on the maps c_map3 and NIC NIS barnwYYMMDDcolor. Despite greater number of concentration levels provided by the AARI map, AARI seemed to be less precise in detail than the NIC MIZ map. Due to the limitation of the precision of the scale of concentration or precision of position, the above maps seemed to be useful for the preliminary voyage planning and routeing of the vessel. They were not useful for optimizing the routeing of the vessel in accordance with the criteria specified by Kjerstad (2011), Arikajnen and Tsubakov (1987) and CCG (1992).

NCEP maps named ice5min.YYYYMM showed general consistency between the data. It was assumed that the average position offset of ice fields shapes corresponded to the dimensions of the grid of 9.4 Nm. However, lack of continuity (consistency) of the data on the NCEP map was noted. This raised concerns that the concentration field visualized by the map may not accurately reflect the actual ice conditions in a particular place and thus lead to an incorrect assessment of navigational situation or prevent proper determination the routeing of the vessel.

NIC maps named masie_ice_r00_v01_2009-222_4km were related to ice floe concentration of 40%. Average offset of ice field shapes on MASIE map in relation to ice fields observed from the vessel was 20.6 Nm. IFREMER maps named YYYYMMDD showed a slight similarity with the observations of ice cover made from the vessel, with NIC maps barnwYYMMDDcolor and NIS c_map3 maps. However, the shape of the edge of the ice floe field of concentration above 11% on IFREMER map most closely corresponded to the shape of ice edge on the IUP map. It was assumed that the average position offset of the characteristic shapes of ice edge fields was as much as a side of the grid (cell) of IFREMER map equal to 6.7 Nm. IUP maps named asi-n3125-YYYYMMDD showed average position offset of the ice field shapes equal to 19.5 Nm. Therefore it was assumed that the fields of ice floe on IUP map did not reflected directly the ice edge observed from the vessel. The edge of 0-10% concentration on the IUP map corresponded to the edge of concentration of 40% on c_map3 NIS map, to the edge of concentration of 30-50% on NIC barnwYYMMDDcolor map, to the edge of concentration of 40% on NIC MASIE map, to the edge of concentration of 81% on the NIC nic_mizYYYYDDDnc_pl_a map, to the edge of concentration of 13-46% on the AARI aari_bar_YYYYMMDD_pl_a map and to edge of concentration of 11-44% on the IFREMER map. It was assumed that the NIC MASIE maps and ice limits on the IFREMER and IUP maps reflect the concentration limits of the ice floe of 30-40% due to the weather filters applied. Full scale of ice concentration on both of the above mentioned maps may be misleading. All three maps cannot be used for routeing of the lowest ice class vessels in the ice or in the vicinity of ice. However, they appear to be useful for vessels of the lowest classes (see Table 1) as they indicate the limits of the region with average 30-40% concentration of ice floe. These vessels can navigate in this area with icebreaker assistance. The use of IFREMER and IUP maps for vessels with higher ice classes routing does not seem to be appropriate because they do not reflect the actual concentration of ice floe.

7 CONCLUSIONS

The results of analyses of different maps obtained by remote sensing methods showed that the information presented is not the same. An example of the edges and limits of the concentration of ice floe that are close to concentration of 40% for each map is shown in Figure 4. Discrepancies among the analysed data on the maps certainly arise from error of the measurement method. According to Rodrigues (2009) based on Comiso (1999) the error may be up to 5-10% and even 15% of the concentration of ice floe. This error depends on the remote sensing measurement method used. The error of the measurement method should be taken into account when planning the route of the vessel in ice-covered areas. It should also be borne in mind that the more generalized information about the state of the ice cover, the less likely becomes detection of ice floe patches of high concentration and theirs spatial extent. It should also be taken into account that the vessel may be trapped in the ice floe of a high concentration while overcoming larger ice floe field. Then icebreaker assistance will be required. In both these cases, the vessel will be able to sail at very low speed and thus lose a lot of time to go through the ice and the travel time becomes significantly longer. Thus, the current schedule and the next expedition schedule will be disrupted.

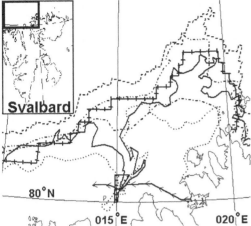

Figure 4. The spatial distribution of edges or borders close to 40% ice concentration for individual maps for the day 15 August 2009: ———— edge of 40-70% concentration on the NIS map, • • • edge of 30-50% concentration on the NIC (EGG) map, — • — edge of 81% concentration on the NIC (MIZ) map, — • • — edge of 46% or 91% concentration on the AARI map, — — — edge of 40% concentration on the NIC (MASIE), +++ edge of concentration 40% on the NCEP map, —◄— route of the vessel, ⊥⊥⊥ edge of concentration 25-65% of ice floe field observed from the vessel.

Now the question should be answered which maps may be used to assess the ice conditions from the navigation point of view. The MASIE, IFREMER and UIP maps seem to be useless for vessels with low ice class, as they relate to concentration of 30-40%. Such concentration edges might be useful for average ice class ships. Low ice class vessels are looking for concentration limit of 15%. The NCEP maps also appear to be useless for low ice class vessels, because they may mislead the user during the preliminary voyage planning and scheduling the next expedition. The NIC (MIZ) and AARI maps seem to meet the needs of preliminary voyage planning and scheduling the next expedition of vessels with low ice classes as these maps depict the lower limits of 13-18% concentration. NIS maps named c_map3 and NIC maps named barnwYYMMDDcolor have a satisfactory scale concentration and precision to present the edge of each concentration. They allow planning the routeing of the vessel and avoiding areas with higher concentration of ice floe. When analysing the ice floe concentration on a map for planning the route and schedule of the vessel, the errors of remote sensing methods used to estimate concentration of ice floe should always be taken into consideration.

Finally, we must answer the question whether the above mentioned vessel of L2 (L1) class could determine more favourable route at an initial stage of routeing. The answer is positive. At the initial stage the crew could follow data on reliable daily NIS maps. The NIC barnwYYMMDDcolor maps were not available for required day. The NIC (MIZ) maps could be used but their content does not advise the vessel about higher concentration of ice patches. Thus the vessel probably would follow standard route to encounter ice floe patches of high concentration and make route around the ice field.

ACRONYMS

AARI - Arctic and Antarctic Research Institute in St.Petersburg
ALOS - Advanced Land Observing Satellite
AMSR-E - Advanced Microwave Scanning Radiometer - Earth Observing System
AMSR2 - Advanced Microwave Scanning Radiometer 2
AMSU - Advanced Microwave Sounding Unit
ASA - Advanced Synthetic Aperture Radar
ASI - ARTIST Sea Ice
AVHRR - VIS Advanced Very High Resolution Radiometer
AVHRR-VIS - Advanced Very High Resolution Radiometer - Visible Band
CERSAT - French ERS Processing and Archiving Facility
CDOP - Continuous Development and Operations Phase by Meteo-France
CIS - Canadian Ice Service
CT - Concentration Total
DMI - Danish Meteorological Institut
DMSP - Defense Meteorological Satellite Program
DMSP OLS - Defense Meteorological Satellite Progam – Operational Linescan System
ENVISAT - ENVIronmental SATellite
EUMETSAT - EUropean Organisation for the Exploitation of METeorological SATellites
GDSIDB - Global Digital Sea Ice Data Bank
GMM - Geometrical Mathematical Model
GOES - Geostationary Operational Environmental Satellite
IFREMER - Institut Français de Recherche pour l'exploitation de la Mer
IMS - Interactive Multisensor Snow and Ice Mapping System
IUP - Institut für Umweltphysik, Universität Bremen
MASIE - Multisensor Analyzed Sea Ice Extent
Met.no - Norwegian Meteorological Institute
MIZ - Mariginal Ice Zone
MMAB - Marine Modeling and Analysis Branch
MODIS - Moderate Resolution Imaging Spectroradiometer
MTSAT - Multi-functional Transport Satellite
NAVO - Naval Oceanographic Office

NCEP - National Centers for Environmental Prediction
NESDIS - National Environmental Satellite, Data, and Information Service
NGDC- National Geophysical Data Center
NIC - National Ice Center, US National Ice Service, US Naval Ice Service
NIS - Norwegian Ice Services
NOAA - National Oceanic and Atmospheric Administration
NOMADS - National Operational Model Archive & Distribution System
NSIDC - National Snow and Ice Data Center
NWP - Numerical Weather Prediction
NWS - National Weather Service
OSISAF - Ocean And Sea Ice Satellite Application Facility, High Latitude Centre
PALSAR - Phased Array type L-band Synthetic Aperture Radar
QUIKSSCAT- "quick recovery" mission from the NASA Scatterometer
RADARSAT - RADAR SATellite system equipped with a powerful synthetic aperture radar
RGB - additive Red, Green, Blue color model
SAF- Satellite Application Facilities
SAR - Synthetic Aperture Radar
SEVIRI - Spinning Enhanced Visible and Infrared Imager

SMHI - Swedish Meteorological and Hydrological Institute
SSM/I - Special Sensor Microwave Imager
UKHO - United Kingdom Hydrographic Office
URL - Uniform Resource Locator

REFERENCES

Arikajnen A.I., 1990. Sudokhodstvo vo l'dakh Arktiki. Moskva "Transport": 247 p.

ArikajnenA.I., TsubakovK.N., 1987. Azbuka ledovogo plavanija. Transport, Moskva: 224 p.

Erzaty R., Girard-Ardhuin F., Croize-Fillon D., 2007. Sea ice drift in the central arctic using the 89 GHz brightness temperatures of the advanced microvawe scanning radiometer, Users manual. Laboratoire d'Océanographie Spatiale Département d'Océanographie Physique et Spatiale, IFREMER: 20 p.

Timco G.W., Gorman B., Falkingham J., O'Connell B., 2005. Scoping Study: Ice Infor-mation Requirements for Marine Transportation of Natural Gas from the High Arctic. Technical Report CHC-TR-029, Canadian Hydraulics Centre, Ottawa: 124 p.

Rodrigues J., 2009, The increase in the length of the ice-free season in the Arctic. Cold Regions Science and Technology 59 (2009): 78-101.

Comiso, J.,1999. Bootstrap Sea Ice Concentrations from Nimbus-7 SMMR and DMSP SSM/I. National Snow and Ice Data Center, Boulder, CO. Digital Media (updated 2005).

Anti-Collision and Collision Avoidance

Apprisal of the Coordinability of the Vessels for Collision Avoidance Maneuvers by Course Alteration

A. Volkov, E. Pyatakov & A. Yakushev
Odessa National Maritime Academy, Odessa, Ukraine

ABSTRACT: Manoeuvres of two interacting vessels should be coordinated or adjusted for collision avoidance, i.e. increase in distance to the Closest Point of Approach (CPA). The article dwells on coordinability of collision avoidance actions by the give-way vessel whilst the stand-on vessel is keeping her motion variables as well as simultaneous course alteration manoeuvre by both vessels. The rate of CPA is selected as evaluation of the vessels' coordinability for collision avoidance. The problem examined in this study is choosing the effective system structure for coordination of interaction of two vessels dangerously approaching each other. The article offers the assessment of the COLREGs-72 regarding coordinability of the vessel manoeuvres and de-scribes the results from the synthesis of the binary coordination system, alternative to COLREGs-72, in which the efficiency of joint coordinated manoeuvres exceeds the COLREGs-72 system.

1 INTRODUCTION

We shall introduce a few notions for mathematical formulation of collision risk assessment for two vessels and for choosing coordinated collision avoidance manoeuvres. These notions have been used previously in research papers [5, 6].

We shall assume that each moving vessel is connected with two-dimensional area $\bar{S}_{nd}$ known as predicted area of danger (PAD), where presence of other moving vessels is unadvisable. PAD shall be arranged in such a way, so that its boundary conformed to zero collision probability. Each point in this region corresponds to a certain nonzero collision probability, and increase in collision probability occurs along with decrease in distance between the vessels. PAD corresponds to a subset of dangerous situations S_ω.

The notion of situational disturbance is introduced to describe collision risk ω_{ij}. Situational disturbance ω_{ij} occurs in case of anticipated entry of vessels into the PAD (the first vessel in the second vessel's region or vice versa – the second vessel in the first vessel's PAD). Situational disturbance exposes impending dangerous position in good time in view of anticipated change in relative position of two vessels based on their motion variables – speed and course.

Intensity of situational disturbance $\tilde{\omega}_{ij}$ is proportional to the vessels' collision probability

connected with the PAD. Applying a certain rule, intensity of situational disturbance $\tilde{\omega}_{ij}$ may be ranged in whole-number values from 0 (no situational disturbance) to a certain maximum value.

The second parameter of situational disturbance is time allowance for each of two interacting vessels, which is equal to the time interval of the vessel's entry into the PAD, i.e. time interval, during which the vessel can change her course to avoid entry into the PAD. This time parameter is more informative than time for reaching the closest point of approach (CPA) since it characterizes possibility to avoid dangerous approach situation.

In case of dangerous approach of vessels and situational disturbance ω_{ij} interaction Bz occurs, calling forth definition of targeted and coordinated strategies for each interacting vessel.

Thus, Bz can be formulated as follows:

$$D = Bz(\omega_{ij}, F)$$

where

D – strategy vector for the vessels' collision avoidance manoeuvres, the components of which are set of strategies of the interacting vessels;

F – essential parameters vector, relating to a specific situation and necessary for targeted and coordinated strategies of collision avoidance manoeuvres.

So, Bz characterizes interrelation between a set of situational parameters and a set of strategies of the vessels' collision avoidance manoeuvres. In other words, collision avoidance manoeuvres are a process of situational disturbance compensation, i.e. conversion of a subset of dangerous situations S_o into a subset of safe situations S_S, based on interaction mechanism Bz, whereas collision avoidance strategy D is an algorithm for performing collision avoidance manoeuvres.

To ensure coordinated collision avoidance manoeuvres of two vessels, both of them must follow the same procedure to expose occurrence of situational disturbance ω_{ij}.

The main binary interaction parameter Bz, coordinating collision avoidance strategies D_1 and D_2 of two vessels, is coordinability, i.e. consistency of strategies, which is denoted by $Coor(D_1, D_2)$

Strategies can be coordinated, if simultaneous application thereof leads to increase in distance to the CPA (D_{min}), in which case $Coor(D_1, D_2) = 1$. If D_{min} decreases as a result of application of strategies D_1 and D_2, then $Coor(D_1, D_2) = -1$.

Thus, the problem solution includes:
- mathematical formulation of the PAD S_{nd};
- definition of relationship between changing the distance D_{min} and alteration of the vessel's course;
- assessment of coordinability of manoeuvres of the vessels both avoiding collision by course alteration;
- suggesting an alternative system for coordination of manoeuvres of two vessels by course alteration.

2 MATHEMATICAL FORMULATION OF THE PAD

Let us analyze two figures of the PAD: circle and ellipse, where dimensions of the region S_{nd} depend on deterministic $\tilde{S}_{nd}$ and stochastic S_{nd} components, i.e. $\bar{S}_{nd} = \tilde{S}_{nd} \cup S_{nd}$. Deterministic component $\tilde{S}_{nd}$ allows for the vessel size, her stopping characteristics, and includes force-majeure distance allowance. Stochastic component S_{nd} is determined through precise obstacle position finding relative to the vessel's position.

The research paper [1] suggests that in order to define stochastic component of the PAD S_{nd}, it is necessary to determine the area marginal dimensions, where the obstacle is located within the area with the given probability P_d close to 1.

So, dimensions of the stochastic component of the given PAD S_{nd} are determined by solving an equation:

$$\iint\limits_{Snd} f(x,y) dydx = P_d, \quad (1)$$

where $f(x, y)$ - is a two-dimensional density of distribution of vector position error.

If position error is governed by normal law of distribution, and the PAD is circle-shaped, then the equation (1) becomes:

$$\frac{4}{2\pi\sigma_x\sigma_y} \int\limits_0^R \int\limits_0^{\sqrt{R^2-x^2}} \exp[-(\frac{x^2}{\sigma_x^2} + \frac{y^2}{\sigma_y^2})] dydx = P_d,$$

or for numerical integration, denoting $A = 4/2\pi\sigma_x\sigma_y$:

$$A \sum_{i=0}^R \sum_{j=0}^{\sqrt{R^2-i^2}} \exp[-(\frac{i^2}{\sigma_x^2} + \frac{j^2}{\sigma_y^2})] = P_d. \quad (2)$$

In the equation (2) it is necessary to determine a value R, for which purpose successive approximation method is applied.

If the PAD S_{nd} is ellipse-shaped, then its boundary equation becomes:

$$\frac{x^2}{a^2} + \frac{y^2}{b^2} = 1,$$

where a and b – ellipse semi-axes of the area S_{nd}, and ratio $\gamma = \frac{a}{b}$ is given.

Dimensions of the area S_{nd} are determined by the equation:

$$\frac{1}{2\pi\sigma_x\sigma_y} \iint\limits_{Snd} \exp[-(\frac{x^2}{\sigma_x^2} + \frac{y^2}{\sigma_y^2})] dxdy = P_d,$$

or allowing for its elliptical shape

$$\int\limits_0^a \int\limits_0^{\frac{1}{\gamma}\sqrt{a^2-x^2}} \exp[-(\frac{x^2}{\sigma_x^2} + \frac{y^2}{\sigma_y^2})] dydx = \frac{P_d\pi\sigma_x\sigma_y}{2}.$$

This integral can be computed using Simpson method.

Thus, the circle-shaped area PAD is characterized by the parameter R, and the ellipse-shaped area – by parameters a and b.

Value of the minimum permissible distance minD for the circle-shaped PAD is a sum of radius ^{dop}R of the stochastic and $\tilde{R}$ deterministic components, i.e. $\bar{R} = R + \tilde{R}$. Thus, situational disturbance occurs, when $minD < \bar{R}$.

In case of ellipse-shaped PAD, minimum permissible distance has two values $\min D_{dop1}$ and $\min D_{dop2}$:

$$\min D_{dop1} = \sqrt{b^2 + x_1^2(1 - \frac{b^2}{a^2})},$$

$$\min D_{dop2} = \sqrt{b^2 + x_2^2(1 - \frac{b^2}{a^2})},$$

where

$$x_1 = -\frac{a^2cb}{a^2+c^2r^2} + \sqrt{(\frac{a^2cb}{a^2+c^2r^2})^2 - \frac{a^2c^2(b^2-r^2)}{(a^2+c^2r^2)}},$$

$$x_2 = -\frac{a^2cb}{a^2+c^2r^2} - \sqrt{(\frac{a^2cb}{a^2+c^2r^2})^2 - \frac{a^2c^2(b^2-r^2)}{(a^2+c^2r^2)}},$$

$$c = \frac{a^2}{b(\overline{Y}_0 \sin K - \overline{X}_0 \cos K)}$$

and

$$r = (\overline{Y}_0 \cos K + \overline{X}_0 \sin K),$$

$$\overline{X}_0 = D\sin\alpha, \quad \overline{Y}_0 = D\cos\alpha,$$

D and α - distance and target bearing, K - target course.

In this case situational disturbance occurs, when inequality $\min D < \min\{\min D_{dop1}, \min D_{dop2}\}$ is valid.

3 RELATIONSHIP BETWEEN CHANGING THE DISTANCE TO THE CPA AND ALTERATION OF THE VESSEL'S COURSE

We shall consider a situation when the vessel changes her course to avoid collision, the target keeps her course and speed, and shall analyze patterns in changes of D_m, depending on the vessel's course K_1. In other words, it is necessary to determine analytic expression of the first-order derivative of the distance to the CPA with respect to the vessel's course $\frac{\partial D_m}{\partial K_1}$, and analyze it changing K_1.

To that effect, we shall first determine an expression of the distance to the CPA D_m, see the figure below.

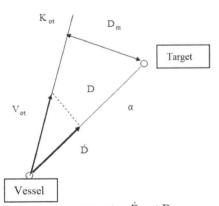

Figure 1. For determining values $\dot{D}$ and D_m

According to the figure above, the value D_m is determined by the expression:

$$D_m = D|\sin(\alpha - K_{ot})|, \tag{3}$$

where
D and α - distance and target bearing;
K_{ot} - relative course, which is a course function K_1.

Dangerous approach situation occurs only in case of shortening the current distance, whenever $\dot{D} < 0$. From this point on, we will need an expression for current distance rate $\dot{D}$, which can also be derived from the figure above. In view of the fact that in relative motion the target remains static, then:

$$\dot{D} = -V_{ot}\cos(\alpha - K_{ot}) \tag{4}$$

If we differentiate the expression (3), we shall obtain:

$$\frac{\partial D_m}{\partial K_1} = -D\cos(\alpha - K_{ot})\frac{\partial K_{ot}}{\partial K_1},$$

or

$$\frac{\partial D_m}{\partial K_1} = -\frac{V_{ot}}{V_{ot}}D\cos(\alpha - K_{ot})\frac{\partial K_{ot}}{\partial K_1}.$$

In view of the expression (4) the last equation becomes:

$$\frac{\partial D_m}{\partial K_1} = -\frac{D}{V_{ot}}\dot{D}\frac{\partial K_{ot}}{\partial K_1}.$$

The derivative sign $\frac{\partial D_m}{\partial K_1}$ also depends on the ratio of values α and K_{ot}, denoted by q. The value sign q is determined by $\sin(K_{ot} - \alpha)$, i.e. $q = \text{sign}[\sin(K_{ot} - \alpha)]$.

When the vessels approach one another, $\dot{D} < 0$, so:

$$\text{sign}(\frac{\partial D_m}{\partial K_1}) = q\,\text{sign}(\frac{\partial K_{ot}}{\partial K_1}). \tag{5}$$

We shall analyze the derivative $\frac{\partial K_{ot}}{\partial K_1}$ from the research paper [2]:

$$\frac{\partial K_{ot}}{\partial K_1} = \frac{V_1[V_1 - V_2\cos(K_1 - K_2)]}{V_{ot}^2}.$$

Note that the sign of the expression $\frac{\partial K_{ot}}{\partial K_1}$ matches the sign of the expression $V_1 - V_2\cos(K_1 - K_2)$.

Analysis of the expression for the first-order derivative $\frac{\partial K_{ot}}{\partial K_1}$ shows that, whenever $V_1 > V_2$, the first-order derivative is positive for all values K_1. If $V_1 < V_2$, the first-order derivative can be both positive and negative.

That is to say that in the first case $(V_1 > V_2)$ relative course K_{ot} may take on any values from 0 to 2π. And when $V_1 < V_2$, the area of values K_{ot} is limited by extreme values, denoted by $K_{ot\,min}$ and $K_{ot\,max}$. These relative courses correspond to the vessel's courses $K_{1\,min}$ and $K_{1\,max}$, obtained in the research paper [3]:

$$K_{1\,min} = K_2 + arccos\rho$$

and

$$K_{1\,max} = K_2 - arccos\rho,$$

where $\rho = V_1 / V_2$.

Extreme values of the relative courses $K_{ot\,min}$ and $K_{ot\,max}$ in the same paper are:

$$K_{ot\,min} = \pi + K_2 - arcsin\rho,$$
$$K_{ot\,max} = \pi + K_2 + arcsin\rho.$$

Thus, whenever $\rho < 1$ for all the vessel's true courses K_1 within the range of 0 to 2π, the relative course shall take on values within the range $[K_{ot\,min}, K_{ot\,max}]$. If $\rho \geq 1$, the relative course shall change from 0 to 2π, whenever the vessel's true course K_1 changes in the same range.

So, whenever $\rho \geq 1$, the sign of the derivative $\dfrac{\partial D_m}{\partial K_1}$ is determined by the expression:

$$sign\left(\frac{\partial D_m}{\partial K_1}\right) = q.$$

If $\rho < 1$, the derivative sign $\dfrac{\partial D_m}{\partial K_1}$ is:

$$sign\left(\frac{\partial D_m}{\partial K_1}\right) = \begin{cases} q, K_1 \in (K_2 + arccos\,\rho, K_2 - arccos\,\rho), \\ -q, K_1 \in (K_2 - arccos\,\rho, K_2 + arccos\,\rho). \end{cases}$$

4 RESULTS OF SIMULATION MODELLING

To analyze relationship between the sign of changing the distance to the CPA D_{min} and alteration of the vessel's course K_1, simulation modelling was performed.

In the course of simulation modelling various vessel-target speed ratios have been analyzed. Modelling results, when the vessel speed is greater than the target speed $V_1 > V_2$, are given in the table below. The following speed values were selected for

modelling: $V_1 = 25$ knots, $V_2 = 15$ knots, $D = 5$ miles, $\alpha = 45°$.

Table. Relationship between changes of D_{min} and K_1 for various K_2, whenever $V_1 > V_2$

K_2	Increasing D_{min} $\left(\dfrac{\partial D_{min}}{\partial K_1} > 0\right)$	Decreasing D_{min} $\left(\dfrac{\partial D_{min}}{\partial K_1} < 0\right)$
0	$K_1 \in (20..110)$	$K_1 \in (340..20)$
90	$K_1 \in (70..110)$	$K_1 \in (340..70)$
180	$K_1 \in (70..160)$	$K_1 \in (290..70)$
270	$K_1 \in (20..160)$	$K_1 \in (290..20)$
	$D_{min} \in (0,0..5,0)$	$D_{min} \in (5,0..0,0)$

According to the table, in the given case the derivative $\dfrac{\partial D_{min}}{\partial K_1}$ has an area of positive values and an area of negative values, depending on the sign of the parameter q.

5 ASSESSMENT OF COORDINABILITY OF MANOEUVRES THROUGH COURSE ALTERATION

To assess coordination of simultaneous alteration of the vessel and the target's courses, we obtained an expression for the mixed derivative of the distance to the CPA with respect to the vessel's course K_1 and the target's course K_2:

$$\frac{\partial D_{min}}{\partial K_1 \partial K_2} = G_s \Bigg\{ \left\{ \left[\mp \frac{DV_1}{V_0^3} V_1 V_2 sin(K_1 - K_2) \right] \times \right.$$
$$\times \frac{[V_1 - V_2 cos(K_1 - K_2)]}{V_0^2} + \frac{DV_1}{V_0} \times$$
$$\left. \times \frac{\mp V_2 sin(K_1 - K_2)\{V_0^2 - 2V_1[V_1 - V_2 cos(K_1 - K_2)]\}}{V_0^4} \right\} \times$$
$$\times [V_2 cos(K_2 - \alpha) - V_1 cos(K_1 - \alpha)] +$$
$$+ \frac{DV_1}{V_0} \frac{[V_1 - V_2 cos(K_2 - K_2)]}{V_0^2} [-V_2 sin(K_2 - \alpha)].$$

In the last formula there is a sign « $\mp$ », before certain members, when means the respective signs "+" or "-" are used, depending on simultaneous turning direction of the vessel and the target.

Thus, the vessels' strategies D_1 and D_2 can be coordinated, i.e. $Coor(D_1, D_2) = 1$, when only the vessel changes her course, if $\dfrac{\partial D_m}{\partial K_1} > 0$, and when both vessels change their course, if $\dfrac{\partial D_{min}}{\partial K_1 \partial K_2} > 0$.

We have analyzed coordinability of collision avoidance manoeuvres stipulated by the COLREGs-72, with the results as follows. According to the research paper [3], the COLREGs-72 provide for three regions of mutual obligations of vessels dangerously approaching each other, where the give-way vessel in the second region may take action to avoid collision by her manoeuvre alone, and in the third region she must take immediate action to avoid collision.

In the first region of mutual obligations collision avoidance manoeuvres are governed by the Rules 13, 14, 15 and 18, and as tests proved all those manoeuvres can be coordinated. In the second and the third regions of mutual obligations, stipulated by the Rule 17, collision avoidance manoeuvres are uncoordinated: in the second region it is not possible to anticipate behaviour of the give-way vessel, which may keep her course and speed, or may take action to avoid collision at any time; and in the third region behaviour of both vessels is impossible to predict. Lack of coordination of collision avoidance manoeuvres in the second and in the third regions of mutual obligations is the COLREGs-72 drawback contributing to accident conditions.

6 ALTERNATIVE SYSTEM OF BINARY COORDINATION OF MANOEUVRES BY COURSE ALTERATION

The research paper [4] reviewed a method for establishing a generalized system of binary coordination and for determining its efficiency. This method was used to assess the quality criterion Q_s of the binary coordination system, applied by the COLREGs-72. Its value is $Q_{COLREG} = 1,45$. The same paper dwelt on a binary coordination system alternative to the COLREGs-72, where the principle of assigning priority to either vessel is different from the one adopted by the COLREGs-72.

In the suggested alternative system of binary coordination it is of crucial importance to define situational disturbance risk. According to this principle, if the initial perturbed situation can be converted into a subset of unperturbed situations by each interacting vessel while keeping motion variables of another vessel unchanged, then the situational disturbance risk is assigned the value 1.

If under the same conditions conversion of the perturbed situation into the subset of unperturbed situations can be carried out only by one of the interacting vessels on her own (while the other keeps her motion variables unchanged), then the situational disturbance risk is assigned the value 2. If under the same conditions conversion of the perturbed situation into the subset of unperturbed situations cannot be carried out by either interacting vessel on her own, and when there is a necessity to change motion variables of both of them, then the situational disturbance risk is assigned the value 3.

In the alternative system of binary coordination priority grade of vessels when $\varsigma = 1$ is determined by two criteria. Firstly, in the perturbed situation when vessels have different priority grades due to their belonging to different subsets of vessels based on control capability, priority grade of the vessels is determined similarly to the COLREGs-72 as in the Rule 18. In other cases, when $\varsigma = 1$, the vessel running at a lesser speed shall have priority, i.e. the vessel running at a greater speed shall keep out of her way. By the way, the same applies to overtaking. In cases of collision risk when $\varsigma = 2$ and $\varsigma = 3$, no priority grade is assigned to the vessels.

Quality criterion of the suggested system of binary coordination is $Q_s = 1,96$. It is higher than the quality criterion of the COLREGs-72 by 36%, due to fuller use of data on navigation situation and application of joint coordination patterns by the vessels dangerously approaching each other. COLREGs-72 system of binary coordination uses just minimum data on the initial navigation situation.

REFERENCES

1. Yakushev A.O. Choosing Optimum Shape of Safe Navigation Area / Yakushev A.O. // Ship Navigation: Collection of research papers./ Odessa National Maritime Academy, Issue 23. – Odessa: IzdatInform, 2013 - p.157-162.
2. Pyatakov E.N. Performance Assessment of Pair Strategies of Vessels Avoiding Collision / Pyatakov E.N., Zaichko S.I. // Ship Navigation. – 2008. – No. 15. – p. 166 –171.
3. Tsymbal N.N. Flexible Collision Avoidance Strategies / Tsymbal N.N., Burmaka I.A., Tyupikov E.E. – Odessa: Odessa Municipal Publishing House, 2007. – 424 p.
4. Pyatakov E.N. Improving Methods for Coordination of the Vessels' Collision Avoidance Maneuvers. Synopsis of the thesis of the Candidate of Technical Sciences: 05.22.13/ Odessa National Maritime Academy. – Odessa, 2008. – 23 p.
5. Tsymbal M.M. Method of synthesis of flexible strategies for preventing collisions. / Odessa National Maritime Academy, Odessa, Ukraine
6. Tsymbal M., Urbansky I. Development of simulator systems for preventing collision of ships / The Interanational Marine Simulator Forum – IMSF 2008

Comparison of Anti-collision Game Trajectories of Ship in Good and Restricted Visibility at Sea

J. Lisowski
Gdynia Maritime University, Poland

ABSTRACT: The paper introduces selected methods of a game theory for ships collisions avoidance at sea. The application of game control algorithms to determine the own ship safe trajectory during the passing of other encountered ships in good and restricted visibility at sea is presented. Methods of comparison of the safe ship control in collision situation: multi-step matrix non-cooperative and cooperative games, multi-stage positional non-cooperative and cooperative games have been introduced. The considerations have been illustrated with examples of computer simulation of the algorithms to determine safe of own ship trajectories in a navigational situation of good and restricted visibility in the Skagerrak Strait.

1 INTRODUCTION

One of the major problems in maritime transport is to ensure the safety of navigation. In order to ensure the marine safety, the ships are obliged to comply with the International Regulations for Preventing Collisions at Sea (COLREGS). However, these Rules refer only to two ships and under the conditions of good visibility.

For the situation of a restricted visibility the Regulations only specify recommendations of a general nature and are not able to consider all the necessary conditions which determine the passing course.

Consequently, the actual process of a ship passing other objects very often occurs in conditions of uncertainty and conflict accompanied by an inadequate co-operation of the ships within COLREGS. It is, therefore, reasonable to investigate, develop and represent the methods of a ship safety handling using the rules of theory based on dynamic game (Bist & Jain 2001, Isaacs 1965, Millington & Funge 2009, Osborne 2004, Straffin 2001).

2 SAFE SHIP CONTROL PROCESS

The process of handling a ship as a multidimensional dynamic object depends both on the accuracy of the details concerning the current navigational situation obtained from the Automatic Radar Plotting Aids ARPA anti-collision system and on the form of the process model used for the control synthesis (Bole et al. 2006, Cahill 2002, Gluver & Olsen 1998).

The ARPA system ensures monitoring of at least $20\ j$ encountered ships, determining their movement parameters (speed V_j, course ψ_j) and elements of approaching to own ship moving with speed V and course ψ to satisfy $D_{min}^j = DCPA_j$ - Distance of the Closest Point of Approach, and $T_{min}^j = TCPA_j$ - Time to the Closest Point of Approach and also assess the risk of collision r_j (Fig. 1).

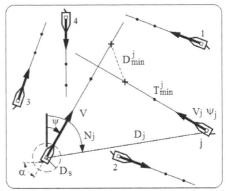

Figure 1. Navigational situation of ships passing – the own ship moving at V speed and □ course with the j ship met moving at Vj speed and □j course.

The model of the process consists both of the kinematics and the dynamics of the ship's movement, the disturbances, the strategies of the encountered ships and the quality control index of the own ship (Clarke 2003, Fang & Luo 2005, Fossen 2011, Kula 2014, Perez 2005).

The diversity of possible models directly affects the synthesis of the ship's control algorithms which are afterwards affected by the ship's control device, directly linked to the ARPA system and consequently determines the effects of safe and optimal control.

Figure 2 shows a set of compromises of a ship safety control measured in terms of a collision risk and time-optimal strategy of the own ship control.

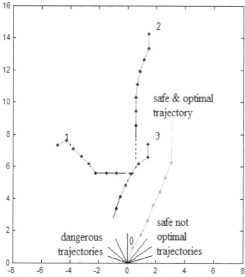

Figure 2. The possible trajectories in collision situation.

3 GAME MODELS OF A SAFE SHIP CONTROL

The way of controlling a ship - which is a multi-dimensional and non-linear dynamic object - depends on the range and accuracy of information on the prevailing navigational situation and on the adopted model of the process (Fig. 3).

The most accurate model is the game differential j objects, which, however, is very complex mathematically and serves as a simulation model for testing control algorithms in real time.

For the practical synthesis of control algorithms are used simplified models of the process of safety steering of the ship in collision situations.

Description of the dynamics of an ship using the differential state equations is replaced in the form of the advance time to the manoeuvre t_m, with element $t_m^{\Delta\psi}$ during course manoeuvre $\Delta\psi$ or element $t_m^{\Delta V}$ during speed manoeuvre ΔV.

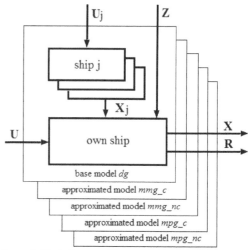

Figure 3. Block diagram of the models for safe control process: U - vector of the own ship control, Uj - control vector of the j ship, Xj - state vector of the j-th ship, Z - disturbance vector, X - state vector of the process, R - vector of safety constraints.

The variety of the models to be adopted directly influences the synthesis of various algorithms supporting the navigator's work, and then on the effects of a safe control of the own ship's movement (Fadali & Visioli 2009, Fletcher 1987, Keesman 2011, Landau et al. 2011, Lazarowska 2014, Mohamed-Seghir 2014, Szłapczyński 2014, Tomera 2014).

3.1 Base differential game model

The most general description of the own ship passing j other encountered ships is the model of a differential game dg of j moving control objects (Fig. 4), (Baba & Jain 2001, Mesterton-Gibbons 2001).

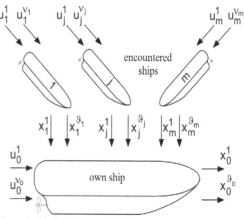

Figure 4. Block diagram of the basic model of differential game.

The properties of the process are described by the state equation:

$$\dot{x}_i = f_i[(x_{0,\vartheta_0}, \ldots, x_{j,\vartheta_j}, \ldots, x_{m,\vartheta_m}),$$

$$(u_{0,v_0}, \ldots, u_{j,v_j}, \ldots, u_{m,v_m}, t] \quad (1)$$

$$i = 1, 2, \ldots, n; \ j = 1, 2, \ldots, m$$

where $\vec{x}_{0,\vartheta_0}(t)$ - ϑ_0 dimensional vector of the process state of the own ship determined in a time span $t \in [t_0, t_k]$; $\vec{x}_{j,\vartheta_j}(t)$ - ϑ_j dimensional vector of the process state for the j-th met ship; $\vec{u}_{0,v_0}(t)$ - v_0 dimensional control vector of the own ship; $\vec{u}_{j,v_0}(t)$ - v_j dimensional control vector of the j-th met ship.

The constraints of the control and the state of the process are connected with the basic condition for the safe passing of the ships at a safe distance D_s in compliance with COLREGS Rules, generally in the following form:

$$g_j(x_{j,\vartheta_j}, u_{j,v_j}) \le 0 \quad (2)$$

Rules and the condition to maintain a safe passing distance as per relationship:

$$D_{\min}^j = \min D_j(t) \ge D_s \quad (3)$$

The closed sets U_0^j and U_j^0, defined as the sets of acceptable strategies of the participants to the game towards one another:

$$\{U_0^j[p(t)], U_j^0[p(t)]\} \quad (4)$$

are dependent, which means that the choice of steering u_j by the j-th ship changes the sets of acceptable strategies of other ships (Fig. 5).

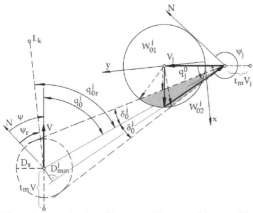

Figure 5. Determination of the acceptable areas of the own ship strategies $U_0^j = W_{01}^j \cup W_{02}^j$.

A set U_o^j of acceptable strategies of the own ship when passing the j-th encountered ship at a distance D_s - while observing the condition of the course and speed stability of the own ship and that of encountered ship at step k is static and comprised within a half-circle of a radius V_r.

Area U_0^j is determined by an inequality (Fig. 6):

$$a_0^j u_0^x + b_0^j u_0^y \le c_0^j \quad (5)$$

$$\left(u_0^x\right)^2 + \left(u_0^y\right)^2 \le V_r^2 \quad (6)$$

$$a_0^j = -\chi_0^j \cos(q_{0r}^j + \chi_0^j \delta_0^j)$$

$$b_0^j = \chi_0^j \sin(q_{0r}^j + \chi_0^j \delta_0^j)$$

$$c_0^j = -\chi_0^j \begin{bmatrix} V_j \sin(q_j^0 + \chi_0^j \delta_0^j) + \\ V_r \cos(q_{0r}^j + \chi_0^j \delta_0^j) \end{bmatrix} \quad (7)$$

$$\chi_0^j = \begin{cases} 1 & dla \ W_{01}^j \ (Starboard \ side) \\ -1 & dla \ W_{02}^j \ (Port \ side) \end{cases}$$

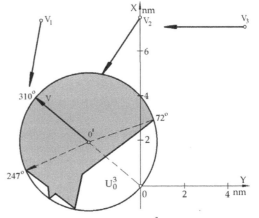

Figure 6. Example of summary set U_0^3 of acceptable manoeuvres for three encountered ships.

The value χ_0^j is determined by using an appropriate logical function Z_j characterising any particular recommendation referring to the right of way contained in COLREGS Rules.

The form of function Z_j depends of the interpretation of the above recommendations for the purpose to use them in the steering algorithm, when:

$$Z_j = \begin{cases} 1 & then \ \chi_0^j = 1 \\ 0 & then \ \chi_0^j = -1 \end{cases} \quad (8)$$

Interpretation of the COLREGS Rules in the form of appropriate manoeuvring diagrams enables to formulate a certain logical function Z_j as a semantic interpretation of legal regulations for manoeuvring.

Each particular type of the situation involving the approach of the ships is assigned the logical variable value equal to one or zero:

- A – encounter of the ship from bow or from any other direction,
- B – approaching or moving away of the ship,
- C – passing the ship astern or ahead,
- D – approaching of the ship from the bow or from the stern,
- E – approaching of the ship from the starboard or port side.

By minimizing logical function Z_j by using a method of the Karnaugh's Tables the following is obtained:

$$Z_j = A \cup \overline{A}(\overline{B}\,\overline{C} \cup \overline{D}\,\overline{E}) \qquad (9)$$

The resultant area of acceptable manoeuvres for m ships:

$$U_0 = \bigcap_{j=1}^{m} U_0^j \quad j = 1, 2, ..., m \qquad (10)$$

is determined by an arrangement of inequalities (5) and (6).

A set for acceptable strategies U_j^0 of the encountered j-th ship relative to the own ship is determined by analogy.

Goal function has form of the payments – the integral payment and the final one:

$$I_{0,j} = \int_{t_0}^{t_k} [x_{0,9_0}(t)]^2 dt + r_j(t_k) + d(t_k) \rightarrow \min \qquad (11)$$

The integral payment represents additional distance traveled by the own ship while passing the encountered ships and the final payment determines the final risk of collision $r_j(t_k)$ relative to the j ship and the final deflection of the own ship $d(t_k)$ from the reference trajectory (Engwerda 2005, Mehrotra 1992, Nisan et al. 2007, Pantoja 1988).

3.2 Approximated models

For the practical synthesis of safe control algorithms various simplified models are formulated:

- dual linear programming model mmg_c of cooperative multi-step matrix game,
- dual linear programming model mmg_nc of non-cooperative multi-step matrix game,
- triple linear programming model mpg_c of cooperative multi-stage positional game,
- triple linear programming model mpg_nc of non-cooperative multi-stage positional game.

The degree of model simplification depends on an optimal control method applied and level of cooperation between ships (Tab. 1).

Table 1. Algorithms of determining of own ship safe trajectory.

Approximate model	Method of optimization	Support algorithm
mult-step cooperation matrix game	dual linear programming	mmg_c
mult-step non-cooperation matrix game	dual linear programming	mmg_nc
multi-stage cooperation positional game	triple linear programming	mpg_k
multi-stage non-cooperation positional game	triple linear programming	mpg_nc

4 COMPUTER SUPPORT ALGORITHMS

In practice, methods of selecting a manoeuvre assume a form of approximate control algorithms supporting navigator decision in a collision situation. Algorithms are programmed in the memory of a Programmable Logic Controller PLC. This generates an option within the ARPA anti-collision system or a training simulator (Fig. 7).

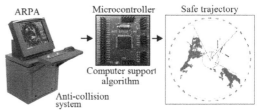

Figure 7. The structure of computer support system of navigator decision.

4.1 Algorithm mmg_c of multi-step cooperative matrix game

The matrix game $\boldsymbol{R} = [r_j(\mathrm{v}_j, \mathrm{v}_0)]$ includes the value a collision risk r_j with regard to the determined strategies v_0 of the own ship and those v_j of the j-th encountered ships (Cockcroft & Lameijer 2006, Luus2000, Modarres 2006, Zio 2009).

The value of the risk of the collision r_j is defined as the reference of the current situation of the approach described by the parameters $D_{\min}^j$ and $T_{\min}^j$, to the assumed assessment of the situation defined as safe and determined by the safe distance of approach D_s and the safe time T_s – which are necessary to execute a manoeuvring to avoid collision with consideration actual distance D_j between own ship and encountered j-th ship (Fig. 8).

$$r_j = \left[\varepsilon_1 \left(\frac{D_{\min}^j}{D_s} \right)^2 + \varepsilon_2 \left(\frac{T_{\min}^j}{T_s} \right)^2 + \varepsilon_3 \left(\frac{D_j}{D_s} \right)^2 \right]^{-\frac{1}{2}} \qquad (12)$$

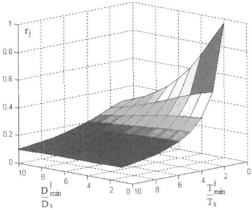

Figure 8. The space of ship collision risk.

As a result of using the following form for the control goal:

$$I_{0,j}^* = \min_{v_0} \min_{v_j} r_j \qquad (13)$$

the probability matrix $P=[p_j\,(v_j,\,v_0)]$ of using particular pure strategies may be obtained.

The solution for the control goal is the strategy of the highest probability:

$$u_{0,v_0}^* = u_{0,v_0}\left\{[p_j(v_j, v_0)]_{max}\right\} \qquad (14)$$

Using the function of lp – $linear\ programming$ from the Optimization Toolbox Matlab, the matrix multi-step game manoeuvring mmg_c program has been designed for the determination of the own ship safe trajectory in a collision situation.

4.2 Algorithm mmg_nc of multi-step non-cooperative matrix game

Goal function (13) for non-cooperative matrix game has the form (Basar & Olsder 1982, Gałuszka & Świerniak 2005, Isil-Bozma & Koditschek 2001):

$$I_{0,j}^* = \min_{v_0} \max_{v_j} r_j \qquad (15)$$

4.3 Algorithm mpg_c of multi-stage cooperative positional game

The optimal control of the own ship $u_0^*(t)$, which is equivalent to the optimal positional steering $u_0^*(p)$ for the current position $p(t)$, is determined from the condition:

$$I_{0,j} = \min_{u_0 \in U_0 = \bigcap\limits_{j=1}^{m} U_o^j}\left\{\min_{u_j^m \in U_j}\min_{u_0^j \in U_0^j(u_j)}\int_{t_0}^{t_k} u_0(t)\,dt\right\} = D_0^*(x_0) \qquad (16)$$

D_0 refers to the continuous function of the manoeuvring goal of the own ship, describing the distance of the ship at the initial moment t_0 to the nearest turning point on the reference route of the voyage.

The optimal control of the own ship is calculated at each discrete stage of the ship's movement by applying the Simplex method to solve the problem of the triple linear programming, assuming the relationship (16) as the goal function and the control constraints (2).

Using the function of lp – $linear\ programming$ from the Optimization Toolbox Matlab, the positional multi-stage game non-cooperative manoeuvring mpg_c program has been designed for the determination of the own ship safe trajectory in a collision situation (Rak & Gierusz 2012).

4.4 Algorithm mpg_nc of multi-stage non-cooperative positional game

Goal function (16) for non-cooperative positional game has the form:

$$I_{0,j} = \min_{u_0 \in U_0 = \bigcap\limits_{j=1}^{m} U_o^j}\left\{\max_{u_j^m \in U_j}\min_{u_0^j \in U_0^j(u_j)}\int_{t_0}^{t_k} u_0(t)\,dt\right\} = D_0^*(x_0) \qquad (17)$$

5 COMPUTER SIMULATION

Computer simulation of mmg_c, mmg_nc, mpg_c and mpg_nc algorithms was carried out in Matlab/Simulink software on an example of the real navigational situation of passing $j=34$ encountered ships in the Skagerrak Strait in good visibility $D_s=0.5$ nm and restricted visibility $D_s=3.0$ nm (nautical miles), (Fig. 9 and Tab. 2).

The situation was registered on board r/v HORYZONT II, a research and training vessel of the Gdynia Maritime University, on the radar screen of the ARPA anti-collision system Raytheon (Fig. 10).

Figure 9. The place of identification of navigational situation in Skagerrak Strait.

Table 2. Movement parameters of own ship and encountered ships.

j	Dj nm	Nj deg	Vj kn	ψj deg
0	-	-	11	130
1	8.0	65	10.0	205
2	5.0	10	1.7	150
3	4.0	300	10.0	100
4	4.0	165	5.0	50
5	5.0	140	4.0	350
6	6.0	48	5.0	150
7	4.0	105	5.0	120
8	10.0	120	3.0	20
9	4.0	140	2.0	350
10	3.0	20	8.0	140
11	3.0	47	5.0	220
12	10	165	6.0	50
13	1.0	200	5.0	100
14	5.0	85	1.0	150
15	8.0	145	7.0	0
16	9.0	120	4.0	330
17	4.0	140	3.0	350
18	11.0	90	2.0	90
19	9.0	250	11.0	300
20	9.0	320	11.0	300
21	12.0	79	8.0	211
22	8.4	170	9.9	32
23	10.2	39	6.8	279
24	11.1	276	8.9	14
25	0.9	27	9.2	182
26	12.0	340	2.1	9
27	8.8	98	4.5	32
28	9.4	29	6.0	326
29	6.0	129	11.0	350
30	10.1	15	8.2	178
31	6.9	87	7.0	123
32	8.7	290	1.4	222
33	5.6	231	6.5	112
34	10.1	48	8.8	228

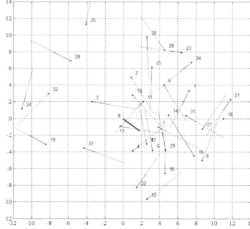

Figure 11. The 6 minute speed vectors of own ship 0 and j=34 encountered ships in navigational situation in Skagerrak Strait.

Computer simulation tests in the software Matlab/Simulink control algorithms were four algorithms of ship steering in collision situations at sea. Each algorithm was tested in conditions of good visibility at sea at a reference value a safe distance passing ships D_s=0.5 nm and in restricted visibility conditions at sea at a predetermined value a safe distance passing ships D_s=3.0 nm.

Fig. 12-15 shows the safe and optimal trajectory of the own ship in collision situation, which is determined using the algorithm of cooperative and non-cooperative matrix game.

Fig. 16-19 shows the safe and optimal trajectory of the own ship in collision situation, which is determined using the algorithm of cooperative and non-cooperative positional game.

Figure 10. The research-training ship of Gdynia Maritime University r/v HORYZONT II.

Examined the navigational situation, illustrated in the form of navigation velocity vectors of own ship and 34 met ships is shown in Figure 11.

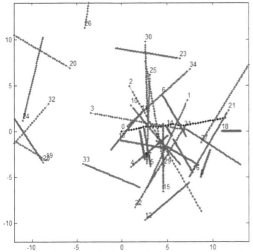

Figure 12. Computer simulation of multi-step cooperative matrix game algorithm *mmg_c* for safe own ship control in situation of passing 34 encountered ships in good visibility at sea, D_s=0.5 nm, $d(t_k)$=0.57 nm (nautical mile).

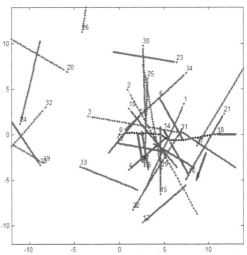

Figure 13. Computer simulation of multi-step non-cooperative matrix game algorithm *mmg_c* for safe own ship control in situation of passing 34 encountered ships in restricted visibility at sea, D_s=3.0 nm, $d(t_k)$=1.88 nm (nautical mile).

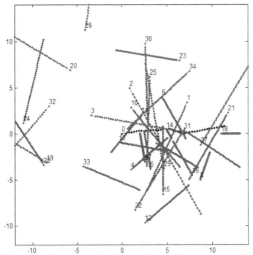

Figure 14. Computer simulation of multi-step cooperative matrix game algorithm *mmg_nc* for safe own ship control in situation of passing 34 encountered ships in good visibility at sea, D_s=0.5 nm, $d(t_k)$=1.19 nm (nautical mile).

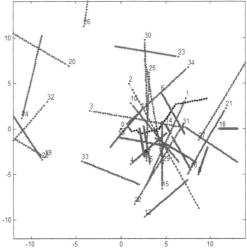

Figure 15. Computer simulation of multi-step non-cooperative matrix game algorithm *mmg_nc* for safe own ship control in situation of passing 34 encountered ships in restricted visibility at sea, D_s=3.0 nm, $d(t_k)$=2.18 nm (nautical mile).

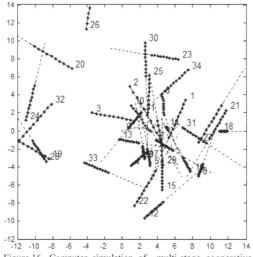

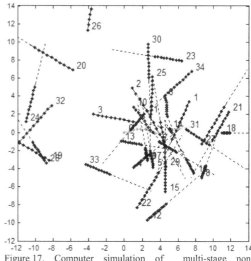

Figure 16. Computer simulation of multi-stage cooperative positional game algorithm *mpg_c* for safe own ship control in situation of passing 34 encountered ships in good visibility at sea, D_s=0.5 nm, $d(t_k)$=0.69 nm (nautical mile).

Figure 17. Computer simulation of multi-stage non-cooperative positional game algorithm *mpg_c* for safe own ship control in situation of passing 34 encountered ships in restricted visibility at sea, D_s=3.0 nm, $d(t_k)$=1.56 nm (nautical mile).

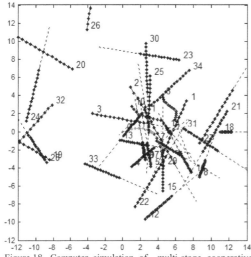

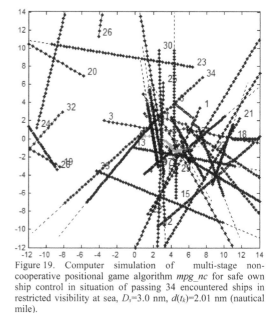

Figure 18. Computer simulation of multi-stage cooperative positional game algorithm *mpg_nc* for safe own ship control in situation of passing 34 encountered ships in good visibility at sea, D_s=0.5 nm, $d(t_k)$=0.71 nm (nautical mile).

Figure 19. Computer simulation of multi-stage non-cooperative positional game algorithm *mpg_nc* for safe own ship control in situation of passing 34 encountered ships in restricted visibility at sea, D_s=3.0 nm, $d(t_k)$=2.01 nm (nautical mile).

Figure 20 shows a comparison of the own ship safe trajectory, designated four algorithms dynamic game, in good visibility at sea if D_s=0.5 nm.

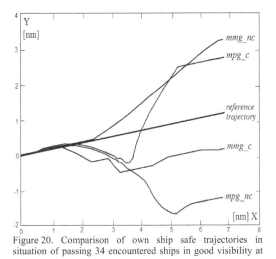

Figure 20. Comparison of own ship safe trajectories in situation of passing 34 encountered ships in good visibility at sea.

The biggest impact on the size of the final game payment, as the final deviation from the reference trajectory, has a degree of cooperation of ships in avoiding collision.

Figure 21 shows a comparison of the own ship safe trajectory, determined by the four algorithms of dynamics games, in restricted visibility at sea if D_s=3.0 nm.

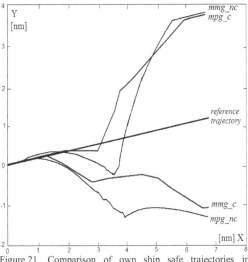

Figure 21. Comparison of own ship safe trajectories in situation of passing 34 encountered ships in restricted visibility at sea.

The biggest impact on the size of the final game payment, as the final deviation from the reference

trajectory, has a degree of cooperation of ships in avoiding collision.

6 CONCLUSIONS

Analysis of the computer simulation studies of game algorithms for good and restricted visibility at sea allows to draw the following conclusions.

The synthesis of an optimal on-line control on the model of a multi-step matrix game and multi-stage positional game makes it possible to determine the safe game trajectory of the own ship in situations when she passes a greater j number of the encountered objects.

The trajectory has been described as a certain sequence of manoeuvres with the course and speed.

The computer programs designed in the Matlab also takes into consideration the following: regulations of the Convention on the International Regulations for Preventing Collisions at Sea, advance time for a manoeuvre calculated with regard to the ship's dynamic features and the assessment of the final deflection between the real trajectory and its assumed values

The essential influence to form of safe and optimal trajectory and value of deflection between game and reference trajectories has a degree of cooperation between own and encountered ships.

It results from the performed simulation testing this algorithm is able to determine the correct game trajectory when the ship is not in a situation when she approaches too large number of the observed ships or the said ships are found at long distances among them

In the case of the high traffic congestion the program is not able to determine the safe game manoeuvre, this sometimes results in the backing of the own ship which is continued until the time when a hazardous situation improves.

REFERENCES

Baba, N. & Jain, L.C. 2001. *Computational intelligence in games.* New York: Physica-Verlag.
Basar, T. & Olsder, G.J. 1982. *Dynamic non-cooperative game theory.* New York: Academic Press.
Bist, D.S. 2000. *Safety and security at sea.* Oxford-New Delhi: Butter Heinemann.
Bole, A., Dineley, B. & Wall, A. 2006. *Radar and ARPA manual.* Amsterdam-Tokyo: Elsevier.
Cahill, R.A. 2002. *Collisions and thair causes.* London: The Nautical Institute.
Clarke, D. 2003. The foundations of steering and manoeuvering, *Proc. of the IFAC Conference on Manoeuvering and Control Marine Crafts, Girona*: 10-25.
Cockcroft, A.N. & Lameijer, J.N.F. 2006. *The collision avoidance rules.* Amsterdam-Tokyo: Elsevier.
Engwerda, J.C. 2005. *LQ dynamic optimization and differential games.* West Sussex: John Wiley & Sons.

Fadali, M.S. & Visioli, A. 2009. *Digital control engineering.* Amsterdam-Tokyo: Elsevier.

Fang, M.C. & Luo, J.H. 2005. The nonlinear hydrodynamic model for simulating a ship steering in waves with autopilot system. *Ocean Engineering* 11-12(32):1486-1502.

Fletcher, R. 1987. *Practical methods of optimization.* New York: John Wiley and Sons.

Fossen, T.I. 2011. *Marine craft hydrodynamics and motion control.* Trondheim: Wiley.

Gałuszka, A. & Świerniak, A. 2005. Non-cooperative game approach to multi-robot planning. *International Journal of Applied Mathematics and Comuter Science* 15(3):359-367.

Gluver, H. & Olsen, D. 1998. *Ship collision analysis.* Rotterdam-Brookfield: A.A. Balkema.

Isaacs, R. 1965. *Differential games.* New York: John Wiley & Sons.

Isil Bozma, H. & Koditschek, D.E. 2001. Assembly as a non-cooperative game of its pieces: Analysis of ID sphere assemblies. *Robotica* 19:93-108.

Keesman, K.J. 2011. *System identification.* London-NewYork: Springer.

Kula, K. 2014. Cascade control systems of fin stabilizers. *Proc. XIX Conference Methods and Models in Automation and Robotics, MMAR 2014, Miedzyzdroje*, 2014: 38-48.

Landau, I.D., Lozano, R., M'Saad, M. & Karimi, A. 2011. *Adapive control.* London-New York: Springer.

Lazarowska, A. 2014. Safe ship control method with the use of Ant Colony Optimization. *Solid State Phenomena* 210:234-244.

Luus, R. (2000). *Iterative dynamic programming*, CRC Press, Boca Raton.

Mehrotra, S. 1992. On the implementation of a primal-dual interior point method. *SIAM Journal on Optimization* 4(2):575-601.

Mesterton-Gibbons, M. 2001. *An introduction to game theoretic modeling.* Providence: American Mathematical Society.

Millington, I. & Funge, J. 2009. *Artificial intelligence for games.* Amsterdam-Tokyo: Elsevier.

Modarres, M. 2006. *Risk analysis in engineering.* Boca Raton: Taylor & Francis Group.

Mohamed-Seghir, M. 2014. The branch-and-bound method, genetic algorithm, and dynamic programming to determine a safe ship trajectory in fuzzy environment. *Proc.18th International Conference in Knowledge Based and Intelligent Information and Engineering Systems KES 2014, Gdynia*, 2014: 348-357.

Nisan, N., Roughgarden, T., Tardos, E. & Vazirani, V.V. 2007. *Algorithmic game theory.* New York: Cambridge University Press.

Osborne, M.J. 2004. *An introduction to game theory.* New York: Oxford University Press.

Pantoja, J.F.A. 1988. Differential dynamic programming and Newton's method. *International Journal of Control* 5(47):1539-1553.

Perez, T. 2005. *Ship motion control.* London: Springer.

Rak, A., Gierusz, W. 2012. Reinforcement learning in discrete and continuous domains applied to ship trajectory generation. *Polish Maritime Research* S1(74): 31-36.

Straffin, P.D. 2001. *Game theory and strategy.* Warszawa: Scholar (in polish).

Szłapczyński, 2014. Evolutionary sets of safe ship trajectories with speed reduction manoeuvres within traffic separation schemes. *Polish Maritime Research* 1(81): 20-27.

Tomera, M. 2014. Dynamic positioning system for a ship on harbor manoeuvring with different observers, experimental results. *Polish Maritime Research* 3(83): 13-24.

Zio, E. 2009. Computational methods for reliability and risk analysis. *Series on Quality, Reliability and Engineering Statistics* 14: 295-334.

Concretization of the Concept "Nearly Reciprocal Course" in Rule 14 of Colreg-72

V.M. Bukaty
The Baltic State Academy of Fishing Fleet FSBAO HPA KSTU, Kaliningrad, Russia

ABSTRACT: Attempt is made to specify the concept "nearly reciprocal course" in Rule 14 of COLREG-72 based on the criterion difference between course over the ground of the ship-observer and bearing of the ship-target: if this difference is less than its tripled mean-square error, with probability 99,7%, the close quarter situation can be considered as close quarter situation with the apposite courses over the ground. Applied to the definition course over the ground of the ship-observer from GPS and bearing of the ship-target from AIS and GPS this criterion means that if the difference between course over the ground of the ship-observer and bearing of the ship-target is less than 5^0, the close quarter situation should be considered at the situation fallen under the jurisdiction of Rule 14 of COLREG-72.

1 INTRODUCTION

"...When two power-driven vessels are me-eting on reciprocal or nearly reciprocal course so as to involve risk of collision each shell alter her course to starboard so that each shall pass on the port side of the other...", - so says in the Rule 14(a) of COLREGs-72 [1]. In the Rule 14(b) of COLREGs-72 says: "*...Such a situation shall be deemed to exist when a vessel sees the other ahead or nearly ahead and by night she could see the mast-head lights of the other in a line and/or both sidelights and by day she observes corresponding aspect of the other vessel...".* And further in Rule 14 (c): "*... When a vessel is in any doubt as to whether such a situation exists see shall assume that it does exit and act accordingly...".* As we see the very wording of Rule 14 COLREG-72 contains a number of uncertainties referring to the attributing of head-on situation to the jurisdiction of Rule 14 of COLREG-72 as it is far from clear what is meant by the word "nearly reciprocal courses", or "nearly ahead", or nearly on line".

Rule 7(d) of COLREG-72 that applies to head-on situation as well indicates that "...such risk should be deemed to exist if the compass bearing of an approaching vessel does not appreciably change ...". And again it is far from clear what is meant by the word "does not appreciably change". That's why it is very accurately and the at same time very fair for one of the word-class specialists in the field of

navigation safety Professor Lushnikov E.M. to ask: "What is meant by the term "nearly" – is it an intersection course at the angle of 3^0, at the angle of 5^0 and can by 10^0 or 15^0 [2]. There is still no answer to this question. And it means that an up-to-date navigator operating the most up-to-date navigation equipment doesn't have science-based quantitative criteria to assess the head-on situation. It often happens that one and the same situation is qualified as ships' meeting on reciprocal courses by one navigator and by the other on it is qualified as ships' meeting on crossing course. Different interpretation of the same situation is often the cause of ships' collision, as, for example, it happened with the passenger lines "The Stockholm" and the "Andrea Doria" when approaching the port of New-York [3]. Intricacies with differentiation of head-on situation on reciprocal courses and sharply crossing courses is, probably, the main reason that a half of navigation accidents takes place when two ships on meeting on reciprocal courses [4].

As to sailing vessels in sight of one another, when for qualification of ships' meeting situation is easy to observe such evidences as ships' masts or top lights in line or visibility of both sidelights, when in real conditions of sailing when there is drift, Rolland pitch, it is rather questionable to use the above evidences as it was shown in the our publications [5].

Russian comments to COLREG-72 do not at all consider the problems quantitative criteria on navigation safety applied to Rule 14 of COLREG-72 that is, minimum allowable difference of ships' courses in head-on situation. In Russian comments to COLREG-72, following the text of Rule 14 in any doubt it is recommended to use Rule 14 executing mutual maneuvering by changing to starboard "…so that each shall pass on the port side of the other…" In foreign comments to COLREG-72 based on judicial and arbitration it is considered that in case of no course differences go beyond $180^0 \pm 0,5$ rhumb, Rule 14 should be used. If the courses differences is more than $180^0 \pm 0,5$ rhumb, Rule 15 should be used [6]. In our view more correct would be to say not about the difference of courses but about courses over ground (COG) as owing to the influence of wind and current the ship in practice always moves with drift and often the drift is considerable. Thus, we can say that ships' meeting doesn't happen on course line but on COG line.

2 CRITERIA FOR QUALIFICATION OF SITUATION

From the above it follows that the problem of finding quantitative criterion for conclusive attributing the situation of ships meeting to the head-on situation on reciprocal courses over ground or to the situation of ship's meeting on sharply crossing courses over the ground as before remains on actual problem that needs to be resolved.

To qualify the head-on situation as the one within the jurisdiction of Rule-14 of COLREG-72 there are several evidences by which we can conclusively do it. Among these evidences we can mention a relative speed of ships meeting and a relative course of ships meeting : it is evident that it the ships are meeting on reciprocal COG, the relative speed of their meeting must be equal to the sum of ground speed of ships and the difference between a relative speed of ships meeting and COG of the ship-observer must be 180^0 . However, the analysis of possible evidences of ships meeting on reciprocal COG had leg us to the conclusion that the only conclusive and quantitative evidence was the deference between COG-of the ship-observer and a measured bearing of the ship-target.

As there are all the grounds to consider that the difference between COG of the ship-observer and a measured bearing of the ship-target B1 is a normally distributed random variable, then in accordance with a fundamental assumption of the theory of probability about the impossibility of distinguishing these values to more than three standard deviations from their mean square as a criteria condition for adjudical about ships meeting on reciprocal COG we should take an inequality:

$$\frac{1}{n}\sum_{i=1}^{n}\left(q_o - B_t\right) \le 3m_{(q_o - B_t)} \tag{1}$$

where $m_{(q_o - B_t)}$ is a root-meat-square error of COG deference of the ship-observer and a bearing of the ship-target B_t.

Root-mean-square error $m(q_o - B_t)$ of difference $q_o - B_t$ is found from the known equation:

$$m_{(q_o - B_t)} = \sqrt{m_{q_o}^2 + m_{B_t}^2} \tag{2}$$

where $m_{(q_o)}$ and $m_{(B_t)}$ are a standard deviation of the COG- of the ship-observer and a bearing of the ship-target correspondingly.

Substituting equation 2 in equation 1 for the criteria condition we obtain:

$$\frac{1}{n}\sum_{i=1}^{n}\left(q_o - B_t\right) \le 3\sqrt{m_{(q_o - B_t)}^2 + m_{B_t}^2} \tag{3}$$

Thus, producing several simultaneous pairs of measurements q_o and B_t and finding the mean of differences q_o-B_t and comparing it with a priori found or found by the internal convergence tripled of the mean square error $m_{(q_o - B_t)}$ according to the criteria condition we can judge about ships meeting situation type.

As we can see, a specific numerical value of the criterion (3) is determined by precise opportunities of the used navigation facilities for finding ship position at sea. Because of the diversity of these facilities and their precise characteristics it is possible to get information about the COG of the ship-observer and the bearing of the ship-target by several combinations and in this case there will be different numerical results of the criterion (3):

- ship-target bearing is defined by compass satellite data and COG of ship-observer – by Satellite Navigation System (SNS) data;
- ship-target bearing is defined by gyroscopic compass data , COG of ship-observer- by SNS data;
- ship-target bearing is defined by compass satellite data, COG of ship-observer- by binary absolute log data;
- ship-target bearing is defined by gyroscopic compass data, COG of ship-observer- by binary absolute log data;
- ship-target bearing is defined by the positions of the ships with SNS and Automatic Identification System (AIS), COG of ship-observer- by SNS data;
- ship-target bearing is defined by the positions of the ships with SNS and AIS, COG of ship-observer- by the absolute log data.

Our calculations show that used navigation equipment allows to define the difference of COG of the ship-observer and the ship –target bearing with a mean-square error of the order of $1,5^0$ - 2^0 . Thus, in practice, if q_o-B_t<$4,5^0$-6^0 , the situation of ships meeting with the probability of 99,73% should be attributed to the jurisdiction of Rule 14, but it go-B1>4,5⁰-6⁰ the situation must be attributed to the jurisdiction of Rule 15 of COLREG-72.

Thus, for example, it the heading measure-ment at sea is carried out by a modern course position indicator (mean-square error is $0,4^0$ $-0,5^0$ [7], then heading measurements by a radar indicator with an electronic cursor will have a mean-square error of about $1,3^0$ [8].

The accuracy of COG determination by the SNS transceiver does not exceed 1^0 [9]. And in accordance with(2) it means , that $m_{qo}-B_t$ is equal to $1,7^0$. The criteria condition for this case will be:

$$\frac{1}{n}\sum_{i=1}^{n}\left(q_0 - B_t\right) \le 5,1^0$$

Consequently, it the mean value of a measured difference between the COG go of ship-observer obtained by SNS and read from the electronic radar cursor, the bearing of the ship-target does not exceed $5,1^0$, the ships meeting situation with the probability of 99,7% should be attributed to the jurisdiction of Rule 14 of COLREG-72. Otherwise, with the same probability the situation should be considered as the situation of ships meeting on the crossing courses (Rule 15 of COLREG-72)

If we talk about the practical implementation of the discussed ships' meeting situation is, of course, this problem must be solved by the computing devices, which receives the information about COG of the ship-observer and about the bearing of the ship-target required for the numerical representation of the criteria condition.

3 CONCLUSION

As the navigation equipment used ships allows to define the difference of COG of the ship-observer

and the bearing of the ship-target with mean-square error of the order of $1,5^0$-2^0, this means that in practice the difference of mean value q_0-B_t is less that $4,5^0$-6^0, that is, it is less than half rhumb and ships meeting situation should be attributed to Rule 14 of COLREG-72. Doesn't this consideration explain the very half rhumb $\pm180^0$ in the difference COG of ship-observer and COG of the ship-target adopted in the foreign arbitration practice to differentiate the ships meeting situation by reciprocal COG (courses) in head-on situation from the crossing COG (courses)? In our view, the question itself and the results indicate that there is a basis for the proposal of specifying Rule 14 of COLREG-72.

SOURCES USED

1. Preventing Collision at Sea, 1972.–M.: RosKonsult, 2001.- 84 s
2. Lushnikov E.M Navigation safety sailing. - Kaliningrad: BFFSA, 2007 . - 260 p.
3. Moscow E. Collision in the ocean. – Moskwf, Transport, 1989.
4. Lushnikov E.M. Safety sailing and fishering. - Kaliningrad: BFFSA,2003.-156 p
5. Bukaty V. M. On determination of head-on situation uder Rule 14 COLREG-72. International Conference @Marine Navigation and Safety at sea Transportation? Gdynia, Poland< 2009. – London: Taylor&Francis Group, 2009. – p. 277-282
6. Karapuzov A. I., Mironov A. V. Maneuring large vessels. – Novorossijsk: NSMA, 2005. 152 p.
7. Usikov V F., Kirillov N. O. Sovremennye kursoukazateli. – Kaliningrad:BSAFF, 2013.-294 s.
8. PK-2306. RK-2306. Ustroistvo distacionnoi peredachi kursa (sputnikovyi kompas) na baze GNSS GLONASS/GPS serii "Farvater". Rukovodstvo po ekspluatacii. – OOO "Radio Compleks", 2012. - 124 s.
9. Peskov U.A. Morskaia navigacia s GLONASS/GPS/ - M.: Morkniga, 2010-344 s.

Visualization of Holes and Relationships Between Holes and Latent Conditions

K. Fukuoka
Graduate School of Maritime Sciences, Kobe University, Kobe City, Japan
Japan Transport Safety Board, Tokyo, Japan

ABSTRACT: The objective of this research is to find the relationships between holes and latent conditions in collisions. If the relationships can be determined, efforts can be directed to specified latent conditions in order to prevent collisions. 28 collision cases published by the Japan Transport Safety Board were studied. This research considered the Safety Management System and risk management imbedded in the collision avoidance action as defensive layers in the Swiss cheese model. The research found the following results. First, most holes at the organizations opened during the process of Do of the Plan-Do-Check-Act (PDCA) cycle. Second, most holes at the local workplaces opened during the risk analysis of the risk management process. Finally, inadequate conditions of operators were the most common among other latent conditions. Visualization of holes and awareness of latent conditions will show people where efforts are directed for the prevention of collisions effectively and efficiently.

1 INTRODUCTION

The Swiss cheese model theorized by Reason (1990, 1997) indicates that there are a number of defensive layers and their associated holes between hazards and potential losses. These holes are in continuous motion, moving from place to place, opening and shutting. Holes are caused by latent conditions and active failures. No one can foresee all possible accident scenarios. Therefore, some defensive layers with holes will be present from the establishment of the system or will develop unnoticed or uncorrected during system operations. When such holes line up in a number of defensive layers, hazards come into direct contact with potential losses and an accident occurs.

Requirements for avoiding potential losses and accidents are to prevent the holes in Swiss cheese model from lining up. This means that when one of the holes, which had lined up due to latent conditions, is shut, losses and accidents do not occur. Therefore, it is important to find the relationships between holes and latent conditions for preventive measures. The objective of this research is to find the relationships between holes and latent conditions in collisions. If the relationships can be determined, efforts can be directed to specified latent conditions in order to shut holes or prevent collisions.

ISO/IEC GUIDE 51:1999 states that safety does not mean absolute freedom from risk, but it means freedom from unacceptable risk. It is achieved by reducing risks to tolerable levels. This research defined the opening of a hole in a defensive layer as a situation of unacceptable risk that remains in an organization or local workplace. At a local workplace, the research defined the situation of unacceptable risk in a collision as the situation in which two vessels approached within the maximum advance of a give-way vessel (Fukuoka 2015). At an organization, the research defined the situation of unacceptable risk as the situation in which functional requirements related to an accident prescribed by the *International Management Code for the Safe Operation of Ships and for Pollution Prevention* (ISM Code) (IMO 1993) were not satisfied. The research defined 10 latent conditions by modifying the concepts of the Software-Hardware-Environment-Liveware (SHEL) model (Hawkins 1987), and the *IMO/ILO Process for Investigating Human Factors* (IMO 2000a). The 10 latent conditions are as follows: (1) inadequate passage planning, (2) inadequate procedures, (3) inadequate or deviations from rules, (4) inadequate human-machine interface, (5) inadequate conditions of equipment, (6) adverse environment, (7) inadequate conditions of operators, (8) inadequate communication, (9) inadequate team work at local workplace, and (10) inadequate management by an organization.

The research considered the Safety Management System (SMS) and risk management as defensive layers in the Swiss cheese model (Fukuoka 2015). The focus of this research was on the processes both of the SMS and risk management imbedded in the collision avoidance action and on the latent conditions at the time of the accidents. With regard to the process of the SMS, the PDCA cycle described by the ISO (2008a, 2008b) was used. For risk management process the terms defined by the International Standard 31000 (ISO 2009) and 31010 (IEC 2009) were used. Fukuoka (2015) demonstrated how the holes opened and moved in a collision, grounding and occupational casualties by using real cases of three types of marine accidents.

The research studied 28 collision cases of serious marine accidents published from 2010 to 2014 by the Japan Transport Safety Board (2010, 2011, 2013, 2014), 4 of which were investigated as the investigator-in-charge. These investigation reports were reviewed and locations of holes at local workplaces and organizations were identified by using analysis methods to find the opening and moving of holes. The latent conditions related to the collisions were then clarified. Finally, these latent conditions were classified into 10 groups in accordance with the definitions of latent conditions.

The research found the following results. First, most holes at the organizations opened during the process of Do of the PDCA cycle, which indicated that established safety plans related to the accidents were not executed at the local workplaces, followed by Plan, which indicated that safety plans related to the accidents were not established at the organizations. Second, most holes at the local workplaces opened during the risk analysis of the risk management process, which indicated that most vessels did not use all available means to determine if risk of collision existed; that radar equipment was not properly used; and that assumptions were made on the basis of insufficient information. Third, inadequate conditions of operators and inadequate or deviation from rules were common latent conditions that caused holes to open and were closely related to each other in collisions. Finally, in addition to inadequate conditions of operators and inadequate or deviation from rules, other latent conditions such as adverse environment, inadequate communication, inadequate management, inadequate team work, inadequate passage planning, inadequate procedures, inadequate human-machine interface, and inadequate conditions of equipment were also causes of holes.

This research indicated that locations of holes in defensive layers were visualized and that inadequate conditions of operators and inadequate or deviation from rules were common latent conditions and closely related to each other in collisions. People can come to understand the locations of holes and latent conditions that caused holes to open by using the 10 latent conditions and the processes of both the PDCA cycle and risk management. Visualization of holes and awareness of latent conditions will show people where efforts are directed for the prevention of collisions effectively and efficiently.

2 METHODS

2.1 Definition of a Hole

Holes are caused by latent conditions and active failures. No one can foresee all possible accident scenarios. Therefore, some defensive layers with holes caused by latent conditions will be present from the establishment of the system or will develop unnoticed or uncorrected during system operations. Holes caused by active failures are triggered by operators' unsafe acts and appear immediately. An unsafe act is an error or a violation performed in the presence of a potential hazard. An unsafe act can be performed in a hazardous situation (Reason 1990, 1997). According to ISO/IEC GUIDE 51:1999, safety does not mean absolute freedom from risk, but it means freedom from unacceptable risk. It is achieved by reducing risks to tolerable levels. The research defined the opening of a hole in a defensive layer as a situation of unacceptable risk that remained in an organization or local workplace. Unacceptable risk exceeds the limit of the tolerable region defined by the ALARP (as low as reasonably practicable) principle.

When defining a hole in each collision case hazardous situations which resulted in an accident because an unsafe act can be performed in a hazardous situation were considered. A hazardous situation is a circumstance in which people, property, or environments are exposed to one or more hazards defined by ISO/IEC GUIDE 51:1999. According to Honda (2008), when taking collision avoidance action, a hard over turn is more effective than a crash astern stop maneuver and other maneuverings. For a local workplace, the research defined the situation of unacceptable risk in a collision as the situation that two vessels approached within the maximum advance of a give-way vessel (Fukuoka 2015).

The SMS in the marine domain is established by the requirements of the ISM Code, which is based on quality management systems of the International Standard (ISO 2008a, 2008b). Section 1.4 of ISM Code states the functional requirements for the SMS. At an organization, the research defined the situation of unacceptable risk in a collision as the situation that functional requirements related to an accident prescribed by ISM Code were not satisfied.

2.2 Analysis Methods to Find the Opening and Moving of Holes

When investigating collisions as the investigator-in-charge, the research found that almost all accidents were related to the procedures of collision avoidance action prescribed by the *International Regulations for Preventing Collisions at Sea, 1972* (COLREGs) (Maritime and Coastguard Agency 2004). The research then compared the risk management process with collision avoidance action (Fukuoka 2015).

With regard to risk management process terms defined by the International Standard 31000 (ISO 2009) and 31010 (IEC 2009) were used. The process of risk identification is to find, recognize, and record risks. It includes identifying hazards in the context of physical harm. The process of risk analysis involves understanding risk and consists of determining the consequences and probabilities of the harm caused by identified risk events. The process of risk evaluation is to determine the level of significance and type of risk. The process of risk treatment is to select and implement relevant options for changing the probability of occurrence, the effect of risks, or both after completing a risk assessment. The process of monitoring and reviewing risk assessment is to update risk assessment when necessary.

Procedures of collision avoidance action are based on the COLREGs. The research considered that Rule 5 included the process of risk identification; Rule 7 (a), (b), and (c), the process of risk analysis; Rule 7 (d), the process of risk evaluation; Rule 8 (a) and (b), the process of risk treatment; Rule 8 (d), the process of monitoring and review. In restricted visibility, Rule 19 (d) first paragraph included the process of risk analysis; Rule 19 (d) (i) , (ii) and 19 (e), the process of risk treatment.

Portions of the COLREGs are as follows: (1) Rule 5 states that every vessel must always maintain a proper look-out to make a full appraisal of the risk of collision; (2) Rule 7 (a), (b), and (c) state that every vessel must use all available means to determine if risk of collision exists. Radar equipment must be used properly, including long-range scanning to obtain early warnings of risk of collision and systematic observation of detected objects. Assumptions must not be made on the basis of insufficient information; (3) Rule 7 (d) states that the risk of collision exists if the compass bearing of an approaching vessel does not appreciably change; (4) Rule 8 (a) and (b) state that any collision avoidance action must be taken in accordance with this rule and be made in ample time and with good seamanship. Any alteration of course and/or speed to avoid collision must be large enough to be apparent to another vessel; (5) Rule 8 (d) states that the collision avoidance action must result in passing at a safe distance. The effectiveness of the action must be carefully checked until the other vessel is past and clear.

In restricted visibility, first paragraph of Rule 19 (d) states that a vessel which detects by radar alone the presence of another vessel must determine if a close-quarters situation is developing and/or risk of collision exists. Rule 19 (d) (i) and (ii) state that when a vessel detects by radar alone the presence of another vessel she must not take an alteration of course to port for the another vessel forward of the beam other than for a vessel being overtaken, or not take an alteration of course towards a vessel abeam or abaft the beam. Rule 19 (e) also states that every vessel must reduce her speed to the minimum at which she can keep her course when she hears forward of her beam the fog signal of another vessel, or cannot avoid a close-quarters situation with another vessel forward of her beam.

With regard to the process of the SMS, the PDCA cycle described by the ISO (2008a, 2008b) was used.

To find the opening and moving of holes in defensive layers, this research focused on the processes of the SMS and risk management which were observed by an organization and local workplace at the time of an accident. At a local workplace, prior to a system operation, the risks associated with the system must be reduced until they fall into a tolerable region by means of protective measures. In collisions, reducing risk level means taking measures of collision avoidance action prescribed by the COLREGs. When measures of collision avoidance action were not taken by an operator, the research considered this situation to be holes in a defensive layer of risk management opened during one of the processes and moved through the processes until the accident occurred. At an organization, when functional requirements related to an accident prescribed by ISM Code were not satisfied, the research considered this situation to be holes in a defensive layer of SMS opened during one of the processes of the PDCA cycle and moved through the cycle until the accident occurred (Fukuoka 2015).

2.3 Definitions of Latent Conditions

For definitions of latent conditions the research modified concepts of the SHEL model and the *IMO/ILO Process for Investigating Human Factors* (IMO 2000a). Software includes organizational policies, procedures, manuals, checklist layout, and charts. Hardware includes the design of work stations, displays, and controls. Environment includes the internal and external climate, temperature, visibility, and other factors which constitute conditions within which people are working. The central liveware includes capabilities and limitations of an operator. The peripheral liveware includes management, supervision, crew interactions, and communications.

The research defined 10 latent conditions: (1) inadequate passage planning, (2) inadequate procedures, (3) inadequate or deviations from rules which indicates an operator does not take appropriate measures in accordance with such rules as the COLREGs, (4) inadequate human-machine interface, (5) inadequate conditions of equipment, (6) adverse environment, (7) inadequate conditions of operators which includes physical, physiological, psychological, psychosocial conditions, and knowledge, skill, experience, education and training, (8) inadequate communication which includes such communication as among the bridge team, between a pilot and the bridge team, or bridge and Vessel Traffic Services, (9) inadequate team work at a local workplace, and (10) inadequate management by an organization.

Of those 10 latent conditions, (1), (2) and (3) are liveware-software interactions; (4) and (5) are liveware-hardware interactions; (6) is a liveware-environment interaction; (7) is the central liveware; and (8), (9) and (10) are the central liveware-the peripheral liveware interactions. Adams (2006) states that standardized procedures include governmental regulations, checklists, station bills, voyage plans, standing orders of captains, and company rules. When investigating various kinds of marine accidents, the research found that operators were using different kinds of standardized procedures at the time of such accidents, including collisions, groundings, and occupational death or injuries. Therefore, the standardized procedures were divided into three categories: passage planning, procedures, and rules. The research defined procedures as the rest of the standardized procedures defined by Adams which excluded passage planning and rules (governmental regulations). Definitions of passage planning were based on the concept of Swift (2000) and IMO (2000b), which consist of the stages of appraisal, planning, execution, and monitoring. Rules included the COLREGs, STCW, SOLAS, and local navigation rules. Deviations from rules were used in this research because collision avoidance action proscribed by the COLREGs is clear and comprehensive. For adverse environment traffic density and geographical features of the waters were included because proceeding to those waters was decided by organizations. Team work referred to roles and responsibilities of crew, a pilot and others related to an accident. Inadequate management was the situation in which functional requirements related to an accident prescribed by ISM Code were not satisfied. Management also included safety culture defined by Reason (1997).

2.4 Scope of this Study

The research studied 28 collision cases of serious marine accidents investigated from 2008 to 2014 by the Japan Transport Safety Board (Japan Transport Safety Board 2010, 2011, 2013, 2014). These investigation reports were reviewed; locations of holes at local workplaces and organizations were identified by using the analysis methods to find the opening and moving of holes; and the latent conditions related to the collisions were clarified. These latent conditions were classified into 10 groups in accordance with the definitions of latent conditions. The collisions between fishing vessels and/or pleasure boats were not in the scope of this research because the Swiss cheese model applies only to organizational accidents.

3 RESULTS

3.1 Locations of Holes at Organizations and Local Workplaces

The total number of organizations studied was 23. One organization among 23 had two holes in one defensive layer of the SMS. Day and night show the time of an accident occurrence. At 12 of the 23 organizations, holes opened during the process of Do. At eight organizations, holes opened during the process of Plan (Table 1). The total number of local workplaces studied was 54. When a pilot was onboard, one vessel had two defensive layers of risk management at a local workplace: one of the pilot and the other of the master. At 38 out of 54 local workplaces, holes opened during the process of risk analysis. At 10 local workplaces, holes opened during the process of risk identification (Table 2).

Table 1. Locations of holes observed at 23 organizations.

Locations	Day Number	Night Number	All day Total
Plan	2	6	8
Do	5	7	12
Check	2	2	4
Action	0	0	0
Total number	9	15	24

Table 2. Locations of holes observed at 54 local workplaces.

Locations	Day Number	Night Number	All day Total
Risk identification	4	6	10
Risk analysis	16	22	38
Risk evaluation	1	0	1
Risk treatment	1	4	5
Monitoring and review	0	0	0
Total number	22	32	54

3.2 Latent Conditions

The total number of 237 latent conditions was categorized into 10 groups. Two local workplaces did not include inadequate conditions of operators because the investigation reports did not mention such. Inadequate conditions of operators were the most common latent conditions. Inadequate or deviation from rules were the second most common latent conditions (Table 3). Among inadequate conditions of operators, psychological conditions counted for 51 out of 57, which included multiple factors (Table 4). Assumptions accounted for 42 out of 51 psychological conditions. The most frequent pattern of assumptions was that operators assumed that another vessel would take collision avoidance action or alter her course and/or reduce her speed (Table 5).

Table 3. Number of latent conditions observed at 54 local workplaces.

Locations	Day Number	Night Number	All day Total
Inadequate passage planning	5	10	15
Inadequate procedures	1	6	7
Inadequate or deviations from rules	21	30	51
Inadequate human-machine interface	1	6	7
Inadequate conditions of equipment	0	3	3
Adverse environment	14	24	38
Inadequate conditions of operators	21	31	52
Inadequate communication	10	14	24
Inadequate team work	4	13	17
Inadequate management	9	14	23
Total number of latent conditions	86	151	237

Table 4. Subdivisions of 52 inadequate conditions of operators.

Subdivisions of inadequate conditions of operators	Number
Physical conditions	0
Physiological conditions	3
Psychological conditions	51
Psychosocial conditions	0
Skill, experience, education and training	3
Total number of subdivisions of 52 inadequate conditions of operators	57

Table 5. Patterns of assumptions categorized in 51 psychological conditions.

Patterns of assumptions	Number
Another vessel would take collision avoidance action or alter her course and/or reduce her speed	14
Own vessel would pass starboard to starboard, port to port etc. to another vessel	7
Another vessel would not approach to own vessel	5
Another vessel would take certain route	4
Own vessel would overtake another vessel safely	3
Approaching vessel would not exist	3
Others	6
Total number of assumptions	42

4 DISCUSSION

The research found the locations of holes and relationships between holes and latent conditions in collisions in the following manner. (1) At organizations the most common holes opened during the process of Do, which meant that established safety plans related to the accidents were not executed at the local workplaces. The second most common holes opened during the process of Plan, which meant that safety plans related to the accidents were not established at organizations. These holes moved through the PDCA cycle until the accident occurred. Reason states that some defensive layers with holes will be present from the establishment of the system or will develop unnoticed or uncorrected during system operations. This finding accords with his theory. (2) At local workplaces the most common holes opened during the process of risk analysis, which meant that most vessels did not use all available means to determine if risk of collision existed, radar equipment was not properly used, and assumptions were made on the basis of insufficient information. The second most common holes opened during the process of risk identification, which meant that vessels did not always maintain a proper look-out to make a full appraisal of the risk of collision, and then vessels did not notice other approaching vessels until accidents occurred. This research could not clarify the manner in which holes in defensive layers lined up theorized by Reason. However, this research identified the locations of holes in defensive layers in collisions. People can notice holes by using both the PDCA cycle and risk management. (3) Inadequate conditions of operators were the most common latent conditions in collisions, followed by inadequate or deviation from rules. Among inadequate conditions of operators, psychological conditions accounted for 51 out of 57. Assumptions accounted for 42 out of 51 psychological conditions. The relationship between inadequate conditions of operators and inadequate or deviation from rules indicated that operators' assumptions tended to intervene into the process of collision avoidance action and lead to deviation from rules. This finding indicated that inadequate conditions of operators and inadequate or deviation from rules were common causes of holes and closely related to each other in collisions. (4) In addition to inadequate conditions of operators and inadequate or deviation from rules, other latent conditions such as adverse environment, inadequate communication, inadequate management, inadequate team work, inadequate passage planning, inadequate procedures, inadequate human-machine interface, and inadequate conditions of equipment were also causes of holes. This finding indicates that holes in collisions were not always caused by the same latent conditions. Therefore, it suggests that it is necessary for people at organizations and local workplaces to find their own latent conditions, which caused the holes to open by using the 10 latent conditions and the processes both of the PDCA cycle and risk management.

Although this research was restricted to show the locations of holes in two kinds of defensive layers in the Swiss cheese model, the visualization can show the location in which effort toward preventive measures should be directed. Some investigation reports did not mention organizational factors and latent conditions; therefore, the findings of this research are not comprehensive. When these organizational factors and latent conditions are investigated thoroughly, the location of holes and the relationship between the holes and latent conditions will become more accurate.

This research was restricted to collisions; however, the methods can be applied to other kinds of accidents as long as the SMS and risk management are established, and the 10 latent conditions are modified to fit the domain.

5 CONCLUSIONS

This research indicates that locations of holes in defensive layers were visualized and that inadequate conditions of operators and inadequate or deviation from rules were common latent conditions and closely related to each other in collisions. Visualization of holes and awareness of latent conditions which caused the holes to open will show people where efforts are directed and the prevention of collisions effectively and efficiently.

ACKNOWLEDGEMENTS

I thank the Japan Transport Safety Board when I conducted marine accident investigations and this research.

REFERENCES

Adams, M. 2006. *Shipboard Bridge Resource Management.* Eastport : Nor'easter Press.
Fukuoka, K. 2015. *WMU Journal of Maritime Affairs, Visualization of a hole and accident preventive measures based on the Swiss cheese model developed by risk management and process approach.* New York: Springer New York, LCC. Doi:10.1007/s13437-015-0076-2.
Hawkins, F. 1987. *Human Factors in Flight.* Aldershot : Gower Technical Press Ltd.
Honda, K. 2008. *Sousen Tsuron.* Tokyo :Seizando-Shoten Publishing Co., Ltd.(Japanese)
IEC. 2009. *IEC/ISO 31010:2009 Risk Management-Risk Assessment Techniques.*
IMO. 1993. *The International Management Code for the Safe Operation of Ships and for Pollution Prevention (International Safety Management (ISM) Code), Resolution A.741(18).*
IMO. 2000a. *Amendments to the Code for the Investigation of Marine Casualties and Incidents, Resolution A.884(21), Appendix 1 The IMO/ILO Process for Investigating Human Factors.*
IMO. 2000b. *Guidelines for Voyage Planning, Resolution A.893(21).*
ISO/IEC. 1999. *ISO/IEC GUIDE 51:1999(E) (1999) Safety Aspects - Guidelines for their Inclusion in Standards.*
ISO. 2008a. *ISO 9000 Introduction and Support Package: Guidance on the Concept and Use of the Process Approach for Management Systems, ISO/TC 176/SC 2/N 544R3.*
ISO. 2008b. *ISO 9001:2008 Quality Management Systems – Requirements.*
ISO. 2009. *ISO 31000:2009 Risk Management-Principles and Guidelines.*
Japan Transport Safety Board. 2010. *Marine Accident Investigation Report MA2010-4-2: Collision of cargo vessel SHURI and cargo vessel KOUHEI-MARU.* (Japanese) Available from http://www.mlit.go.jp/jtsb/ship/rep-acci/2010/MA2010-4-2_2010tk0003.pdf
Japan Transport Safety Board. 2010. *Marine Accident Investigation Report MA2010-10-1: Collision of cargo vessel MAY STAR and fishing vessel MYOJIN-MARU.* (Japanese) Available from http://www.mlit.go.jp/jtsb/ship/rep-acci/2010/MA2010-10-1_2010tk0004.pdf
Japan Transport Safety Board. 2010. *Marine Accident Investigation Report MA2010-7-1: Collision of cargo vessel SUN GRACE and cargo vessel SEISHIN-MARU.* (Japanese) Available from http://www.mlit.go.jp/jtsb/ship/rep-acci/2010/MA2010-7-1_2010tk0013.pdf
Japan Transport Safety Board. 2010. *Marine Accident Investigation Report MA2010-5-1: Collision of cargo vessel NORD POWER and cargo vessel HAI YING.* (Japanese) Available from http://www.mlit.go.jp/jtsb/ship/rep-acci/2010/MA2010-5-1_2008tk0003.pdf
Japan Transport Safety Board. 2010. *Marine Accident Investigation Report MA2010-10-4: Collision of cargo vessel SUMIRIKI-MARU No7 and fishing vessel DAIGYO-MARU.* (Japanese) Available from http://www.mlit.go.jp/jtsb/ship/rep-acci/2010/MA2010-10-4_2010tk0016.pdf
Japan Transport Safety Board. 2011. *Marine Accident Investigation Report MA2011-2-1: Collision of chemical tanker SANSHUN-MARU and fishing vessel SHIN KISSHO-MARU.* (Japanese) Available from http://www.mlit.go.jp/jtsb/ship/rep-acci/2011/MA2011-2-1_2010tk0029.pdf
Japan Transport Safety Board. 2011. *Marine Accident Investigation Report MA2011-5-7: Collision of cargo vessel DAIO DISCOVERY and cargo vessel AURORA SAPPHIRE.* (Japanese) Available from http://www.mlit.go.jp/jtsb/ship/rep-acci/2011/MA2011-5-7_2009yh0075.pdf
Japan Transport Safety Board. 2011. *Marine Accident Investigation Report MA2011-12-1: Collision of cargo vessel OCEAN SEAGULL and cement carrier SYMISE-MARU No 2.* (Japanese) Available from http://www.mlit.go.jp/jtsb/ship/rep-acci/2011/MA2011-12-1_2011tk0024.pdf
Japan Transport Safety Board. 2011. *Marine Accident Investigation Report MA2011-7-3: Collision of chemical tanker KINYO-MARU and tug boat KAIRYU barge MARUSEN 2.* (Japanese) Available from

http://www.mlit.go.jp/jtsb/ship/rep-acci/2011/MA2011-7-3_2011tk0010.pdf

Japan Transport Safety Board. 2011. *Marine Accident Investigation Report MA2011-11-4: Collision of car carrier CYGNUS ACE and cargo vessel ORCHID PIA.* (Japanese) Available from http://www.mlit.go.jp/jtsb/ship/rep-acci/2011/MA2011-11-4_2009tk0004.pdf

Japan Transport Safety Board. 2011. *Marine Accident Investigation Report MA2011-10-5: Collision of oil tanker TAIYO No23 and cargo vessel KATSU-MARU No38.* (Japanese) Available from http://www.mlit.go.jp/jtsb/ship/rep-acci/2011/MA2011-10-5_2011tk0027.pdf

Japan Transport Safety Board. 2011. *Marine Accident Investigation Report MA2011-3-1: Collision of cement tanker FUYO-MARU No3 and fishing vessel SHOFUKU-MARU No18.* (Japanese) Available from http://www.mlit.go.jp/jtsb/ship/rep-acci/2011/MA2011-3-1_2010tk0027.pdf

Japan Transport Safety Board. 2011. *Marine Accident Investigation Report MA2011-10-3: Collision of cargo vessel MARINE STAR and container vessel TAKASAGO.* (Japanese) Available from http://www.mlit.go.jp/jtsb/ship/rep-acci/2011/MA2011-10-3_2009tk0005.pdf

Japan Transport Safety Board. 2011. *Marine Accident Investigation Report MA2011-6-3: Collision of container vessel SKY LOVE and cargo vessel HAEJIN.* (Japanese) Available from http://www.mlit.go.jp/jtsb/ship/rep-acci/2011/MA2011-6-3_2010tk0030.pdf

Japan Transport Safety Board. 2012. *Marine Accident Investigation Report MA2012-1-6: Collision of cargo vessel WIEBKE and cargo vessel MARINE PEACE.* (Japanese) Available from http://www.mlit.go.jp/jtsb/ship/rep-acci/2012/MA2012-1-6_2011tk0021.pdf

Japan Transport Safety Board. 2012. *Marine Accident Investigation Report MA2012-6-3: Collision of cargo vessel SHIN KENWA-MARU and cargo vessel SHOWA-MARU No8.* (Japanese) Available from http://www.mlit.go.jp/jtsb/ship/rep-acci/2012/MA2012-6-3_2012tk0021.pdf

Japan Transport Safety Board. 2012. *Marine Accident Investigation Report MA2012-7-3: Collision of cargo vessel SEIREI-MARU and cargo vessel GYOREN 1.* (Japanese) Available from http://www.mlit.go.jp/jtsb/ship/rep-acci/2012/MA2012-7-3_2012tk0027.pdf

Japan Transport Safety Board. 2012. *Marine Accident Investigation Report MA2012-9-2: Collision of cargo vessel MEDEA and fishing vessel KOSEI-MARU.* (Japanese) Available from http://www.mlit.go.jp/jtsb/ship/rep-acci/2012/MA2012-9-2_2012tk0028.pdf

Japan Transport Safety Board. 2012. *Marine Accident Investigation Report MA2012-2-1: Collision of cargo vessel HARMONY WISH and cargo vessel SHINKAZURYU.* (Japanese) Available from http://www.mlit.go.jp/jtsb/ship/rep-acci/2012/MA2012-2-1_2011tk0029.pdf

Japan Transport Safety Board. 2012. *Marine Accident Investigation Report MA2012-1-5: Collision of cargo vessel TY EVER and cargo vessel LOFTY HOPE.* (Japanese) Available from http://www.mlit.go.jp/jtsb/ship/rep-acci/2012/MA2012-1-5_2011tk0020.pdf

Japan Transport Safety Board. 2012. *Marine Accident Investigation Report MA2012-5-2: Collision of cargo vessel DAISENZAN and leisure fishing vessel HISA-MARU.* (Japanese) Available from http://www.mlit.go.jp/jtsb/ship/rep-acci/2012/MA2012-5-2_2012tk0009.pdf

Japan Transport Safety Board. 2012. *Marine Accident Investigation Report MA2012-11: Collision of cargo vessel RYUNAN II and leisure fishing vessel KOYO-MARU.* (Japanese) Available from http://www.mlit.go.jp/jtsb/ship/rep-acci/2012/MA2012-11-3_2011tk0009.pdf

Japan Transport Safety Board. 2013. *Marine Accident Investigation Report MA2013-2-3: Collision of cargo vessel MARUKA and fishing vessel KAIRYO-MARU No18.* (Japanese) Available from http://www.mlit.go.jp/jtsb/ship/rep-acci/2013/MA2013-2-3_2011tk0040.pdf

Japan Transport Safety Board. 2013. *Marine Accident Investigation Report MA2013-11-2: Collision of container vessel TIAN FU and chemical tanker SENTAI-MARU.* (Japanese) Available from http://www.mlit.go.jp/jtsb/ship/rep-acci/2013/MA2013-11-2_2012tk0033.pdf

Japan Transport Safety Board. 2013. *Marine Accident Investigation Report MA2013-1-2: Collision of cargo vessel AQUAMARINE and fishing vessel HIRASHIN-MARU.* (Japanese) Available from http://www.mlit.go.jp/jtsb/ship/rep-acci/2013/MA2013-1-2_2011tk0014.pdf

Japan Transport Safety Board. 2014. *Marine Accident Investigation Report MA2014-5: Collision of container vessel KOTA DUTA and cargo vessel TANYA KARPINSKAY.* (Japanese) Available from http://www.mlit.go.jp/jtsb/ship/rep-acci/2014/MA2014-5-1_2012tk0003.pdf

Japan Transport Safety Board. 2014. *Marine Accident Investigation Report MA2014-1-4: Collision of container vessel YONG CAI and fishing vessel SHINYO-MARU No2.* (Japanese) Available from http://www.mlit.go.jp/jtsb/ship/rep-acci/2014/MA2014-1-4_2012tk0023.pdf

Japan Transport Safety Board. 2014. *Marine Accident Investigation Report MA2014-6-5: Collision of bulk carrier NIKKEI TIGER and fishing vessel HORIEI-MARU.* (Japanese) Available from http://www.mlit.go.jp/jtsb/ship/rep-acci/2014/MA2014-6-5_2012tk0037.pdf

Maritime and Coastguard Agency. 2004. *Merchant Shipping Notice MSN 1781(M+F): The Merchant Shipping (Distress Signals and Prevention of Collision) Regulations 1996.* Southampton: Maritime and Coastguard Agency.

Reason, J. 1990. *Human Error.* Cambridge : Cambridge University Press.

Reason, J. 1997. *Managing the Risks of Organizational Accidents.* Surrey: Ashgate Publishing Limited.

Swift, MNI. 2000. *Bridge Team Management: A Practical Guide.* London: The Nautical Institute.

Global Navigation Satellite System (GNSS)

Global Navigation Satellite System (GNSS)
Activities in Navigation – Marine Navigation and Safety of Sea Transportation – A. Weintrit (ed.)

Nominal Unique BeiDou Satellite Constellation, its Advantages and Disadvantages

J. Januszewski
Gdynia Maritime University, Gdynia, Poland

ABSTRACT: Spatial segment is one of three segments of each satellite navigation systems (SNS). In the case of China BeiDou system this segment will consist of 35 satellites, this number is the greatest among all operational (GPS, GLONASS) or under construction (Galileo) SNSs. Additionally BeiDou constellation uses three types of orbits – medium (MEO), geostationary (GEO) and inclined geosynchronous (IGSO). The results of the analysis of BeiDou geometry (visibility of satellites, dilution of precision coefficients) for the users in different regions in the world, in Europe in particular, and the comparison of the geometry in the areas where the BeiDou GEO and IGSO satellites can be used and in the areas where these satellites cannot be used will be presented. The knowledge of space segment is very important for all users because the type of navigation messages transmitted by satellites depends on theirs orbits. The NAV message broadcast by MEO/IGSO and GEO satellites is D1 and D2 respectively. As the information about integrity navigation (message D2) is transmitted by GEO satellites only the identification of the areas where this information can be received is very important for all BeiDou users, the maritime transport users also.

1 INTRODUCTION

In early 1980s, when the first satellite navigation systems (SNS), Transit in USA and Cikada in Soviet Union were already operational, China began actively study the satellite systems in line with Chinas's conditions. This new system, called BeiDou (Big Dippler), earlier Compass, as all other SNSs, consists of three major components: the space constellation, the ground control segment and the user segment. In 2000, the BeiDou Navigation Satellite Demonstration System was established. With 14 operational satellites – 5 GEO (Geostationary Earth Orbit), 5 IGSO (Inclined GeoSynchronous Orbit) and 4 MEO (Medium Earth Orbit) – BeiDou system (BDS) has formally begun to provide services to China and the surrounding areas since December 2012 (Report...2013, www.insidegnss.com).

The functions and performance parameters at this step are as follows (Report...2013):
- position accuracy: better than 10 m;
- velocity accuracy: better than 0.2 m/s;
- time accuracy: 50 ns;
- short message communications: 120 Chinese characters per message.

Global services BDS will provide by around 2017/2018 (Munich 2014, imo.org). The final BeiDou space constellation will consist of 35 satellites:
- 5 GEO (058.75^O E, 080° E, 110.5^O E, 140° E and 160° E), altitude 35,786 km;
- 27 MEO (altitude 21,500 km, inclination 55°, 3 evenly distributed orbital planes, on each 9 evenly distributed satellites), the nominal values of geographical longitude of ascending node and argument of latitude of all these 27 satellites are showed in the table 1. The phase is selected from the Walker 24/3/1 constellation, and the right ascension of ascending node of the satellites in the first orbital plane is 0°;
- 3 IGSO (altitude 36,000 km, inclination 55°, evenly distributed in 3 orbital planes). The tracks of sub-satellite points for those IGSO satellites are coincided while the longitude of the intersection point is 118° E, with a phase difference of 120°. The sub-satellite tracks for the current (February 2015) other two operational IGSO satellites are coincided while the longitude of the intersection point is at 095° E. These two satellites will not be in final constellation.

It means that the final number of operational satellites (35) will be the greatest among all currently operational (GPS – 32, GLONASS – 24) or under construction (Galileo – 30) SNSs. BeiDou constellation, as the only SNS, uses three mentioned above types of orbits, while all others use MEO orbits only (Januszewski J. 2010, Martin H. 2013).

BDS will meet the demands of China's national security, economic development. This new SNS will be committed to providing stable, reliable and quality satellite navigation services for global users.

The problem of the positioning using the BeiDou satellite constellation and the caalculations of GDOP coefficient values from multi-satellite navigation systems constellations are the theme of many publications already, e.g. (Aigong X. 2014) and (Tang Y. 2015), respectively.

2 NAVIGATION MESSAGES AND SATELLITES ORBITS

The BeiDou navigation (NAV) messages are formatted in D1 and D2 based on their rate and structure in superframe, frame and subframe. The first format is transmitted by all 35 satellites, the second by GEO satellites only. The rate of these messages is 50 bps and 500 bps respectively (www.beidou.gov.cn).

D1 NAV message contains basic, named sometimes fundamental also, information about broadcasting 27 MEO and 3 IGSO satellites, almanac information (10 parameters, week number and health) for all these 30 satellites and BDT (BeiDou time) offsets from other global SNSs (GPS, GLONASS and Galileo).

Table 1. BeiDou system, geographical longitude of ascending node (An) and argument of latitude (Al) of 27 MEO satellites

Orbit I, An $= 0^O$		Orbit II, An $= 120^O$		Orbit IIII, An $= 240^O$	
No of satellite	Al [deg]	No of satellite	Al [deg]	No of satellite	Al [deg]
1	0	10	13	19	26
2	40	11	53	20	66
3	80	12	93	21	106
4	120	13	133	22	146
5	160	14	173	23	186
6	200	15	213	24	226
7	240	16	253	25	266
8	280	17	293	26	306
9	320	18	333	27	346

D2 NAV message contains mentioned above basic NAV information of the broadcasting satellite and additionally augmentation service information – integrity and differential correction information, ionospheric grid information. In the case of this format the definition of basic NAV information is the same as that in format D1 except the page number (Pnum) and seconds of week (SOW), which

are different from these in format D1 (BeiDou, Signal. 2013, BeiDou, Open. 2013).

Whereas the constellation of 27 MEO satellites will provide global coverage at any moment, the constellation of 3 IGSO and 5 GEO satellites will can be used in limited area only. It means that the advantages of D2 NAV message, integrity information in particular will be accessible to users in China and the Asia−Pacific region only.

The satellites are launched from the XSLC (Xichang Satellite Launch Center) in the southwestern province Sichuan of China, using China's Long March 3 CZ launch vehicles (www.beidou.gov.cn).

BeiDou satellite parameters depends on its orbit. Some parameters are the same in all satellites, some differ (tab.2), e.g. power output of MEO satellites is less than in case of two other types of satellites considerably (www.directory.eoportal.org).

3 TEST METHODS

In spite of actually (December 2014) space constellation since two years (December 2012) still consists of 14 satellites only (5 GEO, 5 IGSO, 4 MEO) the calculations were made for mentioned above final space segment with 35 satellites.

Table 2. BeiDou system, parameters of GEO, IGSO and MEO satellites (www.isg.org)

Parameter	Satellite		
	GEO	IGSO	MEO
Satellite bus	DFH−3B	DFH−3B	Navigation satellite bus
Launch vehicle	CZ−3C	CZ−3A	CZ−3B
Launch mass [kg]	4600	4200	800
Dry mass [kg]	1550	1900	no details
Body size [m]	1.8 x 2.2 x 2.5		
Solar array size [m]	3 x 2.2 x 1.7 (two pieces)		
Span width [m]	17.7	17.7	17.7
Cross section [m2]	27	27	27
Power output [kW]	6.8	6.2	1.5
Space design line [years]	8	8	5

All calculations based on reference ellipsoid WGS−84 were made on author's simulating program. The interval of the latitude of the observer between 0° and 90° was divided into 9 zones, each 10° wide. In the observer's receiver masking elevation angle H_{min} was assumed to be 0° (horizon), 5° (the most frequently used value of H_{min}), 10°, 15°, 20° and 25°. The angle 25° is representative for the positioning in restricted area where the visibility of satellites can be limited. This problem is very important in road transport (urban canyon) and in maritime transport.

For each zone of latitude and for each masking elevation angle (H_{min}) one thousand (1000) geographic–time coordinates of the observer were

generated by random–number generator with uniform distribution:
- latitude interval 0 – 600 minutes (10°),
- longitude interval 0 −21600 minutes (360°). In the case of geometry of GEO satellites only this interval was divided into 36 zones, each 10° wide.
- time interval 0 – 10,091.48 minutes (13 MEO BeiDou satellite orbital periods, each 12 hours, 56 minutes 16.05 seconds).

For each geographic–time coordinates: the satellite elevation (H), the satellite azimuth (Az), the number of visible satellites (ls) and GDOP (Geometric Dilution of Precision) coefficient values were calculated. Elevation H was divided into 9 intervals, each 10° wide, azimuth (Az) was divided into 8 intervals, each 45° and GDOP value (w) into 8 intervals: w<2, 2≤w<3, 3≤w<4, 4≤w<5, 5≤w<6, 6≤w<8, 8≤w<20, w≥20.

4 GEOMETRY OF MEO SATELLITES ONLY

The lowest (ls_{min}) and the greatest (ls_{max}) number ls of MEO satellites visible by the observer in open area above different H_{min} in all 9 zones of latitudes are presented in the table 3. As for $H_{min} \leq 25^O$ the number ls is greater than 3 for all user's latitudes it means that 3D position can be obtained always and anywhere in the world. Additionally we can say that:
- the number ls depends on observer's latitude, independently of H_{min} value;
- for $H_{min} = 0^O$ the number ls_{min} is the lowest (6) in zone 20−30° and the greatest (9) in zone 0−10° and at latitudes 60−90°, the number ls_{max} is equal 12 always in all 9 latitude zones.

Table 3. System BeiDou, the lowest and the greatest number of MEO satellites visible in open area above H_{min} at different observer's latitudes.

Latitude [^{O}]	H_{min}					
	0°	5°	10°	15°	20°	25°
0–10	9–12	8–12	7–10	5–10	4–8	4–7
10–20	8–12	6–12	6–10	5–9	4–9	4–8
20–30	6–12	6–12	6–10	5–9	4–9	4–7
30–40	7–12	6–12	6–10	5–9	5–9	4–7
40–50	7–12	6–12	6–10	5–10	4–8	4–7
50–60	8–12	6–11	6–10	5–10	4–9	4–7
60–70	9–12	8–12	7–11	6–9	5–9	4–7
70–80	9–12	8–12	8–11	6–9	5–8	4–8
80–90	9–12	9–12	8–10	7–9	6–9	4–8

If the number ls is equal 4 the user must be very careful because the number of satellites which can be used in position determination can decrease at any moment and for any reason. That's why the additional calculations were made for different H_{min} in order to determine the lowest elevation H for which the number ls of satellites visible at different latitudes in open area above this angle is equal 4

(ls4) or 3 (ls3), (table 4). In the case of ls4 the elevation H is the lowest (18) at latitudes 10–30° and 40–60°, the greatest at high latitudes, 60 – 80° (21) and in zone 80–90° (27) in particular. In the case of ls3 the lowest elevation (26°) is in zone 0–10°, the greatest (34) in zone 80–90°.

Percentage of MEO satellites visible by the observer in open area above given angle H in all 9 latitude zones is showed in the table 5. Additional calculations were made for H = 5°, 15° and 25° in all these zones (table 6). We can say that:
- the percentage of satellites visible decreases with angle H in each zone, if this angle is equal 20° this percentage decreases to about 60% or more, if it is 40° to 30% or more;
- in zone 50−60° (latitude of Poland) the percentage of satellites visible above H ≤ 25° is less than in all other zones;
- at latitudes equal or greater than 70° the percentage of satellites visible above H ≤ 20° is greater than in all other latitudes;
- at latitudes less than 70° the percentage of satellites visible above 20° ≤ H ≤ 50° is the greatest in zone 30−40°;
- because of orbit inclination value (55°) and geometrical figure of the Earth the satellites in the zones 70−80° and 80−90° cannot be visible above 70° and 57° respectively. It means that at high latitudes these limitations must be taken into account in area transport in particular.

Table 4. BeiDou system, the lowest elevation H [^{O}] for which the number ls of satellites visible at different latitudes by the observer in open area above this angle is equal 4 or 3.

Latitude [^{O}]	Number ls of satellites	
	4	3
0 – 10	20	26
10 – 20	18	28
20 – 30	18	30
30 – 40	21	29
40 – 50	18	29
50 – 60	18	28
60 – 70	21	27
70 – 80	21	30
80 – 90	27	34

Table 5. Percentage of MEO BeiDou satellites visible in open area above angle (H) at different observer's latitudes (φ), ls_m – weighted mean number of satellites visible above horizon (H = 0°)

Latitude [^{O}]	ls_m	Elevation angle H [^{O}]								
		0	10	20	30	40	50	60	70	80
0 – 10	10.9	100	81.0	59.7	39.7	25.8	15.5	8.2	3.7	0.9
10 – 20	10.7	100	77.0	58.7	43.7	29.0	16.9	9.0	4.0	1.0
20 – 30	10.0	100	78.1	61.5	46.9	34.4	10.2	6.4	3.7	1.3
30 – 40	9.7	100	79.1	62.3	48.0	35.6	25.0	15.2	6.2	1.5
40 – 50	9.9	100	77.7	60.1	45.9	34.2	24.5	15.7	8.2	2.3
50 – 60	10.6	100	75.9	56.4	42.3	30.7	21.2	13.1	6.7	2.0
60 – 70	10.9	100	77.6	60.0	40.5	27.6	17.6	9.2	3.0	0.2
70 – 80	11.0	100	82.9	64.5	42.5	27.7	12.4	3,1	0	0
80 – 90	11.0	100	83.5	66.6	48.6	25.1	3.4	0	0	0

227

Table 6. Percentage of MEO BeiDou satellites visible in open area above angle (H) at different observer's latitudes (φ), ls_m – weighted mean number of satellites visible above horizon (H = 0°)

Latitude [^{O}]	lsm	Elevation angle [^{O}]		
		5	15	25
0 – 10	10.9	90.9	70.3	49.1
10 – 20	10.7	89.1	67.4	51.1
20 – 30	10.0	88.4	69.7	53.9
30 – 40	9.7	88.9	70.1	54.7
40 – 50	9.9	87.6	68.2	52.7
50 – 60	10.6	88.1	65.3	48.7
60 – 70	10.9	90.5	69.5	48.2
70 – 80	11.0	91.3	73.6	54.2
80 – 90	11.0	91.6	74.6	57.7

Distribution (in per cent) of MEO satellites azimuths in open area for H_{min} = 5° and 25° in all 9 latitudes zones is presented in the table 7. Additional calculations were made in zone 50–60° for four other H_{min} values (0°, 10°, 15° and 20°). We can resume that:

– distribution of satellites azimuths depends on observer's latitude and angle H_{min},
– at latitudes 0–20° the number of satellites with azimuths from intervals 045–135° and 225–315° is for H_{min} = 5° less than in all other intervals,
– at latitudes 20–30° the number of satellites in all 8 azimuths intervals is for H_{min} almost the same,
– at latitudes 30–50° the number of satellites in interval 45–90° and interval 270–315° are for both H_{min} values greater than in all other intervals considerably,

– at latitudes 50–70° in all 6 intervals of azimuths 45–315° the number of satellites increases with H_{min} whereas in two other intervals (azimuth 315–045°) decreases significantly. For H_{min} = 25° the number of satellites in intervals 45–90°
– and 270–315° is several dozen times greater than in interval 315–045°,
– at latitudes 70–80° the number of satellites is in all 8 azimuths intervals almost the same, but for H_{min} = 5° only,
– at latitudes 70–80° the number of satellites in all 8 azimuths intervals is for both H_{min} almost the same.

Distribution (in per cent) of GDOP coefficient values in open area for H_{min} = 5° and 25° in all 9 latitudes zones is presented in the table 8. Additional calculations were made in zone 50–60° for four other H_{min} values (0°, 10°, 15° and 20°). We can say that GDOP coefficient value:
– depends on angle H_{min} and observer's latitudes;
– increases with H_{min} in all 9 zones, but the greatest values are at high latitudes;
– for H_{min} = 5° can be less than 2 at latitudes 0–20° only;
– for H_{min} = 5° is less than 4 in zone 0–10° always and less than 5 at latitudes 10–70° always;
– for H_{min} = 25° is at latitudes equal at least 3, in zone 70–80° and 80–90° equal 4 or 6 at least respectively.

Table 7. Distribution (in per cent) of BeiDou MEO satellites azimuths in open area for different masking elevation angles (H_{min}) at different observer's latitudes (φ), l_m – weighted mean number of satellites visible above H_{min}

φ N [^{O}]	H_{min} [^{O}]	l_m	Azimuth [^{O}]							
			0 – 45	45 – 90	90 – 135	135 – 180	180 – 225	225 – 270	270 – 315	315 – 360
0–10	5	9.9	15.2	10.0	10.0	15.0	14.7	10.1	10.3	14.7
	25	5.4	17.3	10.1	10.4	12.7	11.7	10.0	10.5	17.3
10–20	5	9.5	14.8	11.2	10.2	14.1	13.6	10.3	11.5	14.3
	25	5.5	18.7	10.5	10.5	10.7	10.4	9.9	11.0	18.3
20–30	5	8.9	13.3	13.7	10.8	12.2	11.8	10.6	13.9	13.7
	25	5.4	16.9	12.7	10.5	9.8	10.2	10.0	13.3	16.6
30–40	5	8.7	9.6	18.0	11.1	11.4	11.2	11.0	18.1	9.8
	25	5.3	10.4	18.2	11.0	10.0	10.6	10.6	18.7	10.5
40–50	5	8.7	5.7	21.2	11.6	11.4	11.2	11.9	21.4	5.6
	25	5.2	2.9	23.5	11.9	11.6	10.7	12.8	23.7	2.9
	0	10.6	9.0	17.5	12.2	11.2	11.1	12.6	17.0	9.4
	5	9.3	7.4	18.3	12.8	11.5	11.3	13.1	17.8	7.8
50–60	10	8.0	5.1	19.6	13.4	12.1	11.8	13.9	19.1	5.0
	15	6.9	2.9	20.6	14.2	12.7	12.2	14.6	19.9	2.9
	20	5.9	1.2	21.2	14.7	13.0	12.3	15.4	20.6	1.6
	25	5.1	0.3	21.4	14.9	13.3	12.6	16.3	20.8	0.4
60–70	5	9.8	10.1	14.3	13.7	12.0	12.0	13.3	14.2	10.4
	25	5.3	1.5	16.0	17.5	15.0	14.8	17.7	16.1	1.4
70–80	5	10.0	11.3	13.1	12.9	12.7	12.4	12.7	13.1	11.8
	25	6.0	7.7	13.2	14.6	14.4	14.4	14.3	13.5	7.9
80–90	5	10.1	12.1	12.8	12.6	12.4	12.8	12.1	13.0	12.2
	25	6.4	11.4	12.6	12.9	13.0	13.4	12.2	13.0	11.5

Table 8. BeiDou system, distribution (in per cent) of GDOP coefficient values for different masking elevation angles (H_{min}) at different observer's latitudes (φ)

Latitude [°]	H_{min} [°]	GDOP coefficient value − w							
		w<2	2≤w<3	3≤w<4	4≤w<5	5≤w<6	6≤w<8	8≤w<20	w≥20
0–10	5	0.6	91.3	8.1	–	–	–	–	–
	25	–	–	25.8	37.7	15.0	7.0	9.4	5.1
10–20	5	0.4	86.3	12.4	0.9	–	–	–	–
	25	–	–	16.0	41.3	14.7	8.3	16.9	2.8
20–30	5	–	76.1	17.4	6.5	–	–	–	–
	25	–	–	18.7	34.4	16.8	8.2	13.5	8.4
30–40	5	–	95.8	2.7	1.5	–	–	–	–
	25	–	–	11.2	44.7	36.7	1.6	5.6	0.2
40–50	5	–	86.4	7.8	5.8	–	–	–	–
	25	–	–	24.5	27.2	28.5	–	14.9	4.9
50–60	0	–	91.2	8.8	–	–	–	–	–
	5	–	70.4	27.8	1.8	–	–	–	–
	10	–	37.6	49.6	12.8	–	–	–	–
	15	–	8.7	56.6	34.0	0.7	–	–	–
	20	–	0.3	37.2	43.4	15.6	–	2.3	1.2
	25	–	–	13.6	39.0	22.4	0.1	14.7	10.2
60–70	5	–	28.1	69.3	2.6	–	–	–	–
	25	–	0.1	1.8	29.9	49.2	5.7	8.4	4.9
70–80	5	–	–	35.7	49.1	14.4	0.8	–	–
	25	–	–	–	8.8	27.1	27.0	35.5	1.6
80–90	5	–	–	–	0.8	13.1	27.2	37.1	21.8
	25	–	–	–	–	–	10.8	60.6	28.6

Distribution (in per cent) of GDOP coefficient values for the observer at latitudes $50 – 60°$, if the number of visible satellites l_s is known, for $H_{min} = 5°$ and for $H_{min} = 25°$, is presented respectively in the table 9 and the table 10. We can showed that:

- if $H_{min} = 25°$ GDOP value is less than 4 if $l_s = 7$ and can be less than 4 if $l_s = 6$ or 5. If $l_s = 4$ this coefficient is equal at least 8;
- if $H_{min} = 5°$ GDOP value is less than 3 if $l_s = 11$ and can be less than 3 if $l_s = 8$, 9 or 10. If $l_s = 6$ this coefficient is equal or greater than 3 and less than 5;
- there is not a direct relation between a number l_s of satellites visible above H_{min} and GDOP coefficient value, but we can realize "when l_s is greater, GDOP can be less" and vice versa "when l_s is less, GDOP can be greater".

Table 9. BeiDou system, distribution (in per cent) of GDOP coefficient values at observer's latitudes $50 – 60°$, $H_{min} = 5°$, if the number of visible satellites l_s is known

Visible satellites l_s	%	GDOP coefficient value − w		
		2≤w<3	3≤w<4	4≤w<5
6	2.1	–	0.3	1.8
7	3.5	–	3.5	–
8	14.9	3.3	11.6	–
9	28.1	16.8	11.3	–
10	45.0	43.9	1.1	–
11	11	6.4	–	–
	100	70.4	27.8	1.8

Table 10. BeiDou system, distribution (in per cent) of GDOP coefficient values at observer's latitudes $50 – 60°$, $H_{min} = 25°$, if the number of visible satellites l_s is known

Visible satellites l_s	%	GDOP coefficient value − w					
		3≤w<4	4≤w<5	5≤w<6	6≤w<8	8≤w<20	w≥20
4	24.7	–	–	–	–	14.6	10.1
5	39.3	0.1	16.5	22.4	0.1	0.1	0.1
6	32.7	10.2	22.5	–	–	–	–
7	3.3	3.3	–	–	–	–	–
	100	13.6	39.0	22.4	0.1	14.7	10.2

5 GEOMETRY OF GEO AND IGSO SATELLITES

As the longitudes of two extreme GEO satellites are $058,75°$ E and $160°$ E , the trucks of sub-satellite points for 3 IGSO satellites are coincided and the longitude of the intersection points is $118°$ E, BeiDou system currently provides positioning data between longitude $055°$ E to $180°$ E and from latitude $55°$ S to $55°$ N. This area includes China territory entirely; longitude of this country from $073°$E to $135°$ E, latitude from $18°$ N to $53°$ N.

As the augmentation service information is and will be transmitted by 5 GEO satellites only, at the moment of BeiDou Full Operational Capability FOC (space segment with 35 satellites) for all users without the possibility of the reception this information China system will be another SNS for positioning only.

6 CONCLUSIONS

- Full Operational Capability (FOC) of BeiDou system will made China the third nation in possession of independent, global navigation system following the United States and Russia
- China's BeiDou will accelerate the pace of its development, even as the world's other three SNS are experiencing delays and difficulties May 21, 2014 China Satellite Navigation Office announced in Nanjing that with the launch of a new generation of satellites beginning in 2015, BeiDou expects to complete its planned Phase III (final space constellation with 35 operational satellites) several years ahead of schedule – by 2017 rather than 2020
- BeiDou MEO constellation with 27 fully operational satellites will provide global coverage and the possibility of user's 3D position determination at any moment and any point on the Earth as well as in the case when the satellites are visible by the user below 25° only
- BeiDou 35 satellites constellation by D2 navigation message transmitted via 5 GEO satellites will provide augmentation service information to the users but in China and the Asia−Pacific region only. It means that integrity information will be not accessible to users in Europe, in Poland also.

BeiDou has brought China navigation satellite industry into a new era, and will further provide more high-performance positioning, navigation, timing, and short-message communication services for civil aviation, shipping, railways and other industries.

REFERENCES

Aigong X. et al. 2014. Precise Point Positioning Using the Regional BeiDou Navigation Satellite Constellation, Journal of Navigation, vol. 67, pp. 523−537

BeiDou Navigation Satellite System, Signal in Space Interface Control Document Open Service Signal (Version 2.0), China Satellite Navigation Office, December 2013

BeiDou Navigation Satellite System, Open service, Performance Standard (Version 1.0), China Satellite Navigation Office, December 2013

Januszewski J. 2010. *Systemy satelitarne GSP, Galileo i inne.* Warszawa: PWN, (in polish)

Martin H. 2013. U.S. Space-Based Positioning, Navigation and Timing policy and Program Update. 8th International Committee on GNSS, Dubai

Munich Satellite Navigation Summit, Munich, 2014

Report on the Development of BeiDou Navigation Satellite System (version 2.2) 2013. China Satellite Navigation Office

Tang Y., Wang J. 2015. A closed-form formula to calculate geometric dilution of precision (GDOP) form multi-GNSS constellations, GPS Solutions, DOI 10.1007/s 10291−015−0440−x, Springer−Verlag Berlin Heidelberg

www.beidou.gov.cn
www.directory.eoportal.org
www.gpsworld.com
www.imo.org
www.insidegnss.com
www.isg.org

Experiments with Reception of IRNSS Satellite Navigation Signals in the S and C Frequency Bands

J. Svatoň & F. Vejražka
FEE – Dept. of Radioengineering, Czech Technical University in Prague, Czech Republic

ABSTRACT: The satellite navigation is typically considered as the processing of satellite radio signals in the L frequency band (1.151 to 1.214, 1.215.6 to 1.350, and 1.559 to 1.617 GHz). However, in the process of the Galileo signals design the S (2483.5 to 2500 MHz) and C (5010 to 5030 MHz) bands have been taken into account, too. The S band was allocated to the Radiodetermination Satellite Service by the International Telecommunication Union (ITU) at the World Radio Conference 2000 (WRC 2000), whereas the C band signals were intensively studied in the period 1998 – 2004. Both of the bands were associated with the plans for the Galileo but have not been applied yet. A part of the S band has been lately (since summer 2013) used for the Open Service of the Regional Indian Radio Navigation Satellite System (IRNSS). The two bands have both specific advantages and deficiencies which we will analyse in our contribution. We will refer also to our experiments with the S band IRNSS signals reception and their use for determination of ionosphere and troposphere properties.

1 INTRODUCTION

Usage of an additional band for the Radio Navigation Satellite Service purpose has been considered since 1998. The first ideas were connected with the development of the European navigation system Galileo and were associated mainly with utilization of the C-band (Irsigler, 2004), (Schmitz-Peiffer, 2009). A part of the C-band was allocated to the satellite navigation by the ITU in 2000, but has not been used for this purpose yet. The S-band was allocated to the GNSS purpose in 2003. The first papers on the utilization of the S-band for the planned Indian regional navigation system (IRNSS) emerged around the year 2009 (Mateu, 2009). The first IRNSS space vehicle was launched in 2013 and its Signal In Space - Interface Control Document for its Standard Positioning Service was published by the Indian Space Research Organization in June 2014 (ISRO, 2014). The reception of its signal on the L5 carrier (BPSK(1)) was proved and published (Thombre, 2014). Our contribution describes our new experience with its signal reception in this band for a GNSS purpose.

The first paragraph of this paper deals with the general overview of the current GNSS bands and with the comparison of their properties and the properties of the S and C-bands.

The second paragraph describes the IRNSS properties and parameters for IRNSS observations in a part of Central Europe.

The third paragraph is dedicated to the description of our experimental platform for the S-band signal reception, to the solution of potential problems with interference and to the IRNSS S-band signal measurement results.

The last paragraph deals with the recommendation for construction of future S and C-band front-ends as part of multi-constellation receivers.

2 GNSS SIGNALS OVERVIEW

Satellite navigation systems are mostly associated with the usage of the L-band. Our contribution takes a look above all at the C and S bands.

2.1 L-band (1–2 GHz)

A part of the UHF radiofrequency spectrum from 1 to 2 GHz is by the IEEE terminology called the L-band and the D-band according to the NATO terminology. Its advantages are adequate to physical propagation characteristics. The band has tolerable free space loss and the Earth atmosphere signal propagation attenuation. The wave propagation in the ionosphere is quite well predictable and correctable due to existence of many models or with a help of direct measurements at several frequencies. Receivers, especially their front-end parts, are not too demanding as to the implementation, too. The main reason for the new bands research is mainly the overcrowding of the L-band.

2.2 C-band (4–8 GHz)

One part of the radio spectrum is called the C-band according to the IEEE terminology or as G plus H-band in the NATO terminology. The C-band is primarily used for the satellite communication, the satellite TV network broadcast or raw satellite feeds. The World Radio Conference 2000 (WRC 2000) held in Turkey in 2000 allocated the 20 MHz wide band from 5010 to 5030 MHz to the GNSS. A disadvantage of this band is that it is surrounded by the Radio Astronomy Service band on one side (4990 MHz to 5000 MHz) and with the Air Radio Navigation Service (ARNS) band (5030 MHz to 5150 MHz) on the other side. The high signal propagation attenuation is the major drawback because it increases power requirements for the satellite payload. Some studies about the Galileo have suggested appropriate modulations with a non-rectangle modulation pulse, utilization of a GMSK or an OFDM modulation. A navigation message format was even suggested (Schmitz-Peiffer, 2009), but none of these plans have been exploited yet.

2.3 S-band (2–4 GHz)

The second band is called the S band according to the IEEE, or it is called the E plus F-band in the NATO terminology. The 16.5 MHz bandwidth from 2483.5 to 2500 MHz was allocated for GNSS services in 2003. The central frequency 2492.028 MHz was chosen for the IRNSS. The similar one 2491.75 MHz is for the Chinese BeiDou-1 and for the communication system Globalstar (2483.5 to 2500 MHz). However, bigger problems are associated with interferences from very strong signals of terrestrial communications networks in the 2.4 GHz band (Wi-Fi, Bluetooth, and ZigBee).

2.4 L-C-S bands comparison

Below we summarize a few basic parameters concerning the propagation in these three bands. Propagation attenuation and propagation in the ionosphere affect reception and differ significantly among the bands.

$$B = 20\log_{10}\left(d\right) + 20\log_{10}\left(f\right) - 147.55 \qquad (1)$$

$$\Delta t = \frac{40.31}{cf^2}\int_d Ndl = \frac{40.31}{cf^2}TEC \qquad (2)$$

$$\Delta t_L - \Delta t_S = \Delta t_L \frac{f_L^2 - f_S^2}{f_S^2} \qquad (3)$$

The propagation attenuation is given by Equation (1) as the equation for free space loss in a modified form. Compared with the L-band losses the reckoned C-band losses are 10 dB and the S-band 4 dB higher. Troposphere propagation losses in the L-band are a few tenths of decibel. But these losses in the C-band are twice larger (up to 0.5 dB). The biggest losses in the Earth atmosphere are caused by rain attenuation. These losses can be approximately up to 0.5 dB for the L-band, but up to 5 dB for the C-band in extreme cases. Foliage attenuation in vegetation can add 2 dB/m for the C-band compared to 1 dB/m in the L-band (Irsigler, 2004). Assuming that the transmitted power of the C/A signal in the L-band of the GPS satellite is around 30 W, the previous results mean that for the same received power at the Earth's surface it needs to be transmitted by 6, or 20 times more power for the S band, or the C-band respectively. This may be the greatest current problem in using these bands for GNSS services.

The ionosphere has an important effect on a position error. Propagation time differs from a geometric distance between the satellite and the receiver. This physical phenomena is caused by the wave refraction and the Faraday rotation. These phenomena depend on a number of electrically charged particles in a unit volume of the ionosphere (TEC, Total Electron Content [electrons/m^3]). Several models are used for this error suppression (Klobuchar for example). These errors could be mitigated with the knowledge of a difference among measurements at two or more frequencies with a help of Equations (2) (3) (Bradford, 2009).

The GNSS measurements can be used vice versa for the determination of ionosphere parameters. When we use Δt_L-Δt_S=$\delta(\Delta t)$, then such multi-frequency measurements could be used with Equation (3) for the TEC computation. The more distant frequencies, the better results, because the value of $\delta(\Delta t)$ is then greater than the natural variations of the measured pseudorange caused by other factors. Measurements using GNSS signals become an important source of data for monitoring the ionosphere.

Benefits of using the proposed bands together are frequency diversity and better suppression of influence of the ionosphere.

Disadvantages of the suggested conception consist in a more complex antenna system and in interference from nearby communication services.

Also a quality receiver is required to reduce losses due to the worse power balance.

3 INDIAN REGIONAL NAVIGATION SYSTEM

The IRNSS is expected to determine the position more accurately than 20 m in the Primary Service Area. The Primary Service Area is the region covering the Indian landmass and extending up to 1500 km from Indian borders. The Extended Service Area is also defined and is bounded from 50° to 30°N and from 30° to 130°E. Its European part exceeds the Turkey and Ukraine areas. Our experience from Central Europe is described below.

The IRNSS Interface Control Document (ICD) was officially published in June 2014 in its first version (ISRO, 2014). This ICD defines two carrier frequencies: the L5 (1176.45 MHz with 24 MHz bandwidth) and the IRNSS-S (2492.028 MHz with a 16.5 MHz bandwidth). Two different services are defined: the Standard Positioning Service (SPS) and the Restricted Service (RS). Each carrier is modulated by three signals: by the SPS - BPSK (1), the RS - Data Channel BOC (5,2) and the Pilot Channel BOC (5,2). The SPS signal is defined by Equation (4). A PRN code C_{SPS} is similar to the Gold code and is chipped at 1.023 Mcps with 1 ms period. The Navigation message D_{SPS} is modulo 2 added to data which are transmitted by 50 symbols per second rate.

$$S_{SPS}(t) = \sum_{i=-\infty}^{+\infty} C_{SPS}(i) D_{SPS}(i) rect_{T_{c,sps}} \left(t - iT_{c,sps} \right) \quad (4)$$

The following Table 1 specifies the minimum received power level of a signal at the Earth's surface (0 dBi RHCP antenna is used). The power of the IRNSS L5 signal is almost comparable with similar GPS signals, but the signal in the S-band is about 4 dB weaker than the signals usually received in the L-band.

Table 1. IRNSS minimum received signal power level

Signal	Modulation	Min. P_R [dBW]	Max. P_R [dBW]
L5 - SPS	BPSK (1)	-159.0	-154.0
S - SPS	BPSK (1)	-162.3	-157.3

The first three satellites were launched by the end of 2014 (see Table 2). Their signals in both bands were experimentally received. The other two satellites are to be released in early 2015. The complete constellation of seven satellites (see Figure 1) is scheduled to be in the orbits by 2016. The three satellites in the GEO orbit will be located at 32.5°E, 83°E and 131.5°E. Four IGSO satellites (IGSO is a GSO orbit with non-zero inclination) will cross the equator at 55°E and 111.75°E (two in each plane). The satellite lifespan is scheduled for 10 years. The

weight of the satellite is around 1400 kg and its solar panels supply it with power of about 1660 Watts.

Table 2. IRNSS space segment

	Longitude [°E]	Inclin. [°]	RAAN [°]	Launch
1A	55.0	29 (±2)	135	July, 2013
1B	55.0	29 (±2)	310	April, 2014
1C	83.0	± 5	274	October,2014
1D	111.75	29 (±2)	135	? March, 2015
1E	111.75	29 (±2)	310	? April, 2015
1F	32.5	± 5	270	By end 2016
1G	131.5	± 5	270	By end 2016

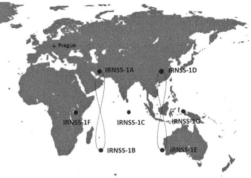

Figure 1. IRNSS world view

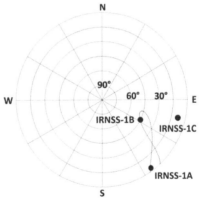

Figure 2. IRNSS sky plot for Prague (50°N, 14°E)

3.1 *IRNSS visibility from Central Europe*

The IRNSS satellites 1A, 1B and 1C were launched by the end of 2014. The geostationary satellite 1C is observable from Prague at low elevation from 2° to 9°. The first two IGSO satellites 1A and 1B reach good elevation of up to 49° and at least one of them is always observable from our laboratory in Prague.

The remaining satellites 1D and 1E will be also observable from our location, but for a shorter period of the day. From the future GEO satellites 1F and 1G only 1F will be visible. It means that we would be able to determine in Central Europe a 2D position with the IRNSS during the whole day from the spring of 2015. Figure 2 illustrates the visibility of the current IRNSS constellation in Prague.

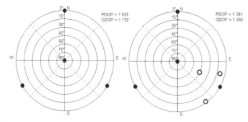

Figure 3. GDOP improvement with IRNSS

Increasing number of satellites in the sky leads to improvements in the positioning accuracy by the dilution factor (DOP). We have analysed the DOP of a simple situation with four GNSS satellites evenly distributed across the sky accompanied by three IRNSS satellites (see Figure 3). The GDOP factor has been improved only by 0.25.

Considerable part of the IRNSS constellation will be observable in the region of Eastern and Central Europe. The system will not be usable for autonomous positioning due to the poor DOP. However but it can be helpful for testing of signal reception in the S-band and for ionosphere propagation models research.

3.2 *IRNSS PRN generator*

The PRN code of the IRNSS SPS comes from the family of Gold codes. The IRNSS PRN generator is based on a similar structure as the GPS C/A code generator, but the outputs of the both LFSR registers with polynomials by Equations (5) and (6) are modulo 2 added directly. The individual codes differ from each other by an initial state of the second 11-bit shift register. This value is specified in the ICD, the first shift register is initialized by "1". The secondary code is not used.

$$G1 = X^{10} + X^3 + 1 \tag{5}$$

$$G2 = X^{10} + X^9 + X^8 + X^6 + X^3 + X^2 + 1 \tag{6}$$

4 EXPERIMENTAL RESULTS

4.1 *The experimental apparatus*

Figure 4 shows a block scheme of the experimental apparatus. The S-band signals were post-processed as recorded samples of the signal complex envelope. Real time processing in our software defined radio receiver (Kovář, 2010) is in preparation.

The signal in L5-band is received by a wideband passive Novatel GNSS 704X antenna. The antenna is placed with a low noise amplifier unit on a mast on the university roof. The L5-band is filtered with the help of the "Amplifying and Splitter Unit". The unit has outputs for eight different GNSS bands altogether. The Unit is described in (Vejražka, 2013).

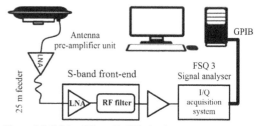

Figure 4. L5/ S-band Front-End block scheme

4.2 *S-band reception*

Reception in the S-band is enabled by the same antenna and by the experimental S-band front-end. This front-end includes the extra-low noise wideband amplifier with high dynamics (G= 15 dB, B> 700 MHz, IP3> 30 dBm) (Dobeš, 2014) and the SAW filter.

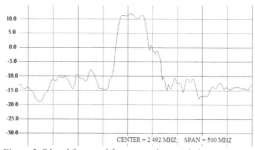

Figure 5. S band front-end frequency characteristic

The frequency amplitude characteristic is shown in Figure 5. The front-end photography is in Figure 6. These parts are followed by the standard line amplifier (20 dB).

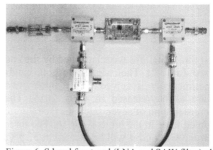

Figure 6. S band front-end (LNA and SAW filter) photography

4.3 Acquisition unit – signal analyser

The Rohde and Schwarz FSQ 3 signal analyser is used as an acquisition system for signal samples with up to 81.6 MHz sample frequency, 16 MSamples acquisition memory (14bit per each I&Q channel) and DANL -157 dBm. The analyser is connected with the PC via the GPIB. We use 44 MHz sampling frequency and a 50 MHz resolution filter. The analyser uses three inter-frequency stages. The first one is for a mirror reception cancellation. The digitalization is implemented in the third stage with a full sampling rate and is down-converted to a required sampling frequency with the help of a cascade of interpolators and decimators.

4.4 Post-processing results from MATLAB

The signal was sampled via the analyser. The ambiguity function of the signal acquisition can be described by Equations (7) and (8).

$$R(\tau, f_d) = \int_{-f_d}^{f_d} s(t) C_{SPS} (t + \tau)^* e^{j2\pi f_d t} dt \qquad (7)$$

$$R(\tau, f_d) = F^{-1} \left\{ F\left\{ s(t) e^{j2\pi f_d t} \right\} F\left\{ C_{SPS}(t) \right\}^* \right\} \qquad (8)$$

where $R(\tau, f_d)$ is the cross-correlation function between the received signal $s(t)$ with Doppler frequency f_d and PRN code replica C_{SPS}. The functions F and F^{-1} (9), (10) are Fourier transformation and the inverse-Fourier transformations, respectively. The presented functions are results of a non-coherent integration process. An integration time 22 ms is used in the L5-band and 360 ms in the S-band. The result is an acquisition of all three IRNSS satellites in both of the bands. The example of the acquired signal ambiguity function of the 1C satellite in the S-band is shown in Figure 7. Figure 10 shows the overview of all satellites.

$$F : S(\omega) = \int_{-\infty}^{\infty} s(t) e^{-j\omega t} dt \qquad (9)$$

$$F^{-1} : s(t) = \frac{1}{2} \int_{-\infty}^{\infty} S(\omega) e^{j\omega t} d\omega \qquad (10)$$

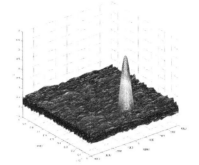

Figure 7. Ambiguity function of IRNSS-1C S-band acquisition

Signals were received with C/N0 from 30 to 39 dB. The S-band signals have worse C/N0 because of the lower gain of the analogue parts and due to the lower received power level. This means that the longer integration time is required. Geostationary satellites signals have better C/N0 than the satellites on inclined-geostationary orbits.

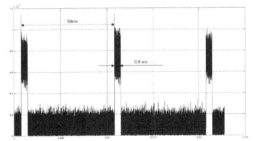

Figure 8. Received signal with interference

In the course of experiments an interference has been observed (Figure 8). It has been probably caused by a Wi-Fi signal in channel 14. Transmission in this channel is not allowed in Europe. Channel 14 exceeds nearly half of the allocated S-band part which can be seen in Figure 9.

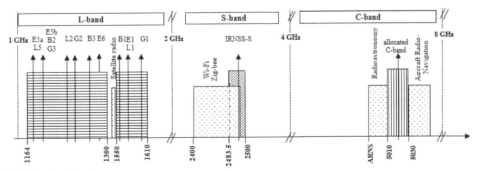

Figure 9. GNSS bands allocation

235

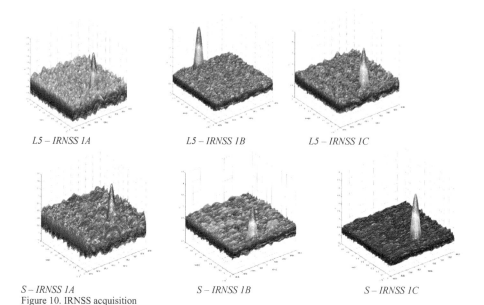

L5 – IRNSS 1A	*L5 – IRNSS 1B*	*L5 – IRNSS 1C*
S – IRNSS 1A	*S – IRNSS 1B*	*S – IRNSS 1C*

Figure 10. IRNSS acquisition

5 CONCLUSION

The IRNSS is in operation. The system covers Indian areas and areas 1500 km far away from its border and as the first GNSS system specifies signals in the S-band. Our contribution is based on the experience with the reception of the signal in the S-band in the area of Central Europe. We are planning to integrate this signal reception to our multi-constellation, multi-frequency GNSS receiver conception.

A quality front-end with low noise figure will be required for the S-band due to the lower received power level and C/N_0 at the antenna output. Receivers will need a good filtration system for the suppression of near-out-band interferences. All active parts will work with a higher dynamic and a reduction of the additional processing loss will be necessary using multi-bit A/D converters.

The IRNSS will be probably useable for 2D positioning in any part of Europe in the first half of the year 2015, however with the poor DOP. It could be useful for the demonstration of the S-band navigation and for ionosphere measurements.

ACKNOWLEDGEMENT

The submitted paper has arisen during the work on the project which has been financially supported by the Technology Agency of the Czech Republic under grant No. TE01020186. This work was also supported by the Student Grant Agency of the Czech Technical University in Prague, grant No. SGS14/150/OHK3/2T/13.

REFERENCES

Bradford, P.W. Spilker, J. and Enge, P. 2009. Global positioning system: theory and applications. *AIAA Washington DC.*

Dobeš, J. et al. 2014. Using the Sensitivity Analysis of the Noise Spectral Density for Practical Circuit Design. *2014 IEEE International Symposium on Circuits and Systems (ISCAS).* p. 1676-1679.

Irsigler, M. et al. 2004. Use of C-Band frequencies for satellite navigation: benefits and drawbacks. *GPS Solutions.* vol. 8. no. 3. pp. 139.

ISRO (Indian Space Research Organisation). 2014. IRNSS Signal In Space ICD. Version 1.0. Bangalore.

Kovář, P. et al. 2010. Universal front end for software GNSS receiver. *Proc. 13th IAIN World Congress.*

Majithiya, P. et al .2011. Indian Regional Navigation Satellite System, Correction parameters for timing group delays. *Inside GNSS* vol.6 no.1: 40 - 46.

Mateu, I. et al. 2009. Exploration of possible GNSS signals in S-band. *Proceedings of the 22nd International Technical Meeting of The Satellite Division of the Institute of Navigation (ION GNSS 2009).*

Schmitz-Peiffer, et al. 2009. Architecture for a future C-band/L-band GNSS mission. *Inside GNSS:* 47-56.

Thombre, S. et al. 2014. Tracking IRNSS Satellites. *InsideGNSS.* online: http://www.insidegnss.com/node/4279.

Vejražka, F. et al. 2013. Study of the RF Front-end of the Multi-Constellation GNSS Receiver. *Marine Navigation and Safety of Sea Transportation: Advances in Marine Navigation.*

Evaluation of a Low Cost Tactical Grade MEMS IMU for Maritime Navigation

R. Ziebold, M. Romanovas & L. Lanca
Institute of Communications and Navigation, German Aerospace Centre (DLR), Neustrelitz, Germany

ABSTRACT: The paper evaluates the performance of a tactical grade MicroElectroMechanical System (MEMS) Inertial Measurement Unit (IMU) with respect to possible application for maritime navigation. The evaluation is based on a measurement campaign on the ferry vessel Mecklenburg Vorpommern, where both a MEMS IMU and a reference Fiber Optical Gyroscope (FOG) based IMU were installed. The evaluation concentrates not only on the provision of a backup functionality for GNSS based heading and position determination, but also on the usage of a MEMS IMU for the integrity monitoring for GNSS faults using a tightly-coupled IMU/GNSS integration scheme.

1 INTRODUCTION

Efficient and safe vessel navigation under all weather conditions nowadays heavily relies on a reliable provision of position, navigation and timing (PNT) data by electronic means. The main source of position, velocity and timing data are currently the Global Navigation Satellite Systems (GNSS), like GPS and GLONASS. Especially due to their small signal power they are vulnerable against natural and artificial interferences. In a survey among mariners within the e-Navigation initiative [1] of the International Maritime Organization (IMO) the main user needs with respect to electronic positioning were (a) an improved reliability and (b) a corresponding indication of reliability. Focusing on these user needs the German Aerospace Center has developed a concept for the whole Maritime PNT System. This system incorporates satellite based services (e.g. GNSS), shore side services (e.g. augmentation and backup services), onboard systems (e.g. GNSS receiver, speed log, gyro compass) and the communication links between them [2,3]. The proposed concept is based on an open architecture and combines all available sources of the navigation information in order to enable resilient PNT data provision. The available redundancy on the measurement level allows, at least to some extent, the determination of an overbound of the actual error, which could serve as the requested indicator of reliability.

In order to support this conceptual work we have developed a prototype of the onboard element of the PNT System, which we call a PNT data processing unit [4], where as an additional onboard sensor we have used an Inertial Measurement Unit (IMU). The usage of inertial sensors is advantageous in order to overcome the GNSS vulnerability due to their complementary noise properties, inherent independence from both the surroundings and possible RF channel contamination. All this allows the IMU data to be integrated synergistically with the GNSS information so that excellent short term performance of the IMU and long term stability of the GNSS can be combined optimally within the hybrid system.

In our previous implementations we have used a rather expensive (~30k€) Fiber Optical Gyro (FOG) based tactical grade IMU (type iMAR FCAI). For practical usage onboard merchant vessels a reduction of Cost of Ownership is clearly required. Here a dramatic progress in performance of lower cost MicroElectroMechanical System (MEMS) sensors had also led to an appearance of affordable MEMS IMUs of tactical grade. Although the main characteristics of the MEMS sensors are still inferior of those of more expensive FOG-based systems, their noise performance can be already considered sufficient for some application scenarios. As the performance of modern MEMS IMUs is rapidly improving both in terms of the sensor noise and bias stability, one can expect the overall characteristics of the PNT Unit with MEMS sensor to become similar to those of systems based on FOGs already in the near future, while still maintaining a 5x - 20x price advantage in the costs of the inertial part. Moreover, the MEMS sensor can be also used as redundant component in order to detect certain inertial failure modes.

In principle, IMU/GNSS fusion had attracted attention already some time ago with multiple works reported [5], mainly for automotive, robotics and aerospace applications. However, most of them were addressing a problem of the estimation accuracy or GNSS gap bridging and GNSS position smoothing, but few works have considered the problem of robustness of the tightly-coupled schemes under the presence of the GNSS observations outliers. Moreover, the adoption of the inertial sensing for maritime application is still not standardized (IMU is not a mandatory sensor), although some recent works have been discussing the possibility of using the MEMS inertial sensors for maritime applications. For example, [5] addressed a problem of detection of slowly growing errors in tightly-coupled IMU/GNSS systems and the associated integrity monitoring algorithms, but not specifically for maritime applications. The potential impact of the IMU/GNSS integration on maritime navigation is discussed in [6]. Here the authors claimed that the main advantage of the INS is to enable the navigation of the ship to revert to alternative navigation techniques in an orderly manner. Unfortunately, the conclusions of the authors have to be, probably, revisited due to extreme performance gain of the MEMS sensor characteristics from the time of the publication and due to the fact that the authors did not assess the sensor quality influence on the Fault Detection and Exclusion (FDE) performance in a tightly-coupled IMU/GNSS integration scheme.

The presented work analyzes the performance of the developed PNT Unit where an expensive FOG sensor is replaced by a tactical grade MEMS-based IMU. The paper is organized as follows. The relevant mathematical methods including the design of the tightly-coupled EKF are presented in Section 2 and are followed by the description of the experimental setup in Section 3. The obtained results are discussed in Section 4 and the conclusions are provided in Section 5.

2 METHODS

For the filter design we follow a classical approach, where the subtle vessel dynamics is tracked using the methods of the strapdown inertial mechanization while the GNSS pseudoranges and Doppler measurements are used to correct the drift of the inertial solution. Additionally, the output of the GNSS Compass (preprocessed heading) is used to

compensate the drift of the heading information. Both the process and measurements models are integrated using the Extended KF (EKF) framework [7,8] with the mathematical details presented below.

Within the EKF both process and measurement models are linearized through a first-order Taylor series expansion around the current state estimate. Here the original nonlinear functions are used for the state transition and the measurement prediction, while the covariances are approximated by calculating the associated Jacobian matrices.

At epoch k, the dynamic model is linearized around the *a posteriori* state estimates from the last epoch (denoted by $\hat{\mathbf{x}}_{k-1}^+$) and the measurement model is linearized around the *a priori* state estimates at current epoch (denoted by $\hat{\mathbf{x}}_k^-$). The EKF routine can be formulated as [7]:

Filter initialization:

$$\hat{\mathbf{x}}_0^+ = E(\mathbf{x}_0)$$

$$\mathbf{P}_0^+ = E\left[\left(\mathbf{x}_0 - \hat{\mathbf{x}}_0^+\right)\left(\mathbf{x}_0 - \hat{\mathbf{x}}_0^+\right)^T\right]$$

Time update:

$$\hat{\mathbf{x}}_k^- = f_{k-1}\left(\hat{\mathbf{x}}_{k-1}^+\right)$$

$$\mathbf{P}_k^- = \mathbf{F}_{k-1}\mathbf{P}_{k-1}^+\mathbf{F}_{k-1}^T + \mathbf{Q}_{k-1}$$

$$(1)$$

Kalman gain calculation:

$$\mathbf{K}_k = \mathbf{P}_k^-\mathbf{H}_k^T\left(\mathbf{H}_k\mathbf{P}_k^-\mathbf{H}_k^T + \mathbf{R}_k\right)^{-1}$$

Measurement update:

$$\hat{\mathbf{x}}_k^+ = \hat{\mathbf{x}}_k^- + \mathbf{K}_k\left[\mathbf{z}_k - h_k\left(\hat{\mathbf{x}}_k^-\right)\right]$$

$$\mathbf{P}_k^+ = \left(\mathbf{I} - \mathbf{K}_k\mathbf{H}_k\right)\mathbf{P}_k^-\left(\mathbf{I} - \mathbf{K}_k\mathbf{H}_k\right)^T + \mathbf{K}_k\mathbf{R}_k\mathbf{K}_k^T$$

where $\mathbf{x}$ is the state vector; f and $\mathbf{F}$ are the original nonlinear process model and linearized dynamic matrix respectively, h and $\mathbf{H}$ are the original nonlinear measurement model and linearized observation matrix, respectively; $\mathbf{P}$ is the state error covariance matrix; $\mathbf{Q}$ is the dynamic error covariance matrix; $\mathbf{K}$ is the Kalman gain; $\mathbf{z}$ is the measurement vector; $\mathbf{I}$ is the identity matrix and $\mathbf{R}$ is the measurement error covariance matrix.

The error states of the Kalman filters in Earth-Centered-Earth-Fixed frame (ECEF) read [8]:

$$\mathbf{x}_{IMU}^e = \begin{bmatrix} \delta\boldsymbol{\psi}_{eb}^e & \delta\mathbf{v}_{eb}^e & \delta\mathbf{r}_{eb}^e & \mathbf{b}_a & \mathbf{b}_g & \delta t & \delta\dot{t} \end{bmatrix} \quad (2)$$

where ψ reflects the attitude angles from the body frame to ECEF frame; $\mathbf{v}$ is the velocity vector expressed in ECEF; $\mathbf{r}$ is the position vector expressed in ECEF; $\mathbf{b}_a$ and $\mathbf{b}_g$ are biases of accelerometer and gyroscope respectively, δt and its time derivative represent the receiver clock offset and receiver clock error rate.

With the strapdown processing the IMU sensor errors are accumulated onto the navigation solutions. In order to quantify the errors brought by the IMU raw data, the system matrix is needed by each strapdown routine, as expressed [8]:

$$\dot{\mathbf{x}} = \mathbf{F}^e_{IMU}\mathbf{x}(-) + \mathbf{w}$$

$$\mathbf{F}^e_{IMU} =$$

$$\begin{pmatrix} -\Omega^e_{ie} & \mathbf{O}_3 & \mathbf{O}_3 & \mathbf{O}_3 & \hat{\mathbf{C}}^e_b & 0 & 0 \\ \left[-\left(\hat{\mathbf{C}}^e_b\hat{\mathbf{f}}^b_{ib}\right)^\wedge \right] & -2\Omega^e_{ie} & \dfrac{2\mathbf{g}}{r^e_{eS}}\dfrac{\hat{\mathbf{r}}^e_{eb}}{\left|\hat{\mathbf{r}}^e_{eb}\right|^2}\hat{\mathbf{r}}^{eT}_{eb} & \hat{\mathbf{C}}^e_b & \mathbf{O}_3 & 0 & 0 \\ \mathbf{O}_3 & \mathbf{I}_3 & \mathbf{O}_3 & \mathbf{O}_3 & \mathbf{O}_3 & 0 & 0 \\ \mathbf{O}_3 & \mathbf{O}_3 & \mathbf{O}_3 & \mathbf{O}_3 & \mathbf{O}_3 & 0 & 0 \\ \mathbf{O}_3 & \mathbf{O}_3 & \mathbf{O}_3 & \mathbf{O}_3 & \mathbf{O}_3 & 0 & 0 \\ 0 & 0 & 0 & 0 & 0 & 0 & 1 \\ 0 & 0 & 0 & 0 & 0 & 0 & 0 \end{pmatrix} \quad (3)$$

$$\mathbf{w} = \begin{bmatrix} \mathbf{w}_\omega & \mathbf{w}_f & 0 & \mathbf{w}_a & \mathbf{w}_g & \mathbf{w}_t & \mathbf{w}_t \end{bmatrix}$$

Here b means the IMU body frame; e denotes the ECEF frame; i indicates the inertial frame; $\hat{\mathbf{C}}^e_b$ is the Direction Cosine Matrix (DCM) from body frame to ECEF frame, Ω is the skew-symmetric matrix for angular rate measurements; $\mathbf{f}^b_{ib}$ is the vector of acceleration measurements from the accelerometers. $\mathbf{F}$ is the system matrix applied in the ECEF frame; r^e_{eS} is the distance from the earth geometric center to the earth surface; $\mathbf{g}$ is the local gravity; $\hat{\mathbf{r}}^e_{eb}$ is the position of the IMU in ECEF. The noise vector $\mathbf{w}$ contains, in the indicated order, gyroscope bias, acceleration bias, acceleration noise, angular rate noise, receiver clock error and receiver clock rate noise. These noise terms are described by the error covariance matrix $\mathbf{Q}$ in the Kalman filter routine:

$$\mathbf{Q}_{IMU} = diag\left(n^2_{rg}\mathbf{I}_3 \quad n^2_{ra}\mathbf{I}_3 \quad \mathbf{O}_3 \quad n^2_{bg}\mathbf{I}_3 \quad n^2_{ba}\mathbf{I}_3 \quad n^2_t \quad n^2_t \right)\tau_s \quad (4)$$

where n represents power spectral densities of components of vector $\mathbf{w}$ given in (3).

In the classic GNSS/IMU tightly-coupled integration the observation vector includes the GNSS pseudorange and Doppler data. In order to enhance attitude determination, the observation vector can also contain the preprocessed heading, pitch and roll angles from the GNSS-compass. This is an extension of the classic tightly-coupled GNSS/IMU integration architecture as we make the attitude fully observable independently of the dynamics of the vessel. The measurement vector $\mathbf{z}$ can be expressed as

$$\mathbf{z} = \begin{bmatrix} y_{GNSS} & p_{GNSS} & r_{GNSS} & p_1 & p_2 & \cdots & p_{np} & d_1 & d_2 & \cdots & d_{nd} \end{bmatrix}^T \quad (5)$$

and is composed of GNSS pseudorange and Doppler data; the subscripts np and nd represent number of available pseudorange and Doppler measurements;

y, p and r with subscripts "GNSS" indicate the attitude angles obtained from GNSS Compass.

The measurement matrix $\mathbf{H}$ projects the measurements onto the state domain. In a classic GNSS/IMU integration, the $\mathbf{H}$ matrix is associated to the measurement vector $\mathbf{z}$ in (5) and the state vector in (2) can be formulated as:

$$\mathbf{H} = \begin{bmatrix} \mathbf{I}_3 & \mathbf{O}_{3\times3} & \mathbf{O}_{3\times3} & \mathbf{O}_{3\times3} & \mathbf{O}_{3\times3} & 0 & 0 \\ \mathbf{O}_{1\times3} & \mathbf{O}_{1\times3} & \mathbf{h}_1 & \mathbf{O}_{1\times3} & \mathbf{O}_{1\times3} & 1 & 0 \\ \mathbf{O}_{1\times3} & \mathbf{O}_{1\times3} & \mathbf{h}_2 & \mathbf{O}_{1\times3} & \mathbf{O}_{1\times3} & 1 & 0 \\ \vdots & \vdots & \vdots & \vdots & \vdots & \vdots & \vdots \\ \mathbf{O}_{1\times3} & \mathbf{O}_{1\times3} & \mathbf{h}_{np} & \mathbf{O}_{1\times3} & \mathbf{O}_{1\times3} & 1 & 0 \\ \mathbf{O}_{1\times3} & \mathbf{h}_1 & \mathbf{O}_{1\times3} & \mathbf{O}_{1\times3} & \mathbf{O}_{1\times3} & 0 & 1 \\ \mathbf{O}_{1\times3} & \mathbf{h}_2 & \mathbf{O}_{1\times3} & \mathbf{O}_{1\times3} & \mathbf{O}_{1\times3} & 0 & 1 \\ \vdots & \vdots & \vdots & \vdots & \vdots & \vdots & \vdots \\ \mathbf{O}_{1\times3} & \mathbf{h}_{nd} & \mathbf{O}_{1\times3} & \mathbf{O}_{1\times3} & \mathbf{O}_{1\times3} & 0 & 1 \end{bmatrix} \quad (6)$$

where $\mathbf{h}$ is the line-of-sight vector describing the satellite position with respect to the GNSS antenna; $\mathbf{O}$ is the zero-vector (matrix). A detailed explanation of the models above can be found in [8].

Finally, the measurement error covariance matrix $\mathbf{R}$ has to be defined and is usually determined from the GNSS error characterization (both pseudoranges and Doppler) as well as the accuracy of the extracted angles from the GNSS Compass. Note that often the values in $\mathbf{R}$ are artificially increased to accommodate the modelling errors, setup errors as well as the linearization inaccuracies within the EKF.

3 EXPERIMENTAL SETUP

The original sensor measurements were recorded using the ferry vessel Mecklenburg-Vorpommern from Stena Lines, which is plying continuously between Rostock and Trelleborg. The vessel was equipped with three dual frequency GNSS antennas and receivers (Javad Delta). The antennas are placed on the compass deck (red circles in Figure 1). The chosen geometry of the antenna placement enables an accurate determination of the vessel 3D attitude by applying the GNSS Compass algorithms (see section 4.2). Furthermore, an IALA beacon antenna and receiver were employed for the reception of the IALA DGNSS code corrections. An additional VHF modem is used for the reception of RTK corrections data from our Maritime Ground Based Augmentation System (MGBAS) station located in the port of Rostock. The MGBAS reference station provides GPS code and phase corrections with 2 Hz update rate for both L1 and L2 frequencies. These correction data are used for a highly accurate RTK positioning onboard the vessel, whenever the vessel is within the service area of the station.

Figure 1. Vessel Mecklenburg-Vorpommern with positions of the GNSS antennas (red circles) and location of the IMU-s (green circle)

The two IMUs: (a) a FOG-based IMU IMAR FCAI and (b) a tactical grade Analog Devices ADIS16485 MEMS IMU were installed in a technic room close to the bridge. Both IMUs were mounted on a joint base plate (see Fig. 2) so that the orientation of the IMUs with respect to each other remains fixed.

Figure 2. Installation of the IMUs (front: ADIS IMU, behind IMAR IMU)

All the relevant sensor measurements are provided either directly via Ethernet or via serial to Ethernet adapter to a Box PC (see Fig. 3) where the observations are processed in real-time and stored in a SQLite3 database along with the corresponding time stamps. The described setup enables a record and replay functionality for further processing of the original sensor data. The system consists of a highly modular hardware platform (here realized with a box PC) and a Real-Time software Framework implemented in ANSI-C++ [9].

Figure 3. Board with GNSS receiver, VHF modem, IALA Beacon receiver, serial to Ethernet adapter and Box PC for data processing and storage

4 RESULTS

In the section the evaluation results are presented. It starts with the characterization of the IMUs and GNSS Compass and later evaluates the performance of the system in several representative scenarios such as GNSS outage bridging, GNSS fault detection and the smoothing performance.

4.1 IMU Characterization

One of the classical approaches to analyze the performance of the inertial sensors is to calculate a so-called Allan Variance (AV) for each of the sensor axes. This analysis technique is able to take into account the long-term noises in the sensor outputs and is a time domain approach which was originally developed to study the frequency stability of the oscillators (clocks) [10]. The AV curve is extremely helpful in identifying the noise terms which affect the observed output of the inertial sensors (see Fig. 4) and is calculated as [12]:

$$\sigma^2(T) = \frac{1}{2T^2(N-2n)} \sum_{k=1}^{N-2n} \left(\theta_{k+2n} - 2\theta_{k+n} + \theta_k\right)^2 \quad (7)$$

where T represents the cluster time – the time associated with a group of n consecutive observed data samples from the set of the complete data of the length N. In the expression above θ is the output velocity (in the case of accelerometers) or the output angle (in the case of the angular rate sensors).

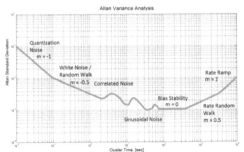

Figure 4. A sketch of the Allan Variance (AV) of a hypothetical inertial sensor (adopted with modifications from IEEE Std. 952-1997

The basic idea behind the AV analysis is to take a long sequence of data (N samples) with the sensor being static (no motion) and integrating the output of the inertial sensor to get θ values with correspondingly removed initial turn-on biases [12] with the experimental data usually plotted as a log-log curve.

The experimental AV analysis for the gyroscope outputs for both MEMS and FOG IMUs is shown in Fig. 5. Clearly, the IMAR FOG gyroscopes have lower bias instability (minimum of curve with slope equal to 0) and possess significantly smaller additive noise which is the source for the angular random walk (segment with slope -0.5). A more detailed calculation would show that the obtained characteristics for both IMUs are close to the manufacturer specifications (IMAR: bias stability < 0.01°/h, ARW < 0.03°/sqrt(h); MEMS: bias stability 6.25°/h (1σ), ARW < 0.3°/sqrt(h) (1σ)). Note that computation of the AV is based on a finite number of sample clusters and the estimation accuracy for a given T depends on the number of independent clusters within the analysis data set. Clearly, the estimation accuracy in the region of short cluster length T is better and is correspondingly worse for long cluster times T. In practice this is visible in Fig. 5 as the AV curves start to be less smooth for larger T values. The analysis confirms that all three axis in both IMUs show similar performance, although a slightly worse X-axis bias stability can be observed for MEMS IMU and better noise

performance can be found for Z-axis of the FOG IMU. The latter is probably a result of a coincidence as otherwise identical sensors are used for all three axis of this costly FOG IMU.

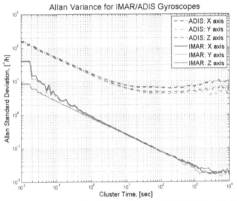

Figure 5. Estimated Allan variance for MEMS and FOG gyroscopes

Similar holds for the accelerometer outputs (not shown) where both IMUs have shown to satisfy the manufacturer specifications, although an increased variability has been observed for FOG IMU accelerometers in the region of bias instability.

Note that there is strong relationship between the AV analysis and the spectral methods and the corresponding conclusions regarding the dominating noise terms can be also drawn from the frequency domain methods. Although in practice it is much easier and convenient to analyze the sensor performacne using the AV methods, the AV analysis does not provide one-to-one matching to the spectrum and this, according to [11], puts a fundamental limitation on how much of the noise information can be extracted from the AV plot. Moreover, while performing the AV analysis one should consider the Nyquist theorem as the recommended sampling rate is suggested to be three to five times the sensor bandwidth to guarantee the validity of the obtained characteristics (we used $F_s = 200\,Hz$ for FOG and $F_s = 205\,Hz$ for MEMS IMUs correspondingly).

For both the accelerometers and the gyroscopes the EKF noise values were selected according to the manufacturer specifications as AV analysis confirmed a close match of the experimentally extracted parameters to those given in the associated datasheets.

4.2 GNSS Compass

As described in section 3 we use three separate GNSS receiver antenna positions for the purpose the attitude determination. The relative baselines between the antennas are determined with a dual frequency RTK algorithm by using both GPS and GLONASS measurements. For the evaluation of the determined attitude angle accuracy we employed a 40 min time period, where the vessel was moored in the port of Rostock.

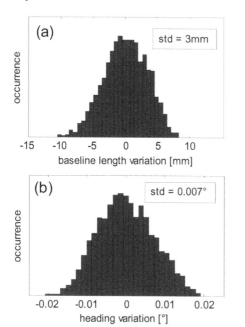

Figure 6. Histogram of (a) baseline length variation and (b) the calculated heading angle variation from the GNSS Compass

The accuracy of the relative RTK positioning can be evaluated by determination of the baseline length. Under the assumption of a rigid body the real baseline length should be constant. Therefore the variation of the calculated baseline length (see Fig. 6(a)) is a measure of the actual measurement accuracy. Under the assumption that the relative positioning error is equally distributed among all three dimensions the resulting accuracy of the attitude angle can be calculated as:

$$\Delta\alpha = a\,sin(\,\Delta l \,/\, l) \qquad (7)$$

where α is the attitude angle and l the baseline length. With a standard deviation of $\sigma_l \sim 3mm$ for the baseline length variation this leads to an attitude accuracy of $\sim 0.17°/l$ [m]. Taking the realized baseline length of $\sim 30m$ into account this would result in a variation of the heading angle of $\sigma_\alpha \sim 0.006°$, which is very close to the measured heading standard deviation of $\sigma_\alpha \sim 0.007°$ (see Fig. 6).

The chosen setup for the GNSS Compass hence enables the determination of 3D attitude with an extremely high accuracy.

4.3 IMU as backup for heading determination

While the GNSS Compass enables a highly accurate and long term stable heading determination, its availability suffers from an obvious dependence on satellite signals especially in harsh harbor environments due to shadowing and multipath effects. Here the IMU could serve as an attitude backup system for a limited time span.

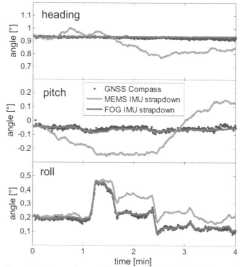

Figure 7. Comparison of FOG and MEMS IMU for the attitude determination in a static scenario

In order to evaluate applicability of both the FOG-based IMAR IMU and the MEMS-based ADIS IMU for this application we have chosen first a static scenario (vessel moored). Within the initialization phase we firstly determined the actual sensor biases and then started the strapdown processing for the orientation only by using the GNSS Compass attitude as a starting point.

From Figure 7 it can be concluded, firstly that the static scenario is not fully static. Especially there are measurable variations in roll angle and slight variations in pitch angle due to loading and unloading activities during the mooring. Secondly, as expected from the AV analysis for the FOG IMU, no significant drift over the timespan of 4min can be seen. The drift for the ADIS MEMS IMU is clearly visible but is limited to maximum error of $\sim 0.2°$ for the presented time interval and is probably dominated by the random walk.

We repeated this analysis for a dynamic scenario with 30min timespan but just focusing on the heading component determination only. In Figure 8 the raw sensor measurements together with the calculated heading angle and the heading angle errors are shown. For the last calculation the GNSS Compass was used as a reference. Here a drift of $\sim 3°$

over the 30min timespan for the MEMS IMU is found under an assumption that an initial sensor offset can be eliminated within the estimation process of the associated EKF.

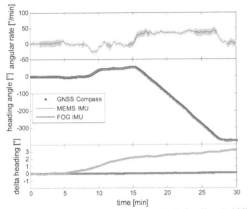

Figure 8. Angular rate measurements (top), heading (middle) and heading error (bottom) with GNSS Compass as reference for FOG and MEMS IMU in a dynamic scenario

Taking into account a required heading accuracy of 1° (see [13]), one can state that the MEMS IMU could serve as a backup sensor for the heading determination for a time period of ~ 10min. In order to bridge possible outages of the GNSS Compass, this time period is for most of the application scenarios clearly sufficient.

4.4 Position accuracy of a tightly-coupled EKF

In order to compare the achieved positioning accuracy of both the MEMS and FOG IMU in a tightly-coupled EKF as described in section 2, we have evaluated a 24h measurement data set of the vessel.

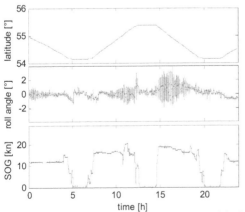

Figure 9. Latitude, roll angle and Speed over Ground for 24h dataset (doy 332 2014)

Figure 9 provides an overview over the vessel movement for this one day. The vessel is plying between Rostock and Trelleborg, being in Rostock ~6 am and 9pm and in Trelleborg ~ 2pm. In order to determine position errors we have calculated a reference trajectory in post processing using the Precise Point Positioning (PPP) approach from the open source RTKLIB software package (http://www.rtklib.com/rtklib.htm). Doing that for all three antennas allows again the estimation of the accuracy by calculation of the baseline length. Here a standard deviation of σ~10cm is reached, which is fully sufficient for the purpose of the following evaluation.

For maritime applications mostly the horizontal position is of interest. Therefore we focus the analysis here also just on the Horizontal Position Error (HPE). In Figure 10 (a) the HPE is plotted for the Single Point Positioning (SPP) together with the HPE from the tightly-coupled EKF using either the FOG or the MEMS IMU. While the SPP solution shows some position jumps with the errors up to 12m, the EKF position solutions for MEMS and FOG IMU are very similar. Figure 10 (b) and (c) show a zoomed data segment spanning ~3min, which confirms, that although the HPE behavior of the EKF solution with FOG and MEMS IMU are not identical, they are very similar and can be considered effectively identical for most of applications. This similarity could be caused by two reasons. Firstly we use in our EKF not only Pseudorange but also Doppler measurements (see equ. (5) Section 2). This already leads to a smoothing of the SPP solutions and therefore possibly reduces the influence of the IMU quality on the final solution. Secondly for the IMU/GNSS integration with a time interval of 0.5s, as we have used it here, not the sensor bias stability but just the sensor noises are mainly relevant and the performance gain of the FOG over MEMS IMU in terms of the additive noise is not as significant as the one caused by the bias stability.

In Figure 10 (d) the histogram and in (e) the cumulated error distribution of the HPE is plotted. While all three graphs in (d) show expected χ distributions with 2 degrees of freedom, again no significant differences are seen between the EKF solutions with MEMS and FOG IMU. The 95% error of ~2.6m for EKF solutions and 2.8m for the SPP solutions are sufficient for most maritime applications. But for safety critical applications the outliers in the SPP solutions might become of the primary importance and, therefore, a trivial definition of just a 95% confidence error bound could be not sufficient for, e.g. vessel navigation scenarios.

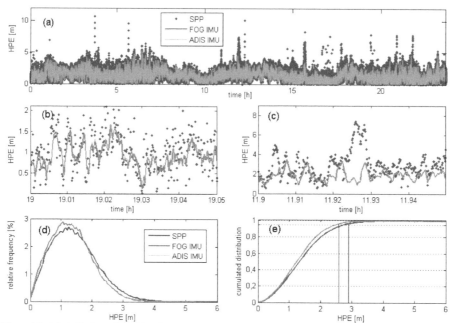

Figure 10. (a,b,c) Horizontal Position Error (HPE) for Single Point Positioning(SPP) and tightly-coupled EKF with FOG and MEMS IMU, (d) error distribution histogram and (e) cumulative error histogram

4.5 Fault detection and exclusion

A more detailed analysis shows that the position jumps in the SPP solution are caused by pseudorange jumps of individual, low elevation satellite signals, most likely caused by the multipath effects. For the integrity monitoring within the tightly-coupled EKF we have implemented a FDE algorithm based on the innovation filtering. The implemented scheme consists of a two-step procedure based on the global and local test. The global test checks the KF innovation consistency among the full set of observations. The detection method is based on a hypothesis testing which compares a test statistics calculated from the available observations with a given theoretical threshold for a required probability of false alarm and redundancy. When the global test detects an inconsistency, local tests are applied to detect and remove the faulty measurements. A detailed description of the applied FDE scheme can be found in [14]. Here we focus on evaluation of this test by using FOG or MEMS IMU within the EKF.

In Figure 10 (c) the HPE is shown for a time period, where such an outlier occurs. In order to distinguish between the effect of inertial smoothing and FDE in Figure 11 additionally the HPE of a loosely-coupled EKF (w/o FDE) is shown. Here the position and velocity of the SPP serve as direct observations instead of using fusion of separate measurements in the tightly-coupled architecture. One can clearly see that although the smoothing

behavior of both tightly- and loosely-coupled EKF is in general similar, but during the measurement fault the non-FDE solution slowly follows the wrong position from the SPP solution, whereas the FDE in the tightly-coupled EKF ensures that the faulty measurements are removed from the navigation solution. A manual inspection of the outliers during that 24h timespan (not shown) leads to the conclusion, that all of these faulty measurements are simultaneously detected by both, the tightly-coupled EKF with FOG and MEMS IMU. Clearly, an in-depth statistical analysis is still necessary in order to confirm that also smaller (not obvious) GNSS faults can be also detected using lower cost tactical grade MEMS IMU.

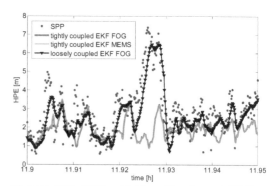

Figure 11. HPE of SPP, and tightly and loosely coupled EKF

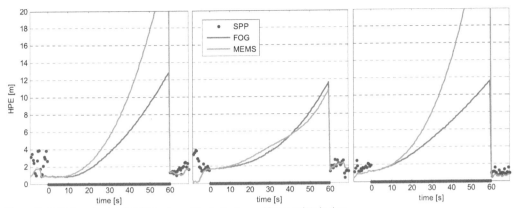

Figure 12: HPE for a simulated GNSS outage of 1min for three arbitrary points in time

4.6 Contingency functionality

In order to check the performance of both IMUs in the classical case of a GNSS outage, we have artificially switched off the GNSS signal for 1min at three arbitrarily chosen points in time. The resulting HPEs for all three outage segments are shown in Fig. 12. As one would expect, the position solution is drifting differently for the three data segments with the time needed to exceed the error bound of 10m shown below:

	a	b	c	minimum
FOG	53 s	55 s	54 s	53 s
MEMS	40 s	58 s	34 s	34 s

Based on this (albeit limited statistics) one can state, that the FOG / MEMS IMU could bridge GNSS outages of ~50s / ~30s while not exceeding a horizontal position error of 10m.

Note that all the presented scenarios were assessed using real-time algorithms without any postprocessing, smoothing or delay compensation. Moreover, differently from [6] we have employed a tightly-coupled IMU/GNSS integration strategy with clear benefits of improved FDE performance and contingency for scenarios with reduced availability of the satellites. Finally, differently from [6], where the RTK and CDGNSS positioning techniques were employed, in the presented work the equivalent performance was achieved using only raw code and Doppler measurements. Obviously, even better performance could be expected assuming measurements discussed in [6]. From the presented results one can conclude that with modern tactical-grade MEMS IMUs one is able to achieve GNSS bridging performance which till recent was only possible for CDGNSS or RTK augmented systems.

5 SUMMARY

In this paper we have evaluated the performance of one of the first available and most popular tactical grade MEMS IMUs (Analog Devices ADIS16485) with respect to possible application for maritime navigation. The results were compared to a FOG-based IMU (IMAR FCAI). The impact of the sensor quality on the performance figures was analyzed for three typical scenarios including GNSS outage bridging (both position and heading), FDE functionality and the accuracy (smoothing), where some significant differences were found only for the former application scenario, while the latter two remained almost not affected by the sensor quality.

For the application as backup sensor for heading determination in combination with a GNSS Compass, the ADIS MEMS IMU could bridge outages of the GNSS Compass up to 10 min while still satisfying an accuracy of ±1°, which should be sufficient for most application scenarios. The usage of both IMUs in a tightly-coupled Extended Kalman Filter shows no significant differences in the resulting position accuracy as well as in the capability to detect and remove faulty satellite signals. Only in the capability of bridging GNSS outages significant differences can be found. While the ADIS MEMS IMU stays within a 10m error bound for ~30s the IMAR FOG IMU could do this for ~50s. Having in mind the factor of ~20 cost benefit of the MEMS compared to the FOG IMU and expected further progress in MEMS technology development, it can be summarized, that tactical grade MEMS IMUs are promising candidates for future maritime navigation applications.

ACKNOWLEDGMENTS

The authors would like to thank Stena Lines and especially the crew of the Mecklenburg Vorpommern with their captains Mr. Watsack and Mr. Franke and our partners from the Hochschule Wismar department of Maritime Studies, Mr. Gluch and Mrs. Schaub for their support during the measurement activities. The authors are also grateful for the assistance of their colleagues Stefan Gewies, Carsten Becker, Anja Hesselbarth and Uwe Netzband.

REFERENCES

[1] IMO, "NAV 55/WP.5 Development of an e-Navigation Strategy Implementation Plan," NAV55/ WP5, Jul. 2009.

[2] Germany IMO, "NAV 58/6/1 IMO input paper: Proposed architecture for the provision of resilient PNT data Submitted by Germany." 30-Mar-2012.

[3] R. Ziebold, Z. Dai, T. Noack, and E. Engler, "Concept for an Onboard Integrated PNT Unit," in *TransNav 2011*, 2011, vol. International recent issues about ECDIS, e-Navigation and safety at sea, pp. 35–42.

[4] R. Ziebold, Z. Dai, L. Lanca, T. Noack, and E. Engler, "Initial Realization of a Sensor Fusion Based Onboard Maritime Integrated PNT Unit," *TransNav Int. J. Mar. Navig. Saf. Sea Transp.*, vol. Journal Vol.7, no. No. 1-March 2013, pp. 127–134, März 2013.

[5] Lee, Young C., O'Laughlin, Daniel G, "A Performance Analysis of a Tightly Coupled GPS/Inertial System for Two Integrity Monitoring Methods," *Navig. J. Inst. Navig.*, vol. Volume 47, no. Number 3, pp. 175–190, 2000.

[6] Terry Moore, Chris Hill, Andy Norris, Chris Hide, David Park, and Nick Ward, "The Potential Impact of GNSS/INS Integration on Maritime Navigation," *J. Navig.*, vol. 61, pp. 221–237., 2008.

[7] D. Simon, *Optimal State Estimation*. John Wiley & Sons, Inc., 2006.

[8] P. D. Groves, *Principles of GNSS, Inertial, and Multi-Sensor Integrated Navigation Systems (GNSS Technology and Applications)*. Artech House Publishers, 2007.

[9] S. Gewies, C. Becker, and T. Noack, "Deterministic Framework for parallel real-time Processing in GNSS Applications," in *NAVITEC 2012*, 2013.

[10] N. El-Sheimy, H. Hou, and X. Niu, "Analysis and Modeling of Inertial Sensors Using Allan Variance," *Instrum. Meas. IEEE Trans. On*, vol. 57, no. 1, pp. 140–149, Jan. 2008.

[11] "IEEE Standard Specification Format Guide and Test Procedure for Single-Axis Interferometric Fiber Optic Gyros," *IEEE Std 952-1997*.

[12] A. G. Quinchia, G. Falco, E. Falletti, F. Dovis, and C. Ferrer, "A Comparison between Different Error Modeling of MEMS Applied to GPS/INS Integrated Systems," *Sensors*, vol. 13, no. 8, pp. 9549–9588, 2013.

[13] IMO, "Resolution MSC.116(73): Performance Standards for Marine Transmitting Heading Devices (THDs)," Dezember 2000.

[14] L. Lanca, M. Romanovas, and R. Ziebold, "Integrity Monitoring In Snapshot And Recursive Estimation Algorithms For Maritime Applications," in *7th ESA Workshop on Satellite Navigation Technologies Era of Galileo IOV*, 2014.

Use of Passive Surveillance Systems in Aviation

M. Džunda, N. Kotianová, K. Holota & P. Žák
Technical University of Košice, Faculty of Aeronautics, Slovakia

ABSTRACT: The submitted contribution analyses the principles underlying the operation of passive surveillance systems also presenting potentials of their use in air transportation. It focuses on their advantages compared to primary radars and the possibilities of integrating passive surveillance systems into operational systems of air transportation. In the contribution there is a more detailed analysis of the most important types of passive systems of surveillance. Main attention is focused on the precision of locating the position of a flying object as well as to design, management and reliability of the Multi-sensor Surveillance System.

1 INTRODUCTION

Airports are strategic points of each country's infrastructure that people are using for travelling around the world. Despite the fact that some people are afraid to fly, according to statistics, air travelling is the safest and it's growing faster than any other type of transportation in the world. There are different types of airports from small, regional to international airports. Their capacity depends on different factors that are very difficult to influence, despite of it every airport have face problem of crowdedness that needs to be solved. (Kršák et al. 2012) Airports are investing more and more money in new technologies of optimal air traffic control system as well as on fixing the problems of economic efficiency, safety and protecting the environment. (Pavolová et al. 2013)

2 PASSIVE SURVEILLANCE SYSTEMS IN AVIATION

Military aviation has been using inactive monitoring systems for many years. Because of their better capability to monitor air traffic in terms of higher accurately, they became a part of many civil airports around the world. (Szabo 2013b)

In this contribution we are presenting technologies, which has been kept in secret for many years. Passive system of surveillance have been used only by reconnaissance units of the army and the civil aviation knew very little about them.

Czechoslovakia was one of the pioneers in the development of inactive surveillance systems. Nowadays, radars are used for more important roles such as air traffic control weather forecasting or ensuring air defense. (Labun 2012)

The very first passive system of surveillance became operational as a radio-technical search system. These were the beginnings of the development of passive systems of surveillance. The advantage of these systems compared radars are in that they do not transmit electromagnetic waves, only receive them from the surrounding environment. The principles underlying the technology of passive systems is not very complex and therefore they became part of civil air traffic control. (Szabo 2013a)

The basic feature of the tracking systems are to guarantee a safe and smooth air traffic at the airport and in the airspace.(EC 2014)

Passive surveillance systems are able to supplement and possibly replace older air traffic control systems. The fundamental principle is based on acquisition of radar signal of all airplanes in system coverage at least by three radio receivers for accurate airplane location. (Kurdel et al. 2014)

Cooperation of active and passive surveillance systems is able to provide air traffic control with all important information.

Passive technologies were originally used under the name as a Precise radio-technical searcher-1. It was a correlation searcher used to jamming guided missiles. The first inactive search device called Ramona was built in the 70s. The upgraded version of Ramona was a system of radio-technical reconnaissance called Tamara. (Koukal 2012)

The beginnings of applying of new technologies in the field passive surveillance systems enabled design of new and more precise systems such as Vera, Borap, MSS a Vera NG and the BCL concept. (Srubar 2005)

3 MULTILATERATION THEORY

3.1 *Time Difference of Arrival (TDOA)*

Multi-sensor Surveillance System (MSS) utilizes multilateration in order to estimate aircraft or vehicle position. Multilateration, or hyperbolic positioning, is the process of locating a target based on the TDOA of a signal emitted from a target to spatially separated sensors. The signal transmitted from a target is received by two spatially separated receiver at different times. This time difference of signal detection is given by the difference in the distances between the target and receivers. Each receiver generates a "time difference of arrival – TDOA". For a pair of receivers with known positions these set of measured distances defines hyperbola (in 2D) or hyperboloid (in 3D) on which target is located. A hyperbola for the TDOA results from the differences in distance of the target and the receivers. A second TDOA measurement from another pair of sensors provides the second hyperbola. The intersection of these two hyperbolas describes a curve on which the target lies. A third TDOA measurement from another pair of sensors provides a third hyperbola. The intersection of the resulting third hyperbola with the curve already found with the other three receivers defines a unique point in airspace, which represents the 3D position of the target. In reality, the position of the target is determined by means of the intersections of the hyperboloids. (ERA 2010)

3.2 *Multilateration (MLAT), Position Accuracy and Dilution of Precision*

The accuracy of the MSS depends on its geometrical configuration. The quality of the TDOA measurements is of crucial influence upon the accuracy MSS accuracy. The differences in the estimated accuracy of position and the position of the target in view of the location of the receivers is termed as the Position Dilution of Precision (PDOP). (ERA 2010) The PDOP is dimensionless coefficient, which is equal to:

$$PDOP(x,y,z) = \sqrt{(\sigma_x^2 + \sigma_y^2 + \sigma_z^2)} / (c * \sigma_T) \qquad (1)$$

Similarly it can be introduced Horizontal Dilution of Precision(HDOP) for horizontal plane or Vertical Dilution of Precision (VDOP) for vertical plane.

Generally it holds that the good HDOP is within the area of coverage defined by location of ground receivers and poor outside of this area of coverage.

HDOP - Horizontal Dilution of Precision is dimensionless coefficient, which is equal to:

$$HDOP(x,y,z) = \sqrt{(\sigma_x^2 + \sigma_y^2)} / (c * \sigma_T). \qquad (2)$$

VDOP - Vertical Dilution of Precision is dimensionless coefficient, which is equal to:

$$VDOP(x,y,z) = \sigma_z / (c * \sigma_T) \qquad (3)$$

The standard deviation of the error in defining the position is then obtained by the formula:

$$\sigma_{x,RN} = \sigma_T * c * PDOP, \qquad (4)$$

where:

σ_T - is the standard deviation of measurement of the time at the receivers of the MSS system, which is equal to 10 ns,
c - speed of light.

The standard deviation of the error in defining the position in the horizontal plane is defined by the formula:

$$\sigma_{x,RN} = \sigma_T * c * HDOP \qquad (5)$$

The standard deviation of the error in measuring the altitude is defined by the relationship:

$$\sigma_n = \sigma_T * c * VDOP \qquad (6)$$

Accuracy of defining the position is the difference between the measured position of the target and real positon of the target in the time of detection.

Accuracy of defining the position by means of the MSS system depends on target position and geometric system layout, which is characterized by the coefficient of PDOP. For the MSS System, an important parameter is mainly the horizontal accuracy. (ERA 2013) Examples of calculation HDOP and VDOP are shown in Figure 1. and 2.

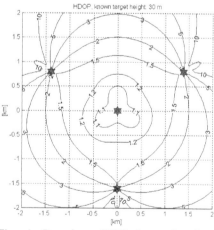

Figure 1. Dependence HDOP for configuration of four receivers

In Fig. 1 the dependence HDOP on configuration of four receivers and for the known target altitude. From the picture it is clear that HDOP depends on the position of the target to the receivers.

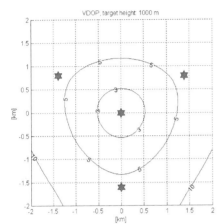

Figure 2. Dependence VDOP for the same configuration and target altitude 1000 m.

Presented in Fig. 2 is the dependence of VDOP for the same configuration and the target altitude of 1000 m. Even in this case the VDOP depend on the position of target with reference to the receivers of the MSS.

3.3 Multi-ranging

Multi-ranging is an new technology that is used in MSS to improve the accuracy in determining the position of targets. Thereby it is possible increase the redundancy of the system or reduce the number of stations required to achieve the same accuracy.

The basic multilateration technique uses time differences between the times of a signal's reception at remote ground stations to calculate hyperboloids in space. A hyperboloid is generated for each pair of ground stations which receive the signal and the position of the target is determined as intersection of three or more hyperboloids.

In addition to this standard technique, use is made of accurate times of interrogation from an active ground station. For each interrogator/receiver pair an ellipsoid is generated, which estimates the target position. The interrogator and receiver form the foci of the ellipsoid. Because the reply to a single interrogation is received at multiple sites, multiple ellipsoids are generated for each interrogation station. This is called multi-ranging. With two receivers the target position will be at the intersection of the ellipsoids.

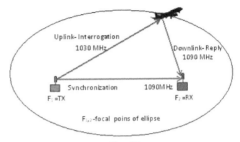

Figure 3. Construction of Ellipsoids

Multi-ranging is cooperating with multilateration. The greatest benefit is derived when the two are used in combination. Multilateration provides high accuracy when the target is located within the area bounded by the sensors.

Multi-ranging improves the both the probability of detection of the target and the operational area of the system. If the target position is erroneously evaluated by the one of the receivers, then it is possible to calculate the 3D position of the target with the help of measuring by the three other receivers. This also means that if one receiver fails to operate then the system can continue in determining the position of the target making use of the other receivers for TDOA combined with the ellipses from multi-ranging. It means an alternative of the system with fewer number of receivers at the same level of determining target position. (Fig.3)

4 MULTI-SENSOR SURVEILLANCE SYSTEM

Further, this paper introduces information about the Multi-sensor surveillance system, the MSS. Described will be a monitoring surveillance system with the Multi-sensor Surveillance System (MMS), which operates on the basis of TDOA.

4.1 MSS Description

The MSS can be a cost effective substitute for conventional radar that monitors cooperative aircraft targets both in the air and on the ground. This new system has higher performance and better reliability. The main difference between this system and the conventional radar is the number of receiver stations. As rule, the radar has just one receiver in the area of coverage, and the MSS makes use as much as 64 of them spread over the coverage area.

MSS is able to operate as both active and passive modes. When in passive mode, the system receives the transmitted information from flying objects. The passive system consists only of receivers. The main benefit of this system is the possibility of using the system as a tool of reconnaissance. Active system, is capable of transmitting and also of obtaining information from aircraft and vehicles. Apart from the receivers it is composed of one or more transmitting units. The most significant benefit of an active mode is its independency on other sources of interrogation.

4.2 *MSS Configurations*

The MSS can be configured based on different requirements. This paper will further describe the selected applications of the system.

Application of Automatic Dependent Surveillance – Broadcast (ADS-B) Based Surveillance is considered to be the simplest MSS implementation. A flying object that is equipped with such a system is able to automatically transmit data about the position derived from the onboard navigation system, together with such information as altitude, velocity, identification and other important data. Making use of the data it is possible to improve safety at air traffic control.

Application of Wide Area Multilateration (WAM) enables location of the position of aircraft equipped with Mode A/C/S transponders with no need to install new avionics aboard. This configuration is used for control and surveillance of aircraft in Terminal Management Areas and En-route for the need of Air Traffic Control. In comparison with radar, the WAM provides higher accuracy, better coverage of the area and improved operational reliability at lower operational costs.

Application of Surface Movement Surveillance is able to provide data such as location and labelling all aircrafts and vehicles in the operational area. It can also determine their manoeuvring area and monitoring their movement over the airport surface. Thereby, the system ensures higher reliability of aviation traffic. Part of the system is an equipment, which identifies potential conflict on the runway. Based on the features presented of the separate MSS system configuration, one can state that the system is contributing to the air traffic safety.

4.3 *Design of the system MSS*

The MSS is made up of three parts:
1 Ground stations are positioned in order to be mutually seen. The module of the ground station is positioned in an alluminium alloy box that meets protection standards. Modules are mounted into a cabinet (made up of a stainless steel and fiber-glass composite) The cabinet, as a whole, can be installed onto a mast, wall or other construction.
2 Central processor station (CPS) is the heart of the product. As a rule, we make use of the central time synchronization or distributed time synchronization. The CPS is housed in a COTS 19-inch rack and the rack type is chosen to allow

easy extension. Moreover, all critical parts of the CPS that ensure signal processing are fully backed-up. The rack is powered through a two-stage power switch, which enables the station to be disconnected from the main electricity supply thereby minimizing the danger of electrical shock to engineering personnel. The CPS includes processors for signal processing, management servers, communication interface, power distributor with optional UPS, monitor for maintenance.
3 Remote Monitoring Terminal (RMT) provides interface between the MSS user and MSS management system. The RMT systems consist of a workstation or of a mobile maintenance notebook. The basic functions RMT include displaying of current system status, providing Graphic User Interface (GUI) for system control, presentation of parameter changes, expert system for failure localization, communicating with the management server, presentation of system failure and printing.

4.4 *MSS Management system*

The Management System allows local and remote operation of the system. It is a web-based and a user friendly interface with separate area for both remote control and for monitoring functions. The management functions fall into following three groups:
– monitoring - diagnostics and on-line monitoring of system status and parameters of the MSS,
– control - modification of MSS parameters and control of MSS operation,
– maintenance - ensuring the functionality of the MSS and the Management System.

The Management System consists of the Management Server, Management Terminals (local and remote), and the management functions of the managed subsystems realized by their Integrated Control Units. The concept is based on unattended operation of the managed subsystems. The architecture is flexible and the system can be extended by further functions such as allows to use multiple management terminals and it can be extended to include other functions like displaying air or ground situation. The management system ensures: remote and local control of the main subsystems, automatic reconfiguration of elements of the system in case of failure of the main equipment, continuous control of quality, and management of the presentation and storage of the operative parameters of the system.

4.5 *MSS Reliability*

The MMS is designed so as to be of high reliability and usable for ATC. The construction of the system involves use of methods, which guarantee high reliability, ease of maintenance and long-term service life. System reliability is determined by method of Mean Time Before Failures (MTBF). The manufacturer calculates that the availability status of the system is achieved the when the system is when the system is able to perform task as required, under given conditions and in real time.

We can sate that the MSS compared to radar is more powerful a less demanding in terms of maintenance. The MSS is a next generation product that enables improvement improves air transport safety.

5 CONCLUSION

Nowadays, there is a big competition between the systems of air traffic control, which operate on the basis of multisensory surveillance. These systems occupy a fixed position in the market air transportation. In view of the ever increasing air traffic and the number of travelers, extension ATC systems is inevitable. The main advantage of Passive Surveillance Systems is potential of changing system architecture according to the needed area operation. Thereby there is a chance to contribute to improved safety and fluency in air traffic. Included among tops system ATC is the MSS that is able supplement and gradually replace older technologies used at air traffic control.

Such systems of MSS have been tested and currently are in use at smaller airports for ATC. Experiences have shown that these system are reliable and enable safe air traffic control. Among the greatest advantages of passive surveillance systems is the possibility of working independently or in cooperation with the existing surveillance systems. Further advantage is in reliable airplane identification and high accuracy of measuring the position of flying objects. Fast information update of MMS systems in real time allows for higher flight efficiency, fluency and higher flight capacity of the airspace. Another advantage is in price accessibility of the MSS systems for small airports with low costs of operation. The object of further research might include application of MSS system in naval transportation.

REFERENCES

EC, 2014. Návrh jednotného európskeho vzdušného priestoru II., Available at: http://ec.europa.eu/transport/modes/air/single_european_sky/doc/ses2_2008_06_25_ses2_citizen_summary_sk.pdf

ERA a.s., 2010. MSS DESCRIPTION. Technical Volume, Customer Reference: N/A, Ere Reference: P110321.

ERA a.s., 2013. MSS-W SK Technical Volume. Document number.: EG0110A00361.

Koukal, M., 2012. Tamara i Věra neměly o nápadníky nouzi. In: 21. STOLETÍ. vol. 5, n. 5: p. 20-21.

Kršák, B. & Tobisová, A.: Prerequisites for the implementation of information technologies in tourism small and medium sized enterprises. - 2012. - 1 elektronický optický disk (CD-ROM). In: SGEM 2012 : 12th International Multidisciplinary Scientific GeoConference : conference proceedings : Volume 3 : 17-23 June, 2012, Albena, Bulgaria. - Sofia : STEF92 Technology Ltd., 2012 P. 205-212. - ISSN 1314-2704

Kurdel, P. & Labun, J. & Adamčík, F.: The estimation method of the characteristics of an aircraft with electromechanic analogue: Naše More. Vol. 61, no. 1-2 (2014), p. 18-21. - ISSN 0469-6255

Labun, J. & Soták, M. & Kurdel, P.: Technical note innovative technique of using the radar altimeter for prediction of terrain collision threats, 2012. In: The Journal of the American Helicopter Society. Vol. 57, no. 4 (2012), p. 045002-1-045002-3. - ISSN 0002-8711

Pavolová, H. & Tobisová, A.: The model of supplier quality management in a transport company. - 2013. In: Naše more. Vol. 60, no. 5-6 (2013), p. 123-126. - ISSN 0469-6255

Srubar, M., 2005. Pasívní radary jako strategická zbraň. Available at: http://www.techblog.cz/technologie/pasivni-radary-jako-strategicka-zbran.html

Szabo, S. & Ferencz,V. & Pucihar, A.: Trust, Innovation and Prosperity. Quality Innovation Prosperity. 2013a-12-27, vol. 17, issue 2, s. DOI: 10.12776/qip.v17i2.224., ISSN 1335-1745. Dostupné z:<http://www.qip/> http://www.qipjournal.eu/index.php /QIP/article/view/224

Szabo, S. &Dorcák,P. & Ferencz,V.: The significance of global market data for smart E-Procurement processes. In: IDIMT 2013b: information, technology human values, innovation and economy. 21st interdisciplinary Information Management Talks, sept. 11-13, 2013, Prague, Czech Republic. 2013, s. 217-224. ISBN 9783990330838.

The Concept of the SWIM System in Air Traffic Management

K. Krzykowska, M. Siergiejczyk & A. Rosiński
Warsaw University of Technology, Warsaw, Poland

ABSTRACT: The paper presents selected issues related to the analysis of network and information exchange systems in air transport. It examines the limitations and drawbacks of current ICT systems used for air traffic management. Analyzing the development of communication systems for the management of general air traffic, it can be concluded that the development of the terrestrial segment of the exchange of information between the parties relating to the air traffic will fluctuate towards a solution based on a service-oriented architecture SOA. This architecture will be the basis for the implementation of the concept of an information exchange system SWIM.

1 INTRODUCTION

The need to increase airspace capacity is a problem existing for traffic engineers for many years. It is not without significance that there is continuous growth of air traffic, particularly over large urban agglomerations in Europe. Any action aimed at streamlining the process in aviation must, however, be consistent with the security policy in aviation, and hence with maintenance an adequate level of that security. Smoothing the process of increasing the number of aircraft in different sectors of airspace may be held only with the improvement of information transmission systems in these sectors (Stańczyk & Stelmach, 2013).

Constantly increasing volume of air traffic entails making modifications to the existing system of air traffic management. These measures concern both organizational changes and procedural and modification of existing ICT systems supporting the management of air traffic safety (Kierzkowski, Kowalski, Magott & Nowakowski, 2012). They provide an opportunity to exchange information between ground services, without which the existence of aviation would not be possible. It is between the ground services exchange bigger than 90% of the information relating to safety of the flight of aircraft (Kierzkowski & Kisiel, 2015). Collection of data about the state of atmosphere and airports, availability of services, the restrictions and the subsequent dissemination of different messages is possible thanks to the fixed network. To ensure safe flight of the aircraft it is essential to ensure communication of all services, such as aeronautical information, weather station, inspectors of particular areas, and many other airspace users. Network technology strongly influenced the present way of exchange of information in air transport (Laskowski, Łubkowski & Kwaśniewski, 2013). Development of aviation networks is focused on the integration of networks and services operating within the national air traffic control systems, and the future extension of these solutions. The paper presents selected issues related to network analysis and information exchange systems in the European air traffic management system. After analyzing the constraints and drawbacks of the existing ICT systems used for air traffic management we indicate the need to change the current "point - point" architecture to the system which exchange information using service-oriented architecture.

2 THE NEED FOR SWIM

Air Traffic Management (ATM) is defined by the International Civil Aviation Organization as a dynamic, integrated management of traffic and airspace - safely, economically and efficiently - through the provision of equipment and uniform services in collaboration with all stakeholders. The proper functioning of the ATM increasingly depends on providing timely, relevant, accurate, accredited and reliable quality of information to make decisions in the process. Sharing best integrated picture of the historical, real-time and planned or anticipated state of the traffic situation on the basis of the whole system will allow the ATM community conducting business in a more secure and efficient manner (Sumiła, 2012). That is how a system SWIM (System Wide Information Management) works, by exchange of information, through the combined set of domains providing or absorbing information. Thanks to SWIM, all information is shared and processed by the service, which must meet the applicable standards and operate in a manner accessible to all users. SWIM aims to improve the management of information, and thus the exchange of information in a wide range, providing support for the ongoing dialogue between the various partners. SWIM meets the safety requirements associated with the exchange of information. Moreover, it provides the exchange of relevant information in a much easier and cheaper in the information exchange process. Aircraft operators will need to constantly update the data on which the ATM service providers and airport operators will have a better knowledge of the intention of flight (Siergiejczyk, Krzykowska & Rosiński, 2014). Thanks to that - controllers, pilots, dispatchers and others will have greater situational awareness with regard to flight status, weather, traffic and other relevant operational information (Sadowski, Siergiejczyk, 2009).

Analysis of the concept of SWIM (System Wide Information Management) enable to confirm that its implementation will require changes in current architecture "point - point" type in area of exchange of information. It is assumed that entities may be geographically dispersed, but should have a valid and uniform information necessary to carry out tasks in specific areas of competence. The figures 1 and 2 show the current structure of the exchange of data (point - point) and a proposal for the future (eg. SWIM).

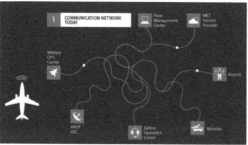

Figure 1. Current structure of the exchange of data

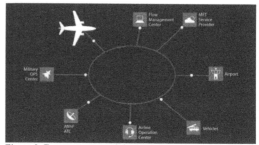

Figure 2. Future structure of sharing information

In addition, the exchange of information between users of fixed network seeks to create a service-oriented architecture SOA (Service-Oriented Architecture), which is the basis for the implementation of the concept of SWIM. SOA is an architecture that uses the definitions of interfaces.

3 ARCHITECTURE OF SWIM

In order to achieve the expected performance of the system - SWIM implementation should be guided by four basic principles:
- separation of the provision of information - distinction between providers of information from the sources of information;
- feedback of the system - each of the elements uses more or less knowledge about other components, in this way - barrier between systems and applications are removed and the interfaces are compatible;
- use of open standards / publicly available;
- use of service-oriented architecture (SOA).

Based on the above principles, the implementation of SWIM will introduce the following elements:
- **AIRM** (ATM Information Reference Model) - will ensure the implementation of each type of information through an ATM conceptually logical data models, among them there will be items such as airports, flight route, airspace, flight procedures and common definition of modeling taking into account the time and space;
- **ISRM** (Information Service Reference Model) - provide a logical division of information services required and their patterns of behavior, it will contain details about service charge, the pattern of information exchange, quality of service (QoS), infrastructure for data exchange system;
- **IMF** (Information Management Functions) - includes functionality such as user identity management, disclosure of resources, aspects of security, including authentication, encryption and notification services;
- **SWIM Infrastructure** as a technical infrastructure (ground/ground and earth/ground).

Technical infrastructure SWIM-TI (System Wide Information Management - Technical Infrastructure) interferes with the termination of the services rendered under the ATM systems supported by SWIM solutions, ensuring their productivity and increasing efficiency and safety. Systems that interact with SWIM cooperate with services specific to ATM systems, and their cooperation is supported by technical solutions offered by SWIM.

SWIM-TI infrastructure is a set of software components distributed in the network infrastructure, providing attributes for the collaboration between systems. These attributes are appropriate for the set of nodes in SWIM (endpoint entities) and common components (ensuring appropriate features in all distributed nodes in SWIM). Therefore, the idea of nodes in SWIM presents a set of features and capabilities of SWIM-TI infrastructure, allowing a given system the use of its solutions.

Examples of common components are:
- register used to enable sharing of information (metadata) about the services within the prescribed time,
- Public Key Infrastructure PKI, which aims to guide the structure of the trust of digital certificates.

Particularly important element of SWIM is AIRM - reference model considered as a model for other developed in SESAR program (Single European Sky ATM Research). The SESAR program was designed to build a modern European air traffic management system. It is a technological and operational component of the initiative of Single European Sky (SES) resulting synchronized and increased capacity of the European airspace. As the founder of SESAR, together with the European Commission, EUROCONTROL plays a key role in all projects (Siergiejczyk, Krzykowska & Rosiński, 2015).

The basic objectives of the program include:
- the introduction of business trajectory, which can be described as the trajectory most common to the set released by performing flight;
- managing trajectory through which it is planned to implement a new approach to design and management of airspace, including:
  - preferred routes of flights;
  - advanced civil – military cooperation (flexible use of space);
  - division of space into controlled and uncontrolled.

The basic principle of the SESAR approach is that all technological achievements should provide the possibility of accomplishment the objectives of which are derived directly from operational requirements and support the growth of the overall air traffic management system performance. In fact, the CNS infrastructure (Communication, Navigation, Surveillance), will have to be more capable, and most importantly, more flexible than ever. The purpose of this is to ensure that technical limitations does not slow down the development of advanced procedures and applications. CNS activities are an important investment in the SESAR program.

Figures 3 and 4 show the structure of the AIRM product and the planned dates for implementation.

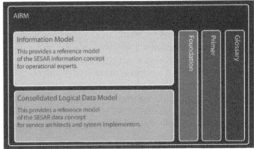

Figure 3. The structure of the AIRM product

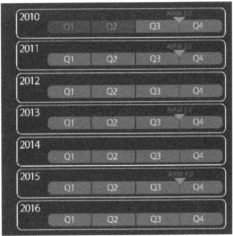

Figure 4. Implementation planned dates

The architecture proposed by SWIM can distinguish a series of mergers, such as shown in Figure 5 below.

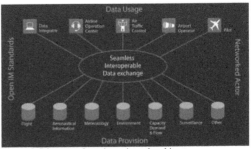

Figure 5. Diagram of service-oriented architecture

SWIM concept is to integrate the various players (ATM providers of air navigation services, airlines, airports, industry, standardization organizations and standardization) in the sense of information, not only in terms of the network but also the system and a B2B interaction.

4 APPLICATION AND BENEFITS FROM SWIM

European air traffic management system operates close to the limit of capacity and has to deal with the challenge of constantly growing demand in the field of air transport. In order to fulfill all the tasks set out by the European Commission and strengthening the air transport value chain, the requirements of airspace users must be better satisfied. Therefore, each flight must be done in strict accordance with the intentions of the owner, while maximizing network performance. This is the main principle driving the future of European ATM system, which represents the airspace user's intent in respect of the flight.

On the basis of the main objectives of the extensive SWIM system architecture the following applications can be provide (Dhas, Mulkerin, Wargo, Nielsen & Gaughan, 2000):

- **synchronized traffic** – aim is to manage arrivals and departures and sequencing of aircrafts during flights and in the airports controlled areas, optimizes the flow of traffic and, consequently, reduces the number of interventions on the part of air traffic control;
- **integration of airports** - is aimed at achieving full integration airports as nodes in the ATM network, ensuring consistency of the process through joint decision-making;
- **support development of 4D trajectory** - is aimed at a systematic breakdown of the trajectory of the aircraft between the different players of ATM process ensuring that all partners have a common perspective of the flight and access to the most current data available for the appropriate performance of their duties;
- **common network management** and balancing liquidity and capacity - improved cooperative network management through dynamic, direct and fully integrated network operating plan NOP (Network Operations Plan).;
- **automation and conflict resolution** - aims to significantly reduce the burden on the flight controller tasks, while meeting the objectives of the SESAR program in the field of safety and environmental benefits, without incurring the provider's significantly higher costs.

In the implementation of SWIM process a number of products associated with the ATM will also be provided, it is shown in Table 1.

Table 1. Available via SWIM selected products for data exchange

Product	Description	Status
AIM FNS	Provides weather telegram NOTAM	After testing, not yet available
AIM SAA	Provides configuration information about airports	After testing, available
AIM	Aeronautical data	After testing,
CSS-Wx	Modernization and centralization of weather data	Not available
ITWS	Integrated Terminal Weather System - provides weather data graphically visualized	After testing
NCR	NAS Common Reference – Aggregation and integration of data depending on the airspace	After testing, partially available
SFDPS	SWIM Flight Data Publication Service - It provides the route data, data about the flight, flight plans, beacon codes	After testing, partially available
STDDS	SWIM Terminal Data Distribution System – Provides data on surveillance from the airports, data from the control tower about the situation on the surface	After testing, available
TBFM	Time Based Flow Management - Provides data on the flow of traffic depending on time	After testing, available
TFMS	Traffic Flow Management- Provides data on Air Traffic Management	After testing, partially available
WARP/EWD	Weather and Radar Processor - The radar data and weather	After testing, partially available

Table 1 shows the set of elements which, according to the FAA organization (Federal Aviation Administration) should be implemented action SWIM products. That is to say, the FAA tested by simulating a range of products, which could be components of SWIM. Some of them, after testing received statute "available" - expressing positive tests.

SWIM system creates a comprehensive solution tailored to the operational policy so as to gradually provide the correct information to specific entities in the right place and time. To the basic benefits of the implementation of the SWIM system we can include:

- availability;
- equality;
- flexibility;
- performance;
- quality, consistency and security of information;
- implementation and development;
- cost;

- orientation services;
- open standards;
- global application.

SWIM system that integrates all the data related to air traffic management can provide a basis for the whole European ATM system. SWIM will support multiple business objectives of high strategic priority through the utilization of shared information. SWIM system may be a key factor in the operation of the SESAR program and can provide direct business benefits, operational and technical.

5 SUMMARY

The paper presents selected issues related to the analysis of network and information exchange systems in air transport. Examines the limitations and drawbacks of the current ICT systems used for air traffic management. Analyzing development of communication systems for the management of general air traffic, it can be concluded that the development of the terrestrial segment of the exchange of information between the parties relating to the air traffic will fluctuate towards a solution based on a service-oriented architecture SOA. This architecture will be the basis for the implementation of the concept of an information exchange system SWIM. SWIM system analysis include the list of advantages and applications of the system and the identification of infrastructure elements. SWIM concept has to integrate the different actors (ATM providers of air navigation services, airlines, airports, industry, standardization organizations and standardization) in the sense of information, not only in terms of network and system. Therefore seems to be the best solution based on service-oriented architecture.

By being able to integrate all the data related to the management and control in the area of air transport, SWIM system seems to be part of supporting the entire European ATM system and its implementation may be necessary to its effective functioning. It includes support for information exchange between terrestrial objects and aircraft, as well as a ground only between objects. As a result of consistent data exchange technology between systems ATM software projects can be a unifying airspace in Europe SES (Single European Sky), and on a global scale.

REFERENCES

Dhas C, Mulkerin T, Wargo C, Nielsen R, and Gaughan T. 2000. Research Report April 2000. *Aeronautical Related Applications Using ATN and TCP/IP*. Springfield, Virginia, Computer Networks and Software, Inc.

Kierzkowski, A., Kisiel, T. An impact of the operators and passengers behavior on the airport's security screening reliability (2015). Source of the Document Safety and Reliability: Methodology and Applications - Proceedings of the European Safety and Reliability Conference, ESREL 2014 pp. 2345-2354.

Kierzkowski, A., Kowalski, M., Magott, J., Nowakowski, T. Maintenance process optimization for low-cost airlines (2012) 11th International Probabilistic Safety Assessment and Management Conference and the Annual European Safety and Reliability Conference 2012, PSAM11 ESREL 2012, 8, pp. 6645-6653.

Laskowski, D., Łubkowski, P., Kwaśniewski, M. 2013. Przegląd Elektrotechniczny 89 (9). *Identyfikacja stanu zdatności usług sieci bezprzewodowych / Identification of suitability services for wireless networks: 128-132.*

Sadowski P., Siergiejczyk M. 2009. *Zastosowanie sieci IP w systemach zarządzania ruchem lotniczym ATM*, XXV Krajowe Sympozjum Telekomunikacji i Teleinformatyki, Warszawa.

Siergiejczyk M., Krzykowska K., Rosiński A. 2014. Proceedings of the Ninth International Conference Dependability and Complex Systems DepCoS-RELCOMEX, editors: W. Zamojski, J. Mazurkiewicz, J. Sugier, T. Walkowiak, J. Kacprzyk, given as the monographic publishing series – Advances in intelligent systems and computing, Vol. 286. *Reliability assessment of cooperation and replacement of surveillance systems in air traffic: 403-411.* The publisher: Springer.

Siergiejczyk M., Krzykowska K., Rosiński A. 2015. Proceedings of the Twenty-Third International Conference on Systems Engineering, editors: Henry Selvaraj, Dawid Zydek, Grzegorz Chmaj, given as the monographic publishing series – „Advances in intelligent systems and computing", Vol. 1089. *Parameters analysis of satellite support system in air navigation*: 673-678. The publisher: Springer.

Siergiejczyk M., Rosiński A., Krzykowska K. 2013. The monograph „New results in dependability and computer systems", editors: W. Zamojski, J. Mazurkiewicz, J. Sugier, T. Walkowiak, J. Kacprzyk, given as the monographic publishing series – „Advances in intelligent and soft computing", Vol. 224. *Reliability assessment of supporting satellite system EGNOS: 353-364.* The publisher: Springer.

Sumiła M. 2012. Telematics in the Transport Environment. *Selected Aspects of Message Transmission Management in ITS Systems: 141-147.* Springer Heidenberg.

SWIM Technical Architecture. 2008. WP 14. SJU_DOW_WP14_V4.0.

Stańczyk P., Stelmach A. 2013. *Wykorzystanie neuronowych modeli do oceny faz wznoszenia różnych typów samolotów*, Prace Naukowe Politechniki Warszawskiej - Transport, Oficyna Wydawnicza Politechniki Warszawskiej, zeszyt nr 100, str. 191 – 200

AUTHOR INDEX

261